C000023045

Solaris® 2 for Managers and Administrators

Second Edition

Curt Freeland, Dwight McKay, and Kent Parkinson

Solaris® 2.x for Managers and Administrators
Second Edition

By Curt Freeland, Dwight McKay, and Kent Parkinson

Published by:

OnWord Press
2530 Camino Entrada
Santa Fe, NM 87505-4835 USA

All rights reserved. No part of this book may be reproduced or transmitted in any form or by any means, electronic or mechanical, including photocopying, recording, or by any information storage and retrieval system without written permission from the publisher, except for the inclusion of brief quotations in a review.

Copyright © Curt Freeland, Dwight McKay, and Kent Parkinson

Second Edition, 1998

SAN 694-0269

10 9 8 7 6 5 4

Printed in the United States of America

Library of Congress Cataloging-in-Publication Data

```
Freeland, Curt.
     Solaris 2.x for managers and administrators / Curt
Freeland, Dwight McKay, and Kent Parkinson. -- 2nd ed.
          p.      cm.
     Revised ed. of: The SunSoft Solaris 2.* for managers and
administrators / Kent Parkinson, Curt Freeland, and Dwight
McKay.
     Includes index.
     ISBN 1-56690-150-2 (alk. paper)
     1. Operating systems (Computers)  2. Solaris (Computer
file) I. McKay, Dwight. II. Parkinson, Kent. III. Parkinson,
Kent.  SunSoft Solaris 2* for managers and administrators.  IV.
Title.
     QA76.76.063P3665      1997
     005.4'4769--dc21
                                                   97-39823
                                                   CIP
```

Trademarks

Solaris is a registered trademark, and SunSoft and OpenWindows are trademarks of Sun Microsystems, Inc. OnWord Press is a registered trademark of High Mountain Press, Inc. Many other products and services are mentioned in this book that are either trademarks or registered trademarks of their respective corporations. OnWord Press and the authors make no claim to these marks.

Warning and Disclaimer

This book is designed to provide information about SunSoft Solaris. Every effort has been made to make the book as complete, accurate, and up to date as possible; however, no warranty or fitness is implied.

The information is provided on an "as is" basis. The authors and OnWord Press shall have neither liability nor responsibility to any person or entity with respect to any loss or damages in connection with or arising from the information contained in this book.

About the Authors

Curt Freeland is Director of Facilities for the Computer Science and Engineering Department at the University of Notre Dame. In addition to serving as system administrator for the department, he teaches senior level system administration and network management courses. Curt earned a degree in systems engineering from Purdue University where he served as systems engineering manager for the Engineering Computer Network from 1979 through 1995. Curt has contributed articles to trade publications such as *Advanced Systems* (formerly *SunWorld*), the *Super-User* newsletter, and *Hardcopy* magazine.

Dwight McKay is system manager for the Markey Center for Structural Biology at Purdue University. Dwight has over 12 years of experience managing Sun systems and integrating them with a variety of other UNIX systems, as well as VMS, Macintosh, and Windows systems. He has a bachelor's degree in psychology from Princeton University.

Kent Parkinson is a principal of Parkinson and Associates, a UNIX systems consulting and training firm located in West Lafayette, Indiana. A former faculty member of Purdue University, Kent has worked with UNIX systems for over 20 years, and with Sun/Solaris operating environments since their respective introductions.

Acknowledgments

We have many people to thank for their assistance with the second edition of this book. Our aim for the second edition "rewrite" was to cover more material at greater depth, and to update the text and illustrations to accommodate Solaris 2.6. These tasks required more time than we originally estimated, but we feel that the resulting publication is much more complete than the first edition.

First, the authors would like to thank Barbara Kohl and the rest of the staff at OnWord Press for their assistance, guidance, and encouragement throughout this project. We would also like to thank all of the people who have read and commented on our manuscripts. This book is much better thanks to your comments and suggestions!

Curt Freeland, Dwight McKay, and Kent Parkinson

Thanks to Dwight McKay and Kent Parkinson for their friendship and patience, and for entrusting me with this edition of the book. The success of the first edition made the production of the second edition much easier to manage!

I would also like to thank the students of CSE498A and CSE498B for their comments and suggestions on early versions of this book. Thanks to Dr. Steven C. Bass for his support and encouragement, and for offering me the opportunity to teach at the University of Notre Dame.

Next, I extend my thanks to my fellow faculty and staff, and the students in the Computer Science and Engineering department for allowing me the time to work on this project. To my wife (Sally) and children (Erin, Shaun): thanks for putting up with me during the development of this book. Finally, thanks to Mindy Canada and her staff at Computer Aided Tools and Service in Indianapolis, and Dr. Kam H. Chan of Tatung Science and Technology, Inc. Without their generosity, my contributions would not have been possible.

Curt Freeland

I would like to extend my thanks to Curt Freeland and Kent Parkinson for sharing their friendship, time, and experience to make this publication possible. Thanks also to Dr. Janet Smith and the members of the Markey Center for Structural Biology at Purdue University for encouraging me to participate in developing the book. And special thanks to my wife, Mary, and my daughters, Catherine and Susan, for their love and support.

Dwight McKay

I am grateful to Curt Freeland and Dwight McKay for sharing their knowledge, time, and experience. I would also like to thank my wife Susan, and children Heidi, Alex, and Nick for their patience, support, and endurance during the production of this book.

Kent Parkinson

OnWord Press...

Dan Raker, President
Dale Bennie, Vice President
David Talbott, Acquisitions and Development Director
Carol Leyba, Associate Publisher
Barbara Kohl, Associate Editor
Cynthia Welch, Production Manager
Michelle Mann, Production Editor
Liz Bennie, Director of Marketing
Lauri Hogan, Marketing Services Manager
Kristie Reilly, Project Editor
Lynne Egensteiner, Cover designer, Illustrator

Table of Contents

Introduction

Welcome to *Solaris 2.x for Managers and Administrators*. This book was written by system administration professionals with over 40 years of combined experience in managing corporate-sized computer installations. Our intent was to approach the topics we cover from a practitioner's perspective. In the following we discuss who should read this book, how the book is organized, and the typographical conventions we use.

Who Should Read This Book

Solaris 2.x for Managers and Administrators was written for beginning Solaris systems administrators and managers who have general UNIX knowledge but little or no Sun systems administration experience. The purpose of this book is to deliver practical information regarding the installation and maintenance of Solaris 2 systems situated in local area networks.

Throughout the book, helpful tips and examples are used to explain and illustrate important concepts. Many examples used in the book derive from the authors' experience as system administrators. Other examples were derived by polling experienced and inexperienced system administrators to identify the tools they used, and how they approach daily system administration tasks.

This text is intended as a handy reference for beginning and experienced Solaris system administrators. The authors hope that readers enjoy learning about the topic of Solaris system administration as much as they enjoyed producing the book. If you have any comments, questions, or recommendations about the book, please contact the authors at OnWord Press, 2530 Camino Entrada, Santa Fe, NM 87505-4835 USA, or e-mail them at *readers@hmp.com.*

Book Organization

Part I, "Fundamentals," covers the concepts of the Solaris distributed environment, Solaris terminology, preparing for a Solaris installation, and basic administration tools.

Part II, "System Management," comprised of the next 12 chapters, is focused on fundamental system administration tasks such as boot and shutdown procedures, managing user accounts, system security maintenance, and working with device names. Separate chapters are devoted to disk subsystem hardware and software, disk and file system maintenance, system software management, adding terminals and modems, managing printers, system backups and disaster recovery, and automating routine administrative tasks.

Part III, "Network Services," the final six chapters, covers the basics of network hardware, use of network utilities such as the Network File System (NFS) and Network Information System (NIS+), diskless and dataless client administration, automating NFS with automount, and network name services. The final chapter consists of an overview on providing World Wide Web services on the Solaris platform.

The book concludes with a detailed index.

Typographical Conventions

A boldface font in regular text is used to highlight command names, daemons, and command options appearing for the first time in a section. Examples appear below.

❐ Use the **chmod** command to execute the changes. Like **chown**, chmod has an **-r** option for handling entire directories.

❐ The **/usr/sbin/sys-unconfig** command allows the administrator to change the network information of the host.

❐ Infrequently required services can be started on demand by a special daemon known as **inetd**.

Italics are used to indicate new terms, file names, directory and path names, and general emphasis in regular text. Selected examples follow.

❐ These logical drives became known as *partitions*.

❐ The files are located in the */sbin* directory, and script directories are located in the */etc* directory.

❐ The *etc/inittab* file contains a series of directives for the init process.

The following monospaced font is used for examples of command statements and computer/operating system responses.

```
glenn% /etc/mount
/ on /dev/dsk/c0t3d0s0 read/write/setuid on Tue Aug 9 05:58:19 1997
/usr on /dev/dsk/c0t3d0s6 read/write/setuid on Tue Aug 9 05:58:19 1997
```

```
/proc on /proc read/write/setuid on Tue Aug 9 05:58:19 1997
/dev/fd on fd read/write/setuid on Tue Aug 9 05:58:19 1997
/tmp on swap read/write on Tue Aug 9 05:58:23 1997
/opt on /dev/dsk/c0t3d0s5 setuid/read/write on Tue Aug 9 05:58:24 1997
/var/mail on mail.astro.com:/var/mail soft/bg/read/write/remote on Tue Aug 9
12:20:52 1997
glenn%
```

Named keys on the keyboard are enclosed in angle brackets, such as <Enter>, <Shift>, and <Ctrl>.

Key sequences, or instructions to press a key immediately followed by another key, are linked with a plus sign.

❏   <Ctrl>+G

➻ **NOTE:** *Information on features and tasks that is not immediately obvious or intuitive appears in notes.*

✓ **TIP:** *Tips on command usage, shortcuts, and other information aimed at saving you time appear like this.*

☛ **WARNING:** *The warnings appearing in this book are intended to help you avoid committing yourself to results that you may not have intended.*

Part I

Fundamentals

An Introduction to Distributed Computing

When the first electronic computer was introduced, few people saw much use for this new technology. In fact, many people were afraid of the early computers. Much of this apprehension resulted from a lack of knowledge about what computers could (and could not) do.

The first computers were slow, difficult to use, extremely large, outrageously expensive, and nearly impossible to maintain. A corporation was "lucky" if it could afford to purchase and operate one computer. All corporate computing was handled at one central computing facility by computer specialists.

As technology marched forward, computers became faster, smaller, and more easily maintained. Suddenly it became feasible for corporations to operate many computers instead of just one. Corporate decision makers were then faced with unanticipated questions such as maintaining a single large computing center versus more diverse and distributed computing resources.

An Abbreviated History of the Microcomputer

Over the past decade corporate computing has undergone many drastic changes. (In this chapter, "corporation" and "corporate" are interchangeable with "organizational" and "organization," or any public or private sector entity using computing power in daily operations.) As the complexity of data processing jobs increased, typical corporate mainframe computers could not keep pace with computing demands. These older *mainframe* computers employed discrete logic circuits (chips) to implement the computer hardware. Many computer manufacturers were having problems making these discrete components operate at the speeds required to produce faster computer.

At about the same time, chip manufacturers were developing single chip computers which could out-perform many of the discrete logic mainframes on the market. These single chip computers were dubbed *microprocessors* or *microcomputers*. Many computer manufacturers began developing a new type of personal desktop computer system based on the microprocessor chips. These high performance systems soon became known as *workstations*.

A Look at Traditional Mainframe Computing

The workstation computing revolution presented business computing with the following quandary: should the corporation rely on one large computer system which was not keeping up with computing demands, or should it "downsize" to the new workstations and distribute computing power among many smaller systems?

Corporations had to examine computing needs and how these needs could be addressed by the systems available on the market. Decisions were often reduced to a choice between the mainframe computing model and the distributed computing model.

Mainframe Model

What is the mainframe computing model? Mainframe computing is the term used when the corporation employs one large (mainframe) computer system for all corporate computing. This mainframe system provides computing power for everything from payroll to materials management and production management.

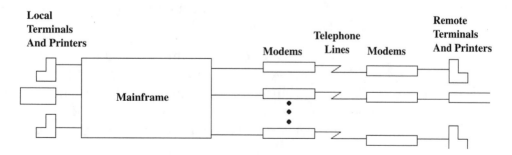

Typical corporate mainframe computing environment.

Access to the mainframe typically occurs via an ASCII terminal. These terminal connections may be local (in the same building or complex as the mainframe) or remote (via telephone lines and modems). Regardless of the type of connection, the data rate between the mainframe and the terminal is usually slow. Viewing a large document or accessing information from a database may take several minutes via ASCII terminal connections.

Hybrid Mainframe Model

A typical extension of the mainframe computing environment produced a tree-shaped computing environment. In this model, the corporation still relied on a mainframe computer (at the root of the tree). Smaller minicomputers were distributed throughout branch offices. These systems were tightly coupled to the mainframe, and in many cases were completely controlled by the mainframe.

Even smaller microcomputers were connected to the branch minicomputers. The microcomputers were used to control machinery and collect data. This information was shipped to the minicomputer for processing. The results of these computations were shipped off to the corporate mainframe for storage and/or further processing.

This version of mainframe computing typically used telephone lines and modems for the interconnection of system elements. Similar to the traditional mainframe computing model, the hybrid mainframe model was characterized by low data rates and lengthy delays in data access.

As minicomputers and microcomputers became more powerful, corporations had to evaluate their computing objectives. Among other things, corporations were forced to weigh the advantages and disadvantages of respective computing models, and make decisions about future directions.

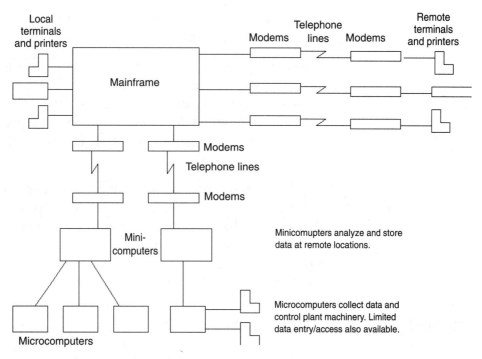

A hybrid mainframe computing environment.

Advantages of Mainframe Computing

The mainframe computing model presents certain advantages over alternative models. One of its major advantages is the centralization of corporate data. In some instances it is desirable to store all corporate data in one place. In other instances, reliance on a single central repository of information became a liability to the corporation.

As an example of mainframe computing advantages, consider the credit card division of a large banking operation. If all credit card customer data are maintained on a single system, corporate headquarters can easily determine the "value" and status of any and all customers. If the same data were distributed among many geographically diverse computing centers, bank management would not be able to track the "value" of customers so easily.

Another advantage of mainframe computing is that the corporation does not have to worry about maintaining large numbers of geographically separated computers. A single computer at a single site presents a (fairly) simple maintenance task for the corporate computing division.

Finally, certain applications simply require enormous amounts of computing power to accomplish basic objectives. These applications are best served with a mainframe system. Examples here might include the operation of international airline ticket offices, and the trading offices of global stock exchanges.

Disadvantages of Mainframe Computing

While the centralization of data allows corporate offices to monitor everyday operations, it could also cause substantial operating expenses and delays. These expenses and delays often appear in conjunction with the distribution of corporate data to branch offices or other entities requiring data access.

Revisiting the credit card division of the bank mentioned above, what happens if customers are serviced by geographically distributed branch offices? If a branch office requires information pertaining to customers serviced by a different branch, tens of gigabytes of information may have to be exchanged between the two branches. The use of a typical ASCII terminal connection over a phone line would require days to access the information required to conduct the branch office's daily business.

Another disadvantage of mainframe computing is the "single point of failure" problem. If the corporation owns a single large computer system and the system fails, all corporate data processing capability is off-line. While many manufacturers provide 24-hour service contracts, problem resolution often requires minutes (or hours) to diagnose the problem and replace the failed module.

Finally, a typical mainframe computer is a very large, very expensive system. As the mainframe becomes obsolete, the corporation must plan well in advance to provide the capital for upgrading the system. Because mainframe computers from different manufacturers rarely run the same operating systems or the same versions of language compilers and other applications, the corporation must also plan for the crossover between the old and new systems.

Distributed Computing Model

The distributed computing environment was originally an offshoot of the modified mainframe computing environment. As the minicomputers and microcomputers positioned at the leaves of the tree became more powerful, they assumed some of the data storage and processing responsibilities. If the mainframe required access to the information on these leaves, it had to request that information from the workstations.

In addition to more powerful processing elements, interconnection technology underwent significant advances. In the distributed computing model, processing elements are typically tied together via *local area* or *wide area* networks. These connections are orders of magnitude faster than a mainframe's ASCII terminal connections. Documents which once required several minutes to access could now be accessed in less than a minute.

These developments eventually led to the distribution of corporate data among the multitude of processing elements owned by the corporation. No single processor accessed all corporate data, but each processor had access to the data required to maintain relevant activities for a particular segment of the corporation. Such data became almost instantly available to other processing elements via local and wide area data networks.

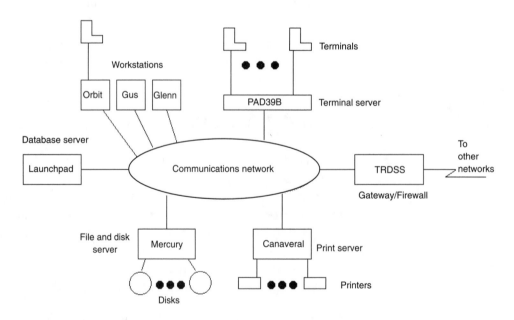

A distributed computing environment.

In today's distributed computing environment, the data processing chores are distributed among many systems. These systems are typically high-speed workstations linked via a local area or wide area network. Each of these processing elements has a specific task assigned as its primary function. But each of these workstations is typically a general purpose computer which can also provide processing capabilities beyond its primary task.

Advantages of Distributed Computing

One of the principal advantages of the distributed computing model is geographic distribution. Systems are distributed by geographic region as required to best serve corporate needs. By linking these systems via high-speed networks, the branch offices have instant access to the data required for everyday operations. In addition, the corporate office computers can still access the branch office data via high-speed networks with minimal delays.

Another advantage of the distributed computing model is that each component of the system is typically less expensive than its equivalent in the mainframe computer model. Consequently, the corporation can often purchase several computer systems for the same capital investment required to purchase one mainframe computer.

As the workstations in the distributed computing model become obsolete, the capital burden of upgrading to new systems can be spread over a longer period of time by upgrading the critical systems first, and the less critical systems at a later date. In addition, because many of the workstations run the same operating system, the corporation should not have as many problems porting software to the new systems.

Another major advantage of the distributed computing model is the "many points of failure" feature. Because no single computer stores all corporate data, no single system failure should be able to shut down corporate data processing. Instead, if one system fails, it is often possible to have another system automatically assume the computing responsibilities of the failed system. This feature allows the corporation to maintain operations and provide maximum access to critical corporate data at all times.

Disadvantages of Distributed Computing

While distributed computing offers many advantages, it also offers a few disadvantages to the corporate data processing center. Because workstations are widely distributed over a broad geographic area, maintenance can become a problem for the corporation. Spatial dispersion of computing equipment may force the corporation to add personnel in order to maintain these systems.

File system backups, installation of new software, system security, and installation of user accounts can also become complicated in the distributed computing environment. When the corporation operates only one computer, the tasks of monitoring the system, allowing access, and installing software are contained within a single location.

In the event of a network failure, corporate office computer systems may not be able to access certain essential data for indeterminate periods of time. This is not an acceptable mode of operation in some applications (e.g., stock trading, control of the space shuttle, and banking transactions).

Evolution of System Software

As minicomputer and microcomputer hardware evolved, so did the system software. Dozens of companies produce workstations. Each company developed operating system software to control its computers. This diversity of hardware and software had to be factored into the corporate decision making process.

Several computer manufacturers offered system software based on traditional mainframe operating systems. Corporate computing centers could easily migrate their applications to such operating systems. Unfortunately, some of these operating systems were incapable of extension to support newer models of system hardware.

At about the same time that minicomputers were introduced on the market, a new operating system that would be portable across several hardware platforms was emerging. The UNIX operating system, in conjunction with workstation technology, changed the direction of corporate computing.

An Abbreviated History of UNIX

The UNIX operating system was originally developed at American Telephone and Telegraph's Bell Laboratories in the late 1960s and early 1970s. Originally intended as a research tool for use within Bell Labs, word of the power and flexibility of UNIX spread quickly. UNIX quickly became the software of choice in many research and higher education institutions.

UNIX Branches Out

As the UNIX operating system offered by AT&T evolved and matured, it became known as System V (five) UNIX. As the developer of UNIX, AT&T licensed other entities to produce their own versions of the UNIX operating system. One of the more popular of these licensed UNIX variants was developed by The University of California at Berkeley Computer Science Research Group. The Berkeley UNIX variant was dubbed Berkeley Software Distribution (BSD) UNIX.

The BSD version of UNIX rapidly incorporated networking, multiprocessing, and other innovations which sometimes led to instability. In an academic environment these temporary instabilities were not considered major problems, and researchers embraced the quickly evolving BSD UNIX environment. In contrast, corporate computing centers were wary of converting to operating systems with a history of instability.

Unlike BSD UNIX, AT&T's System V UNIX offered stability and standardization. New capabilities were introduced at a slower rate, often after evaluating the results of introducing the same capabilities in the BSD UNIX releases. Corporate computing centers tended to favor the stability of AT&T's version of UNIX over that of BSD UNIX.

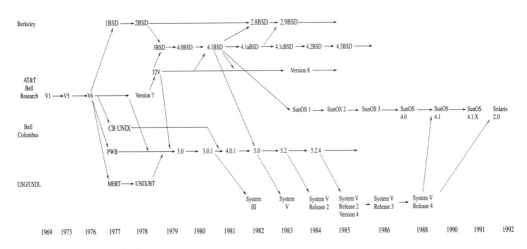

Genealogy of the UNIX operating system.

An Abbreviated History of Sun Microsystems

Sun Microsystems Computer Corporation was one of the manufacturers at the forefront of the distributed computing revolution. The company's first workstations employed microprocessors from the Motorola MC68000 chip family. Sun Microsystems founders believed that a UNIX operating system would be the most desirable option for their new line of workstations. The early Sun systems ran an operating system called SunOS (Sun Operating System, pronounced Sun Oh-ess). SunOS was based on the BSD UNIX distribution.

Enter the SPARC

While the systems based on microprocessor chips were faster than many mainframes, there was still room for improvement. Sun promptly began developing its own microprocessor. In the 1980s Sun introduced a Reduced Instruction Set Computer (RISC) chip called the Scaleable Processor ARChitecture (SPARC) processor. The first implementation of the SPARC chip ran at twice the speed of the fastest MC68000-based systems that Sun was producing at the time.

The SPARC processor chip allowed Sun to produce very powerful, inexpensive, desktop workstations. The SPARC systems also ran the SunOS operating system, thus preserving customers' software development investments. Many other workstation manufacturers were delivering operating systems based on AT&T's System V release 3 operating system standard (also known as the System V Interface Description or SVID). Sun's customer base had to decide between the Sun BSD based operating system, and competitors' System V based offerings.

Sun Develops a New Operating System

In the late 1980s Sun announced plans to develop a new operating system based on the AT&T System V release 4 UNIX. The new operating system is called *Solaris 2*. Because Solaris is a System V based operating system, it is quite different from the BSD based SunOS operating system.

The Solaris 2 operating system presented two problems for Sun customers. The first problem was that many applications available under BSD UNIX were not available under System V UNIX. Software vendors had to expend great effort to recode portions of respective applications in order to make them operate under Solaris 2.

The second problem presented by Solaris to customers was that many BSD-fluent system administrators had no experience with a System V operating system. While most user-level commands are similar under SunOS and Solaris, the system administration commands are very different. Many of the "old reliable" commands from BSD-based UNIX are missing from Solaris. Next, new commands with more functionality replaced these BSD commands. The architecture of the operating system is also very different under Solaris. Access to and management of system devices under Solaris is foreign to BSD system administrators.

Summary

Sun Microsystems is one of the market leaders in the downsizing/rightsizing of corporate computing environments. Recently, Sun introduced Solaris 2, a new operating system for its workstations. While the new operating system has many similarities to the SunOS operating system, there are also many differences between the two offerings.

The primary goal of this book is to deliver practical information regarding the installation and maintenance of Solaris 2 systems. This chapter presented a brief history of the UNIX operating system, and examined selected advantages and disadvantages of the mainframe and workstation computing models. Subsequent chapters of the book explore the Solaris operating system in detail.

Solaris Concepts and Terminology

The previous chapter clarified the differences between a mainframe and a distributed computing environment. The remainder of this book will focus on the Solaris 2 distributed computing environment.

This chapter focuses on concepts and terminology related to installation methods and distributed environments. Installation methods encompass configuration and the role of servers, clients, diskless clients, and diskfull clients.

Solaris Operating Environment

Solaris is best understood as an operating *environment*, rather than an operating system. The environment is comprised of the following components: SunOS (Sun's operating system), OpenWindows (Sun's graphical user interface), OpenWindows DeskSet (a collection of standard OpenWindows applications), and the new CDE (Common Desktop Environment), which can be found on other vendors' workstations as well.

At present, there are two versions of the Solaris operating environment: Solaris 1 and Solaris 2. In Solaris 1, SunOS is based on the Berkeley Software Distribution (BSD), commonly known as Berkeley UNIX, and OpenWindows Version 2 or 3. With Solaris 2, SunOS is based on System V Release 4 (SVR4), also known as AT&T UNIX, and OpenWindows Version 3. If a system is currently running SunOS 4.1.2 or SunOS 4.1.3, it is, in fact, running either Solaris 1.0.1 or Solaris 1.1.

SunOS

SunOS is the heart of the Solaris operating environment. Like all operating systems, SunOS is a collection of software that manages system resources and schedules system operations. Solaris 1 includes a version of SunOS based on the BSD kernel. Solaris 2 has a newer kernel based on AT&T's System V Release 4 with numerous additions.

OpenWindows

OpenWindows, often referred to as Openwin, is Sun's graphical user interface. OpenWindows is based on AT&T's OPEN LOOK graphical user interface functional specification. OPEN LOOK applications adhere to a standard look and feel in the same way the automobile industry has adopted predictable standards. For example, just as drivers know that cars made in the United States will always have a turn signal lever on the left side of the steering column, so the

OPEN LOOK specification ensures OpenWindows users of a standard arrangement of window controls regardless of the application's third party developer. This standardization reduces the time required to learn new applications and improves communications between the applications.

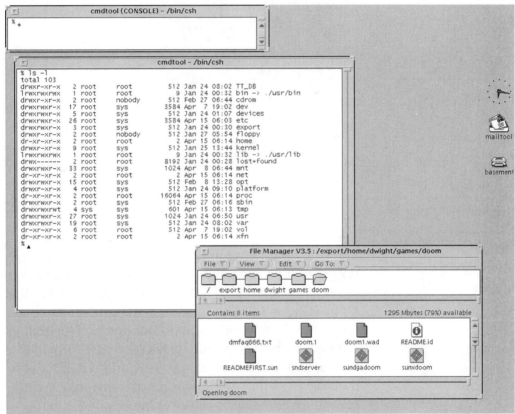

OpenWindows graphical user interface.

DeskSet

OpenWindows DeskSet is a collection of productivity applications including the Mail Tool and Calendar Manager. The Mail Tool is used to compose, send, and receive electronic mail, and the Calendar Manager, to make appointments and schedule resources.

OpenWindows DeskSet tools.

CDE

The Common Desktop Environment (CDE) is a collection of user productivity tools similar to those found in the OpenWindows DeskSet. However, unlike the OpenWindows DeskSet, CDE tools are available on many different vendors' workstations and operating environments. The Solaris version of CDE provides a way to give users a common look and set of tools to their desktops in work environments that may include several different vendors' workstations.

Another feature of CDE is the Motif GUI. Motif differs from OpenLook in that it includes 3D-style window borders and buttons along with graphical controls to adjust environmental parameters such as screen saving, color schemes, and menus. The "dashboard," a thin strip window at the bottom of the display in the following illustration, provides easily modified pop-up menus and common functions such as a clock, calendar, printer control, and e-mail arrival notification.

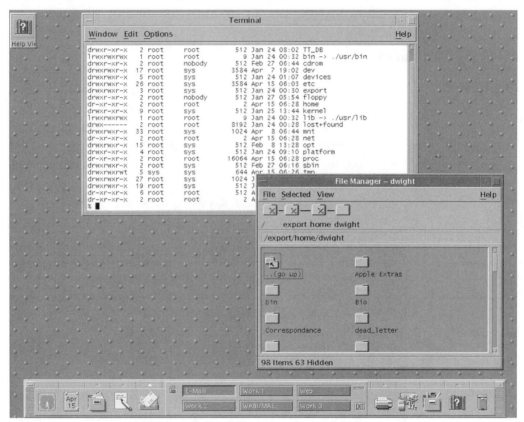

CDE window and selected CDE tools.

Working with Hosts

Host is another word for a computer. The term *host* comes from the notion that the computer hardware hosts the operating system and applications. The hardware is like a house in which the "guests" (programs) live.

In order to distinguish one computer from another, hosts are assigned names, or *hostnames*. To continue the house metaphor, the hostname would be the system's address. In the Solaris operating environment, hostnames are assigned by the system administrator.

Assigning Hostnames

Hostnames usually follow a uniform naming convention. Names do not need to be stuffy or boring. The rocket motor division of an aerospace company, for example, may decide to name its hosts after astronauts, while its aircraft division would use names of famous test pilots. An advantage of the naming convention is to make it easier to keep track of which hosts belong to which organizations. Furthermore, system users like environments with organized hostnames because such systems appear friendlier.

✓ **TIP:** *The* uname *command returns system information about the current host. For example, when the command* uname -n *is entered at the system prompt, the system returns the hostname.*

A typical host naming convention.

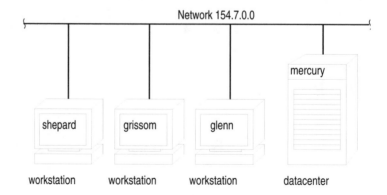

The previous figure illustrates four hosts connected to a network. The host named *mercury* is a large capacity computer that provides network services to other computers connected to the network. These are frequently referred to as *datacenters* or *servers*. The hosts named *shepard, grissom,* and *glenn* are desktop workstations which use the network services provided by *mercury*. Desktop workstations are often referred to as *clients*. The roles of clients and servers in a network are discussed in the section of the same name later in this chapter.

Assigning Network Addresses

The system administrator assigns each host both an Internet and an Ethernet address.

Internet Address

The Internet address, similar to a telephone number, enables hosts to communicate with one another. For example, in the case of a long distance telephone call, the caller dials the area code, exchange number, and line number in order to communicate with a specific telephone location. In the same way, a host's Internet address describes where a host is on the Internet, which in turn allows network traffic to be directed to the host. In the previous naming illustration, the hosts are connected to the Internet network number 154.7.0.0. (See Chapter 17, "Network Configuration and Management," for a detailed discussion of Internet addressing.)

Host Ethernet Address

An Ethernet address functions like a social security number in that it is a unique and permanent hardware address assigned by the hardware manufacturer.

Hosts are identified via such addresses on the Ethernet which in turn enables them to communicate.

↝ **NOTE:** *An Ethernet address uniquely identifies hosts on a network regardless of hostnames, while an Internet address identifies a host's location on the Internet.*

UNIX Operating System Design

Like all UNIX systems, Solaris uses a hierarchical operating system design. Each layer in the hierarchy is responsible for specific system functions and is capable of communicating with adjacent layers.

UNIX hierarchical operating system.

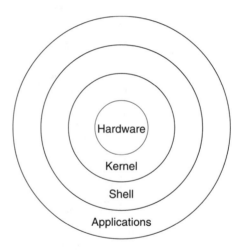

Hardware Core

The core of all computers is hardware. For hardware components such as CPUs, disks, and monitors to work as a team, they require management. The *kernel* is a collection of software components whose primary functions are the control and management of the system hardware. When information is written to a disk drive, the kernel manages the hardware components to make the write occur. Hardware components are directed by the kernel through software components known as *drivers*.

➡ **NOTE:** *The term kernel is used to indicate a part of the operating system found at the core of the UNIX environment.*

Shell Layer

The *shell layer* is the primary user interface with the computer. This layer processes command line instructions issued by a system user and passes them on to the kernel for further processing.

➡ **NOTE:** *Solaris 2 provides the Bourne shell (sh), C-shell (csh), and Korn shell (ksh) versions. The Bourne shell is the default for the root or primary administration user.*

Applications Layer

The *applications layer* holds word processors, spreadsheets, and databases. When the user commands the shell, the kernel starts the application, which in turn calls up the applications layer. This layer interacts with both user and system hardware (via the kernel) to accomplish the desired task.

Modularity

One of the major advantages of the modularity design of the UNIX operating system is portability. In order to port a UNIX operating system to a new hardware platform, only the kernel needs to be changed. The shell and applications layers remain unaffected. This greatly reduces system development time as new hardware technology becomes available. More important, however, is that end users are essentially unaffected by changes in hardware. The applications run on one hardware platform will perform identically on another.

System Architectures

Architecture describes the design and interaction of the various components of a computer system. There are two types of architecture that administrators must be aware of when configuring and maintaining Solaris systems: application and kernel.

Application Architecture

Application architecture is the term used to define the applications and commands that run on a CPU. In the mid-1980s, Sun's dominant architecture was the Sun3, based on the Motorola 68000 series CPU chip set. In the late 1980s, the dominant architecture became Sun4, based on Sun's Scalable Processor ARChitecture (SPARC) chip set.

➙ **NOTE:** *Application binaries built for the Sun3 architecture are not compatible with Sun4 architecture. (See the tables in the next section.)*

Kernel Architecture

Kernel architecture defines the kernel (operating system) version and kernel-specific binaries that run on a

CPU. The following table lists several Sun systems, CPU models, kernel architectures, and applications architectures.

Differences between Sun3 and Sun4 application architecture

Application architecture	CPU architecture
sun3	Motorola 68xxx
sun4	SPARC

Sun systems supporting Solaris 2

System name	CPU model	Kernel architecture	Application architecture
SPARCstation SLC	4/20	sun4c	sun4
SPARCstation ELC	4/25	sun4c	sun4
SPARCstation IPC	4/40	sun4c	sun4
SPARCstation IPX	4/50	sun4c	sun4
SPARCstation 1 +	4/65	sun4c	sun4
SPARCstation 2	4/75	sun4c	sun4
SPARCclassic	4/15	sun4m	sun4
SPARCstation LX	4/30	sun4m	sun4
SPARCstation 10	SS10	sun4m	sun4
Sun - 4/100	4/110	sun4	sun4
Sun - 4/200	4/2xx	sun4	sun4
SPARCserver 300	4/3xx	sun4	sun4
SPARCserver 400	4/4xx	sun4	sun4
SPARCserver 600MP	SS6xxMP	sun4m	sun4
SPARCserver 1000	SS1000	sun4d	sun4
SPARCserver 2000	SS2000	sun4d	sun4
SPARC	engin 1 E	4/E	sun4e

➥ **NOTE:** *Beginning with Solaris 2.6, the Sun 4 architecture and the 4/6XX series are no longer supported. Consequently, these machine types will not be able to run Solaris 2.6 or future versions of Solaris.*

When purchasing applications for a Sun system, check the host application architecture. When configuring the host kernel (O/S), check the host kernel architecture type.

✓ **TIP:** *Use the* uname *command to obtain information on the host's kernel architecture.*

```
% uname -m
sun4c
```

✓ **TIP:** *Similar to the* uname *command, the* showrev *command can be used to display system information.*

```
# showrev
Hostname: mercury
Hostid: 240015cf
Release: 5.5
Kernel architecture: sun4m
Application architecture: sparc
Hardware provider: Sun_Microsystems
Domain: bigcorp
Kernel version: SunOS 5.5 Generic 103093-07 September 1996
```

Roles of Servers and Clients

Servers and clients are the two types of hosts in a distributed computing environment. A *server* is a process or program that provides services to hosts on a network. If a host runs one or more server processes, it can also be referred to as a server. A *client* is both a process that uses services provided by the server and a host which runs a client process.

A *process* refers to the program currently running. It is possible to have multiple instances of a single program, each separately serving different clients. The mail transport agent, *sendmail,* is a good example.

A wide variety of server and client processes can be operating in a network environment. For example, *file servers* share disk storage with hosts, *application servers,* applications, *boot servers,* boot services, and *print servers* attach printers and provide network printing services.

Typical network server/ client configuration.

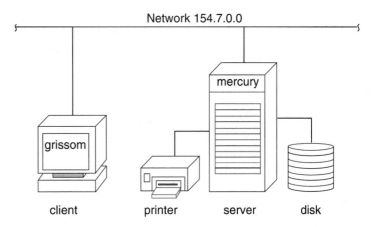

Network 154.7.0.0

mercury

grissom

client   printer   server   disk

The above illustration demonstrates that the host *grissom* (a network client) is able to use resources (i.e., printer and disk) attached to the host *mercury* (a network server). The *mercury* host is *sharing* its resources with *grissom*. A more detailed discussion of resource sharing appears in the chapters on the network file system (NFS), network name services (NIS+, DNS), and managing printers.

Host Configurations

There are two standard host configurations in the Solaris environment: standalone and diskless. These terms seem straightforward, but can be confusing. For example, a diskless client can have a local disk. The important point to remember is that the host's configuration refers to how it boots.

In order for a host to boot to the multi-user state, it must be able to locate three areas on the disk: *root* (/), *swap,* and *usr* (pronounced "user"). The root contains the boot code, system files, and directories necessary for host operations. The usr area holds system software, executable commands, and system libraries. Finally, swap is used for virtual memory which extends the host's total available memory. This area allows currently unused portions of a program (or its data) to be moved out of memory and on to disk, thereby freeing up space to run more programs. All programs and data in the memory are divided into segments known as *pages.* The virtual memory component of the operating system kernel moves unused pages out of memory and onto swap, and then moves the pages out and back into memory when needed again.

Standalone

Standalone hosts boot up independently of network services. When booted, they obtain root, swap, and usr areas from a local disk. The term "standalone" refers to a host *not* connected to a network. "Networked standalone" refers to a standalone host *connected* to a network. One advantage of networked standalone hosts is that in the event of a network shutdown, they can continue to operate (although use of network services may not be possible).

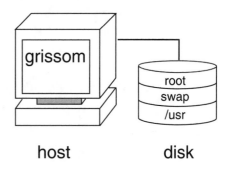

A standalone host configuration.

host   disk

Diskless Client

Diskless clients are wholly dependent on network services to boot and operate. A diskless client's root, swap, and usr are supplied by a client server on the network. Advantages of the diskless client configuration include reduced equipment cost (since no local disk is required) and centralized services. The major disadvantage of using diskless clients is network load. When a diskless client boots, pages (writes) to the swap area, or executes a library routine, all of the I/O must go through the network.

As noted previously, every host must have both a unique root and swap area in order to function properly. A host configured as a boot server has two file systems on its disk which store root and swap for diskless clients (see previous illustration).

The */export* file system on the server contains a directory known as *root*. The */export/root* directory contains a subdirectory for each diskless client supported by the server. For example, */export/root/glenn* located on *mercury's* disk holds the root for the diskless client *glenn*.

The diskless client swap is used in a fashion similar to the root area. Special "swap files," one per supported diskless client, are located under the */export/swap* file system. For example, the */export/swap/glenn* file is the swap area for the diskless client named *glenn*.

✓ **TIP:** *A quick way to identify diskless clients supported by a host is to change directory to* /export/swap *(assuming it exists) and list the files.*

```
cd /export/swap
ls - l
-rw-----T 1 root 285036 Jan 6 1992 glenn
```

Diskless client and boot server configuration.

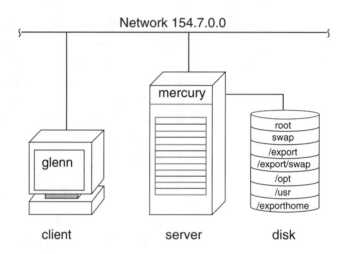

Network 154.7.0.0

client   server   disk

Software Terminology

Solaris 2 bundled and unbundled software are distributed as software packages. These packages are comprised of files which perform diverse functions, such as on-line manual pages or OpenWindows demonstrations. There are more than 80 software packages in Solaris 2. The format of software package names is *SUNWxxx*. For example, the package name for the on-line manual pages is *SUNWman* and the package name for the OpenWindows demos is *SUNWowdem*.

The software packages in Solaris 2 conform to a software standard known as the applications binary interface (ABI). ABI compliant software is easily installed or removed with standard systems administration utilities. These utilities are discussed in Chapter 12, "Managing System Software."

Related software packages are grouped into software clusters. Cluster names use the SUNWC prefix. For example, the OpenWindows Version 3 cluster contains 13 software packages related to OpenWindows version 3.

Example of the relationship between a software configuration, software cluster, and software packages.

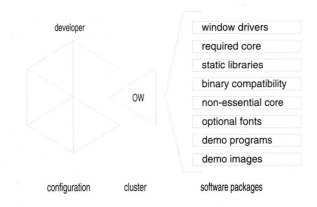

Working with Software Configurations

SunInstall further groups software packages and clusters into categories known as software configurations. The four configurations in Solaris 2 are *core, end user, developer,* and *entire.* Each configuration supports a different level of system sophistication.

❧ **NOTE:** *The configuration space requirements listed in the following sections are approximations based on an early version of Solaris 2.6. The actual size of a configuration depends on system* *architecture, release level of the operating system, and any user customizations performed at the time of installation.*

Core

The core contains the minimum required software to boot and operate a standalone host. It does not include OpenWindows software or the on-line manual pages. It does, however, have sufficient networking software and OpenWindows drivers to run OpenWindows from a server sharing OpenWindows software. A server would not be built from the core configuration, but a dataless client might be built based on the core configuration. Core requires approximately 112 Mb of disk space.

End User

End user contains core configuration software and additional software typically used by end users, including OpenWindows version 3 and the end user version of AnswerBook. It does not contain the on-line manual pages. End user requires approximately 312 Mb of disk space.

Relationship of the four software configurations.

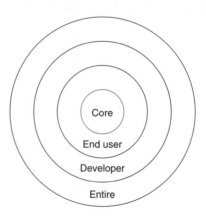

Developer

Developer contains both the core and end user configuration software, and additional software typically used by systems and software developers. It also includes on-line manual pages and the full implementation of OpenWindows and compiler tools. It does not contain compilers or debugging tools. In Solaris 1 and earlier releases of Sun operating systems, compilers were bundled with the release media. Compilers and debuggers are unbundled in Solaris 2 and purchased and installed separately. Developer requires approximately 561 Mb of disk space.

Entire

Entire contains the entire Solaris 2.x release and requires approximately 636 Mb of disk space. The OEM includes additional software provided by the Sun clones manufacturers. Its size varies depending on the specific additional software offered by the clone vendor.

→ **NOTE:** *It is important to note that software configurations can be modified. For example, the on-line manual pages cluster could be added to the end user configuration, or removed from the developer configuration. These modifications can be made when the system is being constructed or afterward.*

Disk space required for different software configurations in Solaris 2.6

Software configuration	Size
Core	112 Mb
End user	312 Mb
Developer	561 Mb
Entire	636 Mb

Obtaining System Access

On a Solaris system, every would-be user must gain access to it through a *log-in* procedure. The log-in process is a security mechanism which helps ensure that only authorized users gain access to system resources and information.

User Account

Before a user can log in, the system administrator must create a user account. The account contains a variety of information about the user, including areas of the system the user can access. In order to log in to a system, users need a name and a password. The creation of user accounts is addressed in Chapter 6, "Managing User Accounts."

The log-in name is a lower-case alphanumeric string up to eight characters long that identifies the user. Sites often employ users' last names. For example, Kathy Smith's log-in could be *smith*. Should more than one person share the same last name, then the last name and first initial serves the purpose. For example, Mark Smith would be assigned *smithm*.

The simpler the log-in convention, the easier it is for system administrators to track and manage users. Simple conventions also facilitate user communication through utilities like electronic mail.

⇢ **NOTE:** *Experience has shown that a simple log-in naming convention will reduce user account management time as well as security problems. This is especially true in large or changing system environments.*

Assigning Passwords

System users must also have a password. The log-in name identifies a user to the system; the password ensures the system that the user is who she claims to be. By default, passwords must be at least six characters in length. In addition, passwords should never use the proper spelling of words, nor have anything to do with the user that could be easily guessed. Choosing secure passwords is discussed in Chapter 6, "Managing User Accounts."

Special Systems Account

A special systems account frequently used in system management is known as *root*. The *root* user account must be carefully guarded because it has unlimited access privileges. In the wrong hands, *root* access allows a user to cause extensive system damage and data corruption.

Logging In

To log in to the system (in this case as *root*), type the log-in name at the log-in prompt and press <Return>. (See the next illustration.) The system then prompts for a password. Type in the appropriate password and press <Return>. Note that the password is not echoed to the terminal screen. This feature prevents a bystander from seeing the password. If the information provided to the system is correct, a command prompt appears on the screen. The command prompt is the shell interpreter that passes commands for execution to the kernel. To log off the system, enter the **exit** command at the shell prompt and press <Return>.

*Example of log
in and log off.*

```
mercury login: root
password:
#
# exit
mercury login:
```

☞ **WARNING:** *Never leave a root log-in unattended!
A clever system cracker could take advantage of
the unattended root log-in to alter critical system
files, thus allowing the cracker undetected root
access. The integrity of an entire corporate computer system can be compromised due to lax root
security.*

Summary

This chapter covered the fundamental concepts and
terminology used in the Solaris operating environment. Discussion topics included the role of servers
and clients in a distributed computing environment;
server, standalone, and diskless host configurations;
and software management under Solaris.

Preparing for a Solaris Installation

One of your principal tasks as a systems administrator is planning for company software needs under Solaris 2 and identifying a hardware configuration to support the software. This chapter focuses on basic pre-installation considerations to keep in mind while reviewing software applications. Requirements may differ depending on whether you are working with a new system installation or an upgrade from the Solaris 1.X (SunOS) to the Solaris 2 operating system. The results of this evaluation are then applied to your application and operating system demands to determine what they mean in terms of hardware configurations.

Evaluating Applications

If you are planning to upgrade a SunOS system to a Solaris 2 system, you should be aware of several important differences between these operating systems. Some of the differences involve changing the names of commands or function calls which perform certain operations. Other changes involve portability issues such as how the compiler links or includes the library calls into the binaries. Because of these differences, you will need to examine the software currently in use on your system to identify mission critical applications from non-mission critical applications.

While many applications that you have been using under SunOS are available under Solaris 2, others may not be available to run under the newer OS. This is the time to evaluate the packages to determine if you need to replace certain existing functionality, and if so, how replacement should be accomplished. In some cases, it may be possible to use existing SunOS binaries under Solaris 2. You will need to understand some of the internals of the applications before they can be used under Solaris 2.

When you are planning a new installation, some of the same compatibility and portability concerns may come into play. Even in this case, it is best for system administrators to understand the differences between the two operating systems so that the differences can be explained to users who have experience with SunOS or other Berkeley-derived UNIX operating systems.

Program Linking

Dynamically and *statically* linked binaries are found in SunOS applications. Linking is the process performed by the link editor (*/usr/ccs/bin/ld*) to resolve references to variables within program binaries. In early versions of SunOS, all binaries were statically linked. Under Solaris 2, the preferred method of linking binaries is dynamic. What are the differences between the two methods of linking a program binary?

A statically linked program links (or includes) all library functions at compile time. If changes are made to a library routine, the program must be recompiled to take advantage of those changes. A statically linked program will usually be quite large because it includes all library functions.

A dynamically linked program is one that loads the library functions at run time. This allows the program to include the latest version of all library functions upon invocation. No recompilation is required to enjoy the advantages of the updated library routines. Dynamically linked programs are usually smaller than their statically linked counterparts, because the library functions are not included in the on-disk binary image.

Native Mode Applications

In Solaris, applications may be run in native mode or binary compatibility mode. There are benefits and costs to both operating modes. For instance, because Solaris is based on industry standards, native mode applications enjoy application portability across all System V Release 4 platforms. Application code developed on one vendor's System V Release 4 system should compile and operate correctly on another vendor's System V Release 4 machine. Applications which contain native mode binaries will operate on Solaris systems, but will not operate on SunOS 4.x systems.

To be considered a native mode application, a program must be compiled under the Solaris environment. This typically means that the binaries are dynamically linked programs.

➥ **NOTE:** *As noted in Chapter 2, compilers are not included in the core Solaris operating system. The compilers must be purchased and installed separately. Keep this in mind for subsequent cataloging of applications software.*

✓ **TIP:** *Developers can save significant time porting applications to Solaris if the applications are developed under SunOS 4.1 using the System V environment, OpenWindows, and dynamic linking.*

Binary Compatibility Mode Applications

What happens if you do not have source code for the application, or if a critical application is not available for Solaris? Would such a situation prevent the corporation from converting to Solaris? Fortunately, in many cases, the answer is no.

Solaris provides a binary compatibility mode package (BCP) for existing SunOS 4.x programs, including SunView applications. It might be possible to run your mission-critical application in binary compatibility mode until a Solaris version is available.

➥**NOTE:** *The binary compatibility package is intended as a short-term transition aid, and should not be used as a long-term solution.*

In order to operate in compatibility mode, the application must strictly adhere to some simple guidelines. The following table summarizes the restrictions on running an application in compatibility mode.

Binary Compatibility Mode Restrictions

Feature	Do not use	Must use
Dynamic linking		X
Traps to operating system	X	
Application-specific IOCTLs	X	
Read/write directly to system files	X	
Access /dev/kmem or libkvm	X	
Use of undocumented SunOS interfaces/ features	X	

Applications run in compatibility mode must dynamically link the C library and other system library routines. Statically linked applications will not work in binary compatibility mode.

Another condition for proper operation in binary compatibility mode includes the absence of traps directly to the operating system, application-specific IOCTLs or drivers, and direct read or write access to the system files. Further conditions require that the application not attempt to access and interpret kernel data structures through the *dev/kmem* or *libkvm* facilities, nor should applications attempt to use any undocumented SunOS interface.

In the case of locally developed software or applications which include source code, it is relatively easy to determine if the code follows the rules of the table. In order to determine if third party or binary-only applications will work in binary compatibility mode, it may be necessary to contact the original software supplier.

Cataloging Current Applications

Planning for application availability under Solaris involves cataloging existing and planned applications, and determining the availability date of these applications under Solaris. The cataloging process requires that you examine your current systems to identify the software in use. Once you develop a catalog of the required applications, you need to determine when these applications will be available for use under Solaris, and how this will affect your corporate computing plans.

In order to facilitate the cataloging process you should develop a form for the tabulation of local application information. This form should allow you to collect as much information as possible about the applications. Some of the more important information about applications includes package size in megabytes (Mb), the number of users who will require this application, and whether the application runs in native or binary compatibility mode. A sample form appears on the following page.

Software Inventory Form

System Model: ————————————————

System Configuration: Server ☐   Standalone ☐   Diskless ☐

System Hostname: ————————————————

System IP Address: ————————————————

Subnet Mask: ————————————————

Domain: ————————————————

Name Service: NIS ☐   NIS+ ☐   NONE (includes DNS) ☐

Name Server Hostname: ————————————————

Name Server IP Address: ————————————————

Time Zone: ————

Disk Layout

Device	Partition Size	Mount Point	Device	Partition Size	Mount Point
Disk 1	a		Disk 2	a	
	b			b	
	c			c	
	d			d	
	e			e	
	f			f	
	g			g	
	h			h	

Locally Developed Software

Application	Language	Native/BCP	When Available	#Users	Size

Third Party Software

Application	Language	Native/BCP	When Available	#Users	Size

SunSoft Products

Application	Language	Native/BCP	When Available	#Users	Size

Sample software inventory form.

Evaluating Hardware Requirements

Once your software requirements are identified, you need to determine hardware dependencies. For instance, do you have enough disk space to support the operating system, users' files, and the necessary application software?

You also need to consider special hardware requirements while the Solaris installation plan is developed. Do you rely on third party or other special hardware to operate your enterprise? If so, is a special device driver required for the operation of this hardware? Which systems have this hardware? Do you know if software is available to allow the operation of this hardware under Solaris 2?

Cataloging Hardware Requirements

When you consider hardware requirements, you need to pay close attention to any non-standard hardware on your system. As with the software, all hardware found on the system is cataloged. Again, you could develop a form to facilitate the cataloging process. A sample form for cataloging hardware appears on the following page.

Hardware Inventory Form

System Name: _____   IP Address: _____   System Memory: _____ MB

Disk 1	Controller:		Unit:		**Disk 2**	Controller:		Unit:	
Partition	Size MB	Mount Point	Notes		Partition	Size MB	Mount Point	Notes	
a					a				
b					b				
c			Entire Drive		c			Entire Drive	
d					d				
e					e				
f					f				
g					g				
h					h				

Third party or locally developed hardware:

Product	Supplier	System(s)	Driver Req'd	Date Available
_____	_____	_____	_____	_____
_____	_____	_____	_____	_____

Network Sketch

Primary Network IP Address: _____

Server   Client   Client

Subnetwork IP Address: _____

Sample hardware inventory form.

Minimal Hardware Requirements

Once you have identified the company's specific software and hardware requirements, you need to determine if your system(s) meet minimum Solaris requirements. To install and run Solaris 2, systems must have a SPARC, Power-PC (PPC), or Intel 486/Pentium processor with a minimum 150 Mb of disk space. The following are examples of the systems that are capable of running Solaris 2.

Solaris architecture information

Kernel architecture	Application architecture	System model	Model name
Sun4c	Sun4	4/20 and 4/25	SLC and ELC
		4/40 and 4/50	IPC and IPX
		4/60 and 4/65	SparcStation 1 and 1+
		4/75	SparcStation 2
Sun4d	Sun4	SS1000, SS2000	SparcCenter 1000 and 2000
Sun4m	Sun4	EC3	SparcEngine EC3
		4/15	SparcClassic
		4/30	SparcStation LX
		SS5	SparcStation 5
		SS10	SparcStation 10
		SS20	SparcStation 20
		S240	SparcStation Voyager
Sun4u	Sun4	UltraSparc 1	
		UltraSparc 2	
		Enterprise 3000, 4000, 5000, 6000	
i86pc	i86pc	486, 586, 686	
		Pentium P5/ P6	
ppc	ppc	Power-PC 603	
		Power-PC 604	

NOTE: *The* ppc *architecture is supported under Solaris 2.5 and 2.5.1 only.*

✓ **TIP:** *To determine if a particular x86 or PPC based system is capable of running the Solaris operating system, consult the Sun Web page at* http://access1.sun.com. *This site maintains hardware compatibility lists specifying which systems have been certified to be compatible with Solaris 2.*

The Solaris operating system installation software defines four typical installation clusters: Core System Support, End User System Support, Developer System Support, and Entire Distribution. The Core cluster provides the minimum software required to boot and run Solaris 2. Each of the other clusters provides more utilities and functionality than the previous cluster in the list. The space requirements for each cluster are summarized below.

Disk space requirements

Software group	Installs	Space required
Core System Support	Minimum software required to boot and run Solaris.	70 Mb
End User System Support	Core group plus recommended end user applications, including OpenWindows and DeskSet.	180 Mb
Developer System Support	End user software and libraries, including files, manual pages, and programming tools. (Compilers and debuggers are not included.)	270 Mb
Entire Distribution	Entire Solaris release. (Compilers and debuggers are not included.)	360 Mb

NOTE: *Some versions of Solaris distribution media include a fifth cluster named "Entire Distribution* *plus OEM support." This cluster should be used if installing Solaris on a SPARC clone system, and requires slightly more space than the Entire Distribution cluster. Consult the manuals distributed with the installation media for more information on this cluster.*

✓ **TIP:** *It is possible to customize any of the predefined clusters at installation time. Customization allows the user to add or delete any optional packages and tailor the installation to individual requirements.*

The Solaris 2 binaries are extracted from the distribution media and placed in predefined file systems on the disk drives by the installation software. These file systems, or disk partitions, must meet certain minimum size requirements. File system space requirements are summarized below.

Typical partition sizes under Solaris

File system	Required on	Minimum size	Typical size
/ (root)	All systems.	32 Mb	64 Mb
swap	Optional on systems with minimum of 32 Mb memory. Recommended for reasonable performance.	0 Mb	2x size of memory
/usr	All systems.	190 Mb	250 Mb
/usr/openwin	Optional. Servers typically mount OpenWindows as a separate partition so that a single copy can serve many clients.	0 Mb	210 Mb
/opt	Optional (licensed) software.	0 Mb	Site dependent.

File system	Required on	Minimum size	Typical size
/export/exec	Required only on servers which support diskless clients.	15 Mb	15 Mb per supported client architecture.
/export/root	Required only on servers which support diskless clients.	10 Mb	10 Mb base plus 20 Mb per diskless client.
/export/swap	Required only on servers which support diskless clients.	0 Mb	24 Mb per diskless client.
/var	Optional. Often mounted as a separate partition to facilitate network log file archival, mail, and printer spooling.	0 Mb	32 Mb typical for small site, and several hundred Mb for larger site. Site dependent.

↩ **NOTE:** *When installing a server system, providing a separate partition for the* /var *directory is often desirable. This directory stores log files and mailboxes and provides temporary storage for print jobs, as well as space for the NIS+ database and user accounting. At a typical large site, this partition may be 128 Mb (or more), and at smaller sites, 32 Mb.*

✓ **TIP 1:** *In order to reduce disk partitioning, the export file systems are sometimes combined into a single file system.*

✓ **TIP 2:** *In the case of a heterogeneous environment, it is usually desirable to have separate* /opt *and* /usr/openwin *file systems for each client architecture. This practice allows each client architecture to mount and access only the binaries that will execute on a particular client architecture.*

Other Important Hardware Considerations

Another requirement which must be met before running Solaris 2 is the availability of hardware used as the software load media when performing the initial Solaris 2 system installation. Unlike SunOS predecessors which were available on several distribution media, Solaris 2 is distributed only on CD-ROM media.

∞ **NOTE 1:** *To install Solaris on x86 systems, a floppy disk drive is also required.*

∞ **NOTE 2:** *Solaris 2 systems must contain (at least) 16 Mb of main memory.*

Installation Type

Each of the following basic types of Solaris systems is designed for a specific function: standalone, diskless, and heterogeneous and homogeneous servers. Each type is discussed below.

∞ **NOTE:** *Solaris 2.5.1 was the last release which supported the dataless client architecture.*

Standalone

A standalone system typically requires a minimum 500 Mb of disk space. Standalone systems run entirely from a local disk, and they do not rely upon other network service providers. Some standalone systems are also servers.

Diskless

These systems typically require (at least) 64 Mb of disk space on the server. Diskless systems are totally dependent on network services. The root, swap, and other file systems are provided by one of the servers on the network.

Heterogeneous and Homogeneous Servers

Servers usually have disk requirements in excess of 1024 Mb. Servers are a special class of standalone system. The servers boot from local disks and provide other network hosts with the file systems that they require. Because servers provide boot and applications services to clients, they require more available software on the local disks than non-server standalone configurations.

A homogeneous server is a system which serves only a single client kernel architecture. For instance, a Sun4m server with all Sun4m clients is considered a homogeneous installation.

In contrast, a heterogeneous server is a system which serves at least one client with a different kernel architecture than that of the server. An installation which includes a Sun4m server with Sun4c, PPC, and x86 clients would be considered a heterogeneous installation.

Example of Installation Evaluation

Now that the terms used in the Solaris installation process are understood, the administrator needs to determine the specific requirements of each system that will be upgraded. For illustration purposes, consider an Enterprise 4000 server (hostname: *mercury*), with one diskless Sun4c client (hostname: *glenn*), and one x86 client (hostname: *grissom*). The *mercury* system contains two 2.1 gigabyte disk drives and 64 Mb of main memory.

The diskless client will receive all services from the server's disks. The x86 client will receive user file space from the server, but will have its own copy of all essential system binaries on its internal disks. Assume that this site will be moving more clients to the *mercury* server as the management and users become more familiar with the new operating system.

For illustration purposes, *mercury* will also be a master Network Information System (NIS/NIS+) name server.

↔ **NOTE:** *NIS+ is the replacement for the original NIS name server distributed under SunOS operating systems. The NIS+ package manages networkwide information such as passwords, file system availability, and names of hosts on the network. NIS+ is discussed in greater detail in Chapter 21, "Network Name Services."*

Configuration of the mercury system.

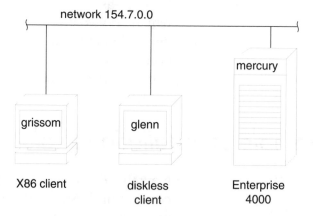

In the example the *mercury* server will be installed as an Entire Distribution cluster. For the purpose of simplicity, there is no locally developed or third party hardware. Of the two applications to be installed on the system, one requires 50 Mb of disk space and runs in binary compatibility mode, and the other requires

75 Mb of disk space and runs in native mode. The inventory forms for the *mercury* system might resemble the following examples.

Software Inventory Form

System Model: ENTERPRISE 4000

System Configuration: Server ☐   Standalone ☐   Diskless ☐

System Hostname: Mercury

System IP Address: 154.7.0.1

Subnet Mask: 255.255.0.0

Domain: astro.com

Name Service: NIS ☐   NIS+ ☐   NONE (includes DNS) ☐

Name Server Hostname: Mercury

Name Server IP Address: 154.7.0.1

Time Zone: US/East Indiana

Disk Layout

Device	Partition Size	Mount Point	Device	Partition Size	Mount Point
Disk 1	a 64 Mb	/	Disk 2	a	
	b 128 Mb	swap		b	
	c			c 2048 Mb	/home
	d 256 Mb	/var		d	
	e 128 Mb	/usr/openwin		e	
	f 256 Mb	/export		f	
	g 256 Mb	/usr		g	
	h 768 Mb	/opt		h	

Locally Developed Software

Application	Language	Native/BCP	When Available	#Users	Size
analyze-cog	C++	Native	now	7	75 mb

Third Party Software

Application	Language	Native/BCP	When Available	#Users	Size
design_a_cog	Binary only	BCP	Now	14	50 mb

SunSoft Products

Application	Language	Native/BCP	When Available	#Users	Size

Software inventory form for mercury system.

Hardware Inventory Form

System Name: _Mercury_   IP Address: _154.7.0.1_   System Memory: _64_ MB

Disk 1	Controller: 0	Unit: 0		Disk 2	Controller:	Unit: 1	
Partition	Size MB	Mount Point	Notes	Partition	Size MB	Mount Point	Notes
a	64	/		a			
b	256	swap		b			
c			Entire Drive	c	2048	/home	Entire Drive
d	256	/var		d			
e	128	/usr/open win		e			
f	256	/export		f			
g	256	/usr		g			
h	768	/opt		h			

Third party or locally developed hardware:

Product	Supplier	System(s)	Driver Req'd	Date Available
N/A				
N/A				

Network Sketch

Primary Network IP Address: _154.7.0_

```
        154.7.0.1         154.7.0.2          154.7.0.3
      ┌─────────┐       ┌─────────┐        ┌─────────┐
      │ Server  │       │ Client  │        │ Client  │
      │ Mercury │       │ Glenn   │        │ Grissom │
      └─────────┘       └─────────┘        └─────────┘
```

Subnetwork IP Address: _N/A_

Hardware inventory form for mercury system.

✓ **TIP:** *Because the* mercury *system is to be the server for client systems, the client inventory information should be included with the* mercury *information to provide a complete picture of installation requirements.*

Determining Disk Partitions

Now that the type of installation and the disk space requirements have been determined, you must decide how to partition (or segment) the disks on the *mercury* system to provide for these needs. This procedure requires a rudimentary understanding of disk terminology.

↪ **NOTE:** *Disk geometry is discussed in more detail in Chapter 9, "Disk System Hardware."*

At a very simple level, the next figure shows a disk drive divided into many data sectors. Each data sector is capable of storing 512 bytes of data. Disk drives are divided into several (up to eight) UNIX file systems or partitions. Each of the UNIX file systems is comprised of numerous data sectors bound together by the **newfs** command

A typical disk layout.

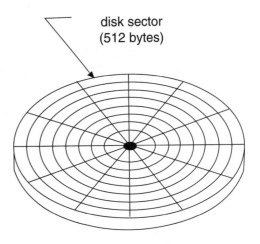

disk sector
(512 bytes)

From the above information it appears that a 100 Mb file system requires a disk partition with at least 200,000 data sectors, or 200,000 x 512 = 102,400,000 bytes. In reality, the file system infrastructure will require extra sectors in order to provide a file system that will hold 100 Mb of data.

A typical rule of thumb when partitioning disks is to allow for ten percent of the sectors in a file system to be consumed by overhead. It is also a good idea to allow room for future growth for each file system.

The *mercury* sample installation requires eight file systems: */, /usr, swap, /opt, /export, /var, /usr/open-win,* and */export/home/mercury.* By examining the cluster size information, the amount of memory in the system, the size of the local packages, and the number of clients supported by the *mercury* server it is possible to develop the following reasonable disk layout for the server.

❒   64 Mb for /

❒   256 Mb for swap

❒   256 Mb for /usr

❒   256 Mb for /var

❒   128 Mb for /usr/openwin

❒   768 Mb for /opt (applications)

❒   256 Mb for /export (root, swap, and exec com-bined into a single file system)

❒   2048 Mb for /export/home/mercury (users' files)

↝ **NOTE:** *The space allocation presented above is very generous. It is possible (and in many cases desirable) to create smaller file systems. Because the* mercury *system may support more users and clients in the near future, the above layout was created to allow for extraordinary expansion. It is easier to make such decisions before installing Solaris than it is to repartition disks after the system has been in use.*

For the example installation one disk drive will be set aside for the user files (*/export/home/mercury*). This leaves one 2.1 GB disk to contain the other seven file systems. Disk partition information and file system mount points appear in the following table.

Example file system sizes

File system mount point	Minimim size (Mb)	mercury system size (Mb)	Anticipated growth
/	32	64	100%
swap	32	256	400%
/usr	190	256	25%
/usr/openwin	210	256	18%
/opt	0	768	
/export	72	256	200%
/var	0	256	
/export/home	0	2048	

At this point pre-planning is complete for the *mercury* installation. If *mercury* is a new installation, the next task would be to load Solaris onto the system disks as per the instructions shipped with the distribution media. If *mercury* were an upgrade installation, a few other tasks should be considered before beginning the Solaris 2 installation.

Saving Critical Data

Before SunOS is shut down for the last time, there are several files you may wish to save. For instance, because you may not have to change the partitioning to load Solaris 2, it would help to know how the disks on the system are currently partitioned. Saving the password file, NIS name server maps, and other critical information before you begin the SunOS shutdown procedure is recommended.

Saving the SunOS Disk Configuration

There are two ways to determine the current disk configuration information on a system. The **dmesg** command will display output from a recent reboot, and the **format** command will display all currently available disks. Because the dmesg information will not be available if the system has been up and running for several days, it is sometimes more convenient to use the format command.

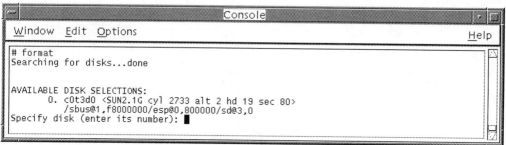

```
# format
Searching for disks...done

AVAILABLE DISK SELECTIONS:
       0. c0t3d0 <SUN2.1G cyl 2733 alt 2 hd 19 sec 80>
          /sbus@1,f8000000/esp@0,800000/sd@3,0
Specify disk (enter its number): █
```

Using format command to identify available disks under SunOS.

Once the device names of all disk drives on the system are known, obtain a hard copy of the current partitioning information for the disks by using the **dkinfo** command as follows:

```
# /etc/dkinfo sd0 sd1 | lpr
```

Another bit of information that may prove useful is a copy of the current **fstab** file. This will help to prevent damage to the users' file systems during the conversion process. Obtain a hard copy of the current file system information by keying in the following command:

```
# lpr /etc/fstab
```

Saving Critical SunOS System Information

It is always wise to have a copy of all critical system information prior to installing a new version of an operating system. A few of the more critical files to be saved are listed below.

- ☐ ./etc
- ☐ ./.cshrc
- ☐ ./.login
- ☐ ./.logout
- ☐ ./.profile
- ☐ ./.rhosts
- ☐ ./var/spool/mail
- ☐ ./var/spool/calendar
- ☐ ./var/spool/cron
- ☐ ./var/spool/uucp
- ☐ ./var/nis

✓ **TIP:** *It is best to perform a full file system backup before installing a new version of the operating system. The backup provides an extra level of security for the critical system information in case something should go wrong during the installation process.*

Creating a file containing the above list of information can be automated with the following shell script:

```
# cd /
# tar cvf tape_drive `cat filename`
```

In the above example, *tape_drive* refers to the device name of the drive to write the files on (e.g., */dev/rst0*). Next, *filename* is the name of the file containing the list of files and directories to be archived.

❖ **NOTE:** *The list of critical files to save is only a suggested list. The actual list is highly dependent on your installation.*

Summary

This chapter explored the prerequisites for installing Solaris 2. Determination of the type of installation, the system type(s) to be supported, and the planned use of the system(s) are factored into Solaris installation planning. An example installation was presented to illustrate typical record keeping and planning required before installing Solaris.

Fundamental System Administration Tools

This chapter addresses the daily tasks involved in Solaris system maintenance. There are two levels to planning day-to-day operations. At one level, the systems administrator will need to work with the corporation to establish procedures for setting aside time and other resources to perform system backups or install new hardware or software. At another level, the administrator must develop methods for performing these tasks, learn about the basic tools required to accomplish them, and even create simple tools to assist in daily chores.

To maintain Solaris, the systems administrator must spend time studying, practicing, and planning basic tasks. Included are the installation of new systems and peripherals, preserving a set of system backups, installing user accounts, allocating disk space, monitoring system performance, and updating system and application software. Each of these tasks is covered elsewhere in this book in more detail. Discussion in this chapter is focused on selected fundamental tools and techniques applicable to maintaining any Solaris system.

Study includes reading books such as this one, familiarizing oneself with system documentation, keeping notes on what works and does not work, and recording how various tasks were accomplished. If the Solaris AnswerBook software is available, take a few moments now to read instructions on how to use it.

✓ **TIP:** *Get a laboratory style notebook and jot down the procedures to use when performing tasks such as installing new hardware or software. Your objective is to build a library of procedures specific to the situation.*

Study also means understanding a few basic UNIX tools and techniques which all systems administrators should know. This is like knowing how to change a flat tire. The knowledge may not be used daily, but it is handy when problems arise.

In the context of Solaris, practice refers to the discipline of carrying out a daily, weekly, or monthly methodology which helps to maintain the functioning and performance of the system and prepares for unexpected problems. A methodology might include a schedule of system backups, routine audits of user accounts and system security, checks to reveal capacity problems such as full disks or overloaded networks, and trimming system log files. Practice also refers to the way basic tasks are performed. A cautious, safe work procedure establishes techniques for easy recovery from problems or missteps. A few rules of thumb for performing system maintenance and administration tasks are listed below.

❒ Avoid using *root* as much as possible. Use a less privileged account where mistakes will be less drastic.

❒ Avoid using wildcard characters such as the asterisk (*) when using *root*. Accidents do happen, and nearly everyone has at least once inadvertently entered a command line such as **rm -r /usr/***.

❒ Create a backup copy of all edited files. Make it a habit to use something like the following command: **cp file file.orig ; vi file**. Using such a command makes starting over easy in the event of an editing mistake.

❒ Allow plenty of time for the first experience of installing a software package or any other new task. System administration tasks can be time-consuming; schedule more down-time than necessary. Users will appreciate it if the administrator has the system back on line ahead of schedule.

A systems administrator needs to know basic UNIX tools as well as specialized tools. The tools discussed in this chapter help supervisors understand the basics, which serve as the foundation for understanding the specialized tools discussed in subsequent chapters. Commands that new systems administrators should know are highlighted, followed by in-depth discussion of commands unfamiliar to many recent UNIX users and administrators. If a command or its usage is unfamiliar to you, take the time to study appropriate pages of the on-line Solaris manual.

Moving Around the UNIX File Tree

The basis of file storage on Solaris and other UNIX systems is a hierarchical directory structure which often looks like a branching, downward growing graph known as a "file tree." The top of the tree is the *root* (/). Files and directories can be referenced by listing names of the directories traversed from the root to the file to the desired directory. This is known as the full path name of a file or directory, such as */etc/inet/hosts* or */usr/local/bin.*

UNIX file tree with selected branches.

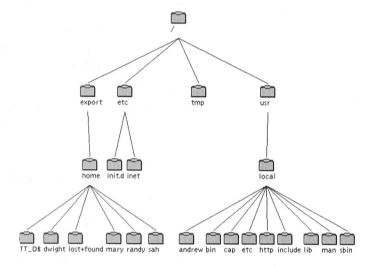

Commands and Graphical Tools

Several commands and graphical tools view, navigate, and manipulate files and directories in the UNIX file tree.

File Manager

The Common Desktop Environment (CDE) provides a graphical tool, known as the File Manager, for browsing the UNIX file tree. The File Manager is similar to the point and click file management tools found in Microsoft Windows and the MacOS finder. Files and directories can be moved by dragging them with the mouse. At the top of the window is a graphical representation of the working directory's full path. This is the current location the file manager is viewing in the UNIX file tree. The lower portion of the window shows the contents of the working directory.

CDE file manager window.

pwd, cd and ls

The UNIX command shells also provide information about a working directory or current location in the file tree. The **pwd** command will print the working directory, and will also display a "relative path name." This term describes the path from the working directory to the file. If, for example, **pwd** returned */etc* as the working directory, the *hosts* file could be referred to by the relative path *inet/hosts* or *./inet/hosts* where the period (.) signifies the name of the working directory. It could also be referred to by the absolute or full path, */etc/inet/hosts*.

All UNIX users should also know the **ls** and **cd** commands. In several formats, ls lists the contents of the current directory or of a directory given as an argument. Although this command has numerous options, the **-l** option (to list the file ownership and protection mode bits) is one of the most common. The **cd** command allows the user to change directories. Examples of using the cd and ls commands appear in the next illustration.

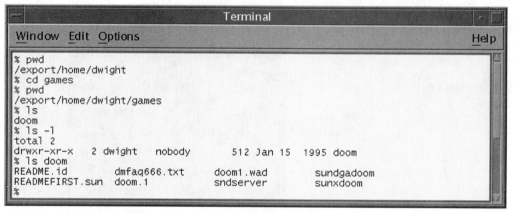

Changing the working directory with cd and viewing directory contents with ls.

✓ **TIP:** *Every directory includes two special entries. First is the single period (.), which refers to the directory itself. Next is the double period entry (..) which refers to the directory one level up, or closer to the root, from the working directory. This feature simplifies navigation within a file system. For example, to move up one directory, type* cd .. *instead of the full path to the directory above the working directory.*

df & du

Two commands which may be unfamiliar to new UNIX users are **df** and **du**. These two commands display the amount of disk space used in two different forms. The df command lists the file systems which make up the UNIX file tree. It shows the mount (start) point of the portion of the file tree on a particular disk or partition as well as the space used and available. The du command lists the amount of space used in a directory. Use df to search for nearly full disks and partitions that require action by the systems administrator. To view the space consumed by a given directory, use du. Examples of the two commands are shown in the next illustration.

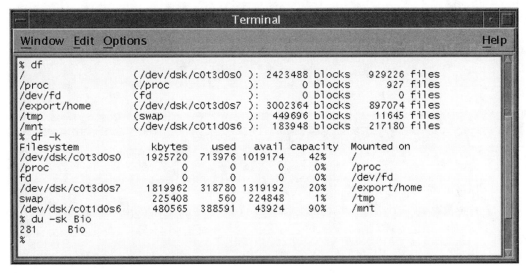

```
                                    Terminal
 Window  Edit  Options                                              Help
% df
/                (/dev/dsk/c0t3d0s0 ): 2423488 blocks    929226 files
/proc            (/proc            ):       0 blocks       927 files
/dev/fd          (fd               ):       0 blocks         0 files
/export/home     (/dev/dsk/c0t3d0s7 ): 3002364 blocks    897074 files
/tmp             (swap             ):  449696 blocks     11645 files
/mnt             (/dev/dsk/c0t1d0s6 ):  183948 blocks    217180 files
% df -k
Filesystem            kbytes    used   avail capacity  Mounted on
/dev/dsk/c0t3d0s0    1925720  713976 1019174    42%    /
/proc                      0       0       0     0%    /proc
fd                         0       0       0     0%    /dev/fd
/dev/dsk/c0t3d0s7    1819962  318780 1319192    20%    /export/home
swap                  225408     560  224848     1%    /tmp
/dev/dsk/c0t1d0s6     480565  388591   43924    90%    /mnt
% du -sk Bio
281      Bio
%
```

Examples of the df and du commands in use.

Editing Files

Numerous files control how Solaris functions. Examples include the password and group files which oversee user accounts, the hosts and networks files for networking, and the configuration files for daemons such as **inetd** and **init**. While special tools exist to manipulate files such as the password, most files are altered via an editor.

Using the Text Editor

On a workstation running CDE, text editing is made easy with the *text editor* found on the CDE "dashboard." The icon looks like a pad and pencil. When accessed, an editor window appears on the screen as shown below.

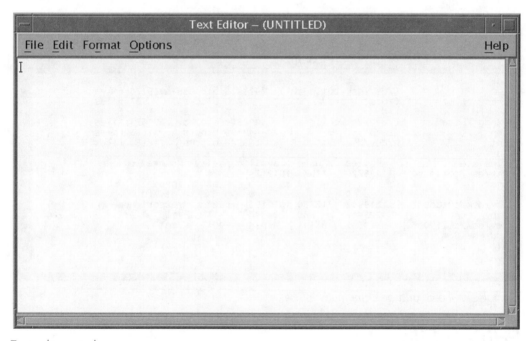

Text editor window.

To select the file to be edited, either drag the file from the File Manager onto the text editor icon on the dashboard or use the Open a File dialog box from the File menu.

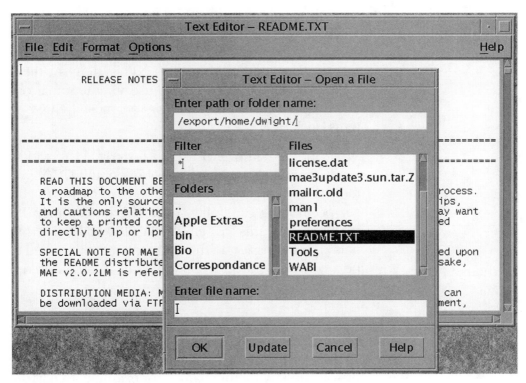

Open a File dialog box.

To edit the file, position the cursor with the keys or the mouse, and then type. To select the text, hold down the right mouse button and sweep the cursor over the text. The Edit menu provides the familiar cut, copy, and paste functions found in other text processing programs. The File menu provides commands to save the edited file and exit the text editor.

vi

Some administrators might prefer using the **textedit** command or the CDE text editor to modify system files. However, there are a surprising number of situations where OpenWindows, CDE, and tools such as textedit are unavailable. Three example situations are working on a server that does not have OpenWindows installed, working over a modem from a PC, or working on another vendor's UNIX system. At these times it helps to know how to use **vi**, a terminal oriented editor which is common on most UNIX systems, including Solaris.

While not as simple to use as textedit, vi is widely available and should be accessible on any UNIX system. The vi editor works easily on terminals with cursor positioning capability. Included are common PC terminal programs that emulate VT100 or ANSI terminals. The discussion in this segment is not intended to be a complete introduction to vi, but rather a description of selected basic functions.

Starting vi

To edit a previously existing file with vi, simply type *vi,* followed by the file name. For example, to edit the file */tmp/test,* type *vi /tmp/test.* If you wish to create a new file, *vi* will present a screen which resembles the following illustration.

Starting up vi.

⊷ **NOTE:** *A very small window is used to make it easy to see the cursor position in the current and following examples. The vi editor adapts to the window size.*

The name of the file is shown at the bottom of the screen. The tilde characters along the left edge of the screen are vi markers indicating the empty region past the end of the file. The markers are not part of the file being created; vi displays them to help identify the boundaries of the screen and the file being worked on.

Entering Text in vi

The vi editor has two modes: command and input. When entering *vi*, begin in the command mode. To switch to the input mode, type either an *A* character (to append input to the line the cursor is on), or an *i* (to insert input at the current cursor position). To see how this works, follow along using vi on a workstation for the remainder of this section.

⊷ **NOTE 1:** *In reality, there are several commands which will place vi in the input mode. The following discussion shows the A and I commands as examples.*

⊷ **NOTE 2:** *The vi editor is modal. It has an input mode and a command mode. Keeping track of the current mode is the key to using vi.*

To edit */tmp/test*, type *A* to put vi into input mode. When inputting the text shown in the next illustration, be sure to press <Return> at the end of each line. Unlike *text edit* and some other text editors, vi does not implement word wrap unless specifically set to do so. Typing in a paragraph without typing <Return> at the end of each line will result in a very long line. After finishing the last line, switch from input mode back to command mode by pressing the <Esc> key.

Entering text in vi.

✓ **TIP:** *When you forget the mode vi is in, press the <Esc> key. The vi editor will switch to command mode if it was in input mode, or cause the terminal bell to beep if it was already in command mode. Either way, vi will end up in command mode.*

In command mode, there are two basic functions: editing using the edit keys, and issuing commands that start with a colon (:).

Moving the Cursor

Moving the cursor is a basic function of the editing keys. The vi editor uses a set of keys to the right of the keyboard's home row.

Cursor movement keys for vi

Key	Cursor action
h	Move left one character
j	Move down one line
k	Move up one line
l	Move right one character

Try moving through the text you input. Finish the motion with the cursor on the *l* in *little* on the second line.

➥ **NOTE:** *On many terminals, the arrow keys can also be used to move the cursor.*

Moving the cursor.

✓ **TIP:** *Tired of moving around one character at a time? The vi editor has motion accelerator keys. The* b *key moves the cursor to the left by one word, and* w *moves the cursor to the right by one word.*

Checking the Terminal Type

Improper cursor motion is a common problem for **vi** users because Solaris does not know what type of terminal or terminal emulator is being used. The vi editor depends on this information to emit the proper sequences of escape and control characters needed to move the cursor around the screen. These sequences vary according to terminal. The vi program determines the type of terminal in use by consulting an environment variable set in the command interpreter or shell.

To check the value of the *TERM* environment variable, exit vi by typing *<Esc> wq* to save the changes and quit the editor. Next, type the following string:

```
% echo $TERM
```

If the terminal listed is the type being used, no changes are needed. If not, the *TERM* environment variable should be set to the correct terminal type. For example, if a VT100 type or VT100 terminal emulator is being used, and you use the Bourne shell (*/bin/sh*), you would input *TERM=vt100*. Export *TERM* to set the *TERM* environment variable.

Check the *TERM* environment variable if vi does not draw the screen correctly, refuses to work (i.e., asserts that the terminal in use does not have the required cursor functions), or moves the cursor incorrectly in response to the motion keys.

   ↝ **NOTE:** *It may also be necessary to use the* stty *command to set the terminal line disciplines. If the user re-sizes a telnet window it may be necessary to use* stty *to set the number of rows and columns in the current window. For example,* stty rows 24 columns 80 *will set the number of lines of text to 24, and the number of characters per line to 80.*

After checking the *TERM* environment variable, restart vi with the same command as shown at the beginning of this section to continue with the following examples.

Changing Text

To change text, use a substitution command or the c edit key. The c key provides a choice of objects to be changed or the number to changed. Command options include **cw** to change the word, and **c** followed by the space bar to change a single character. The cw command will change the characters at the cursor point as far to the right as a space or punctuation mark. Type **c$** to change the remainder of a line, that is, from the cursor position to the right end of the line. The vi program marks the characters to be changed by placing a dollar sign ($) at the right end of the group of characters to be changed and then enters input mode. Try typing cw with the cursor positioned on the *l* in *little* on the second line of the example file.

↬ **NOTE:** *It is also possible to change the entire remainder of the line using the C (upper-case C) command.*

Change command before entering new text.

Type the string, *lot of.* The vi editor inserts the characters in the space formerly occupied by *little.* You can insert as many characters as necessary. Press the <Esc> key to exit input mode and return to command mode.

Change command after entering new text.

Undoing an Action

Typographical errors can easily be remedied. The vi editor has a one-command *oops* buffer. Type **u** to undo the change. Type **u** again to *redo* the change. For the example presented here, the change will be kept.

Deleting Text

The **d** or delete edit key function is similar to the c key. The d key takes an argument to specify what is to be deleted. Try positioning the cursor on the *S* in *Susan.* Type **dw** (delete word). Type **u** to undo the change. The **x** key is an equivalent of d followed by a space bar (d). The x key deletes a single character. To delete the entire line, type **dd**. The following series of illustrations shows the screen changes after using each of these editing key combinations.

Using dw to delete a word.

The dw key sequence deletes the first word and the space following the word. This neatly cleans up a misplaced word without an additional delete command.

Using u to undo.

The undo command, u, pops the deleted text back into place.

↝ **NOTE:** *Remember that* vi *is restricted to a one-item undo buffer. You cannot recover a mistake made two or more commands ago with the* u *command.*

Using x to delete a single character.

The single character delete command, x, deletes the character at the current cursor position. In the example depicted in the previous illustration, the deleted character is the *S* in *Susan.*

✓ **TIP:** *A handy way to fix transposed characters is to place the cursor on the first letter, and then use* xp. *This deletes the character under the cursor, and places the same character after the letter which ends up under the cursor.*

Using dd to delete an entire line.

Review of vi Features

The following table summarizes the basic vi key sequence commands, including the commands used in the preceding examples.

Key(s)	vi action
h	Move left one character
j	Move down one line
k	Move up one line
l	Move right one character
c or r	Change character
cw	Change word
c$ or c	Change remainder of line
x	Delete character
dw	Delete word
d$ or d	Delete remainder of line
dd	Delete entire line

✓ **TIP:** *The dot or period (.) editing key repeats the last edit function. If a word was just inserted or the remainder of a line deleted, or a word is moving to a new cursor location, type a period (.) to repeat the insertion or deletion.*

vi Line Commands

Moving the cursor and making changes with the editing keys works well for small changes. To make changes throughout a large file, **vi** line commands are entered by typing a colon (:).

The vi editor is ready to accept a command.

The colon drops the cursor down to the bottom of the screen to allow input of a line command. Once a line command is entered, press the <Return> key to execute it and vi will automatically return to command mode at the location in the file where the line command finished working. In many instances, line command changes leave the cursor positioned on the last line in the file.

Substituting Text Using a Regular Expression

The most common editing command is a string substitution which takes the form of a *regular expression.* This specialized string substitution and pattern matching language is found in several UNIX commands, including the **grep** family and the stream editor, **sed**. Regular expressions are a rich language with many features. A commonly used feature is their ability to match arbitrary strings when carrying out substitutions. For example, the following command would change the text in the example file from past to present tense:

```
1,$ s/had/has/
```

This command could be translated as: "Starting with the first line and ending with the last line, substitute the string had with the string has once per line."

Using a regular expression to substitute "has" for "had".

There are two parts to a regular expression. The first, *(1,$)*, is an address range. The second, *(s/has/had)*, is the substitution expression. The address range specifies the line numbers over which the expression will be applied. In the example, the entire file is specified by stating the (1,$) range. The dollar sign represents the last line of the file when used in the address range portion of a regular expression.

To restrict a regular expression to the line the cursor is on, type a colon and omit the address range section altogether (e.g., *:s/had/has/*). To specify the range as an offset from the position of the cursor before the colon was typed, use a plus (+) and minus (-) sign as well as line numbers. An example follows:

```
-1,+2 s/foo/bar/
```

The above expression instructs vi to substitute *bar* for the first occurrence of *foo* beginning with the line before the cursor position (-1) through to two lines below the cursor position (+2).

↝ **NOTE 1:** *To substitute for a period (.) or an asterisk (*), the user needs to prevent those characters from being interpreted as special matching characters. This is done by placing a backslash (\ ) before each character. For example, s/./,/ would transform the first character encountered into a comma (,) in the current line, while s/\./,/ would change the first period (.) to a comma (,).*

↝ **NOTE 2:** *Remember that vi has a one-command oops buffer. If a substitution is executed which changes more than is intended, type u and try again.*

Exiting vi

There are several ways to save a file and exit from **vi**. Changes can be saved into the original file or into a different file, or disregarded altogether to leave the original file untouched. A variety of exit commands are described in the table below. The most common exit is to type **ZZ** in command mode, which tells **vi** to save the changes and exit.

Exiting vi

Characters typed	vi action
:w	Write the current file out.
:w!	Write the current file out even if it is marked as *read only*.
:w *foo*	Write the current editor contents to the file *foo*.
:q	Quit the editor.
:q!	Quit even if unsaved changes are present.
ZZ	Save the current file and exit the editor.

Final Comments on vi

The **vi** program is a difficult editor to learn. However, its wide usage and general availability on many types of UNIX platforms makes it worth the effort to learn. Use vi to edit practice files, focusing on the following key points.

❑ Remember to check which mode vi is in. To return vi to command mode, press <Esc>.

❑ Check the context of the character being typed. A dollar sign ($) can mean either move to the end of the line or to the last line of a file in the address part of a regular expression in command mode, or simply the dollar sign character in input mode. Practice using overloaded characters such as the dollar sign in every possible context.

❑ Remember what vi is doing with the characters being typed. Are the characters being inserted into the file or interpreted as commands?

In general, caution is required when editing system files. It is easy to make mistakes in vi, especially for beginners. A good approach to editing a system file is to make a backup copy of any file *before* you edit it.

To change the */etc/inet/hosts* file, which contains system host names and their IP addresses, execute the following commands:

```
# cp /etc/inet/hosts /etc/inet/hosts.990530
# vi /etc/inet/hosts
```

Preceding vi with the **cp** command creates a backup copy, which is invaluable as a quick way out if changes to the file do not work as expected or should problems arise while editing. This is better than retrieving the file from a recent system backup to recover from a typo. The numbers added to the file name are a code for the date. Two digits each represent the year, month, and day, in that order.

Old files are easy to clean up by using a naming convention such as the numerical form of the date used above. For example, backup files from last year can be sorted by typing the following:

```
# rm -i /etc/*.96????
```

The above string will match any file name that contains a *96* date code on the end and will prompt the user for verification before removing any files. The question marks are wildcard characters and will match any single character. In the example, four question marks match the four digits of the month and date. This command should be implemented *after* a system backup so that the files will be on a backup tape in the event they are needed.

✓ **TIP:** *The -i option on* rm *is a good one to use with wildcard characters such as the asterisk (*) or question mark (?) while working as root. The root account has the privilege to remove anything, including things which should not be removed such as critical system files, device entries, and even the system kernel. The -i option asks permission before each deletion. This gives the user a chance to avoid undesirable results from a wildcard character.*

UNIX Power Tools

With file editing in hand, the next collection of basic tools a systems manager needs can best be described as UNIX power tools. The following section demonstrates the basic concepts and uses of these tools.

Modern workstations provide window-based terminals and graphical monitoring tools. The command shell, pipes, and filters in Solaris and other UNIX variants form a flexible and powerful tool set that can be readily applied to common system management problems.

Juggling Tasks with Multiple Windows

Systems administrators will often need to refer to several tasks at once. The CDE terminal window and multiple screen features help by providing a method to organize work and easily switch between tasks. CDE also provides advanced e-mail and performance monitoring tools. Examples of their usage are shown in the next illustration.

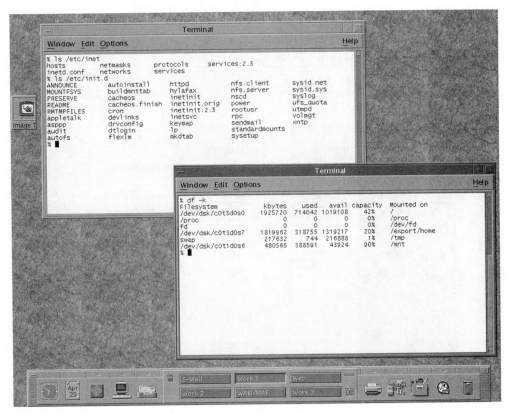

CDE screen with two terminal windows.

The previous illustration shows two CDE terminal windows. The output of the **df** command is in one window and a directory listing is in the other. The terminal window provides a useful scroll bar to review commands or output which have scrolled off the window.

Although the user can open more windows, CDE offers another option. The buttons in the middle of the CDE dashboard allow for switching between several virtual screens, each containing its own set of windows. The next illustration shows another virtual screen which contains only a Web browser. Because the windows can be moved between screens and the screens can be given names, work is organized in logical groupings (e.g., Web browsing, e-mail, and so forth).

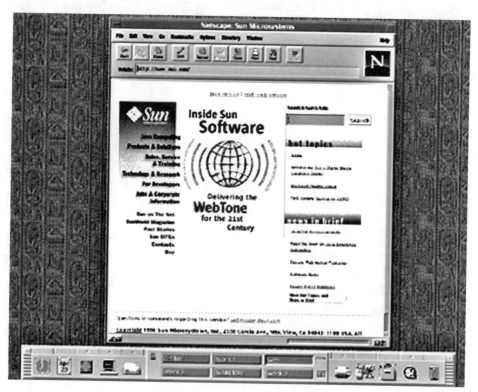

Another CDE screen showing Netscape Navigator.

Electronic mail is another basic tool of systems administration. Many beginner systems administrators are familiar with the shell command, **mailx**, which is used to read e-mail. Systems administrators who receive large volumes of e-mail should consider more powerful mail handling tools such as the CDE mail program. The mail program icon is located next to the screen buttons at the left of the CDE dashboard.

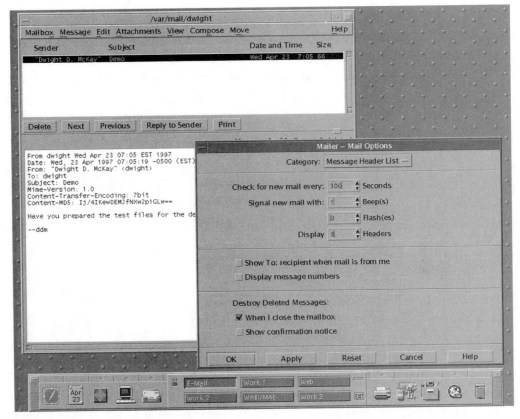

CDE mail program and options dialog box.

The CDE mail program displays a list of the subject lines of e-mail messages, which can be easily scanned to pick out important messages quickly. It has a built-in editor similar to the CDE text editor and option dialog boxes for controlling message and header display, mail boxes, and e-mail address aliases.

Another important window-based tool, the "performance meter," is not on the CDE dashboard. However, it can be started from the program's sub-menu off the workspace menu (which appears when the right mouse button is held down with the cursor point at the screen backdrop). The performance meter provides a graphical display dial or line graph of several system performance metrics. The display style, parameters monitored, and monitoring rate can be changed via the performance meter's Properties dialog box. This is one way an administrator can keep an eye on the health of a system while performing other tasks.

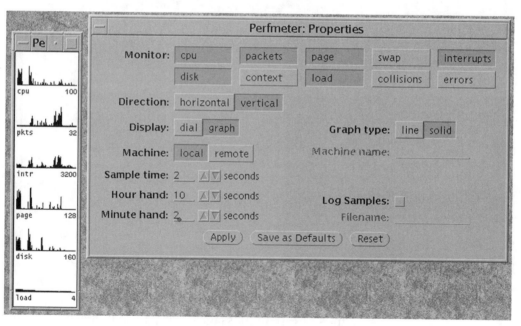

Performance meter and Properties dialog box.

Pipes

The real power of Solaris is found in command combinations. The principal method for combining tools is called a "pipe." A pipe connects a source to a destination. With UNIX command shells, the flow is from left to right along the command line; the pipe connects the output of the command on the left side of the pipe to the input of the command on the right side of the pipe. The following examples will clarify pipe usage.

A common problem on a Solaris system is the identification of a running process by its process ID number which can be used to stop the process or send it a signal. If the Web server process needs to be identified and stopped, the command is **ps -ef**. The -ef options to ps list all processes along with the process IDs, as well as additional information associated with each process.

Filtering with more

Unfortunately, the list of processes, even on a workstation, is too long to fit in a 24-line window. While you can scroll bar backwards, another approach is to filter the output of **ps** to show only items of interest. A good first choice is to use the **more** program, which displays output a page at a time. The more program will read from its standard input if no file name is provided. The output of ps is connected to the **more** input by using a pipe (|).

```
┌─────────────────────────────────────────────────────────────────────────┐
│─                              Terminal                              ·  □ │
├─────────────────────────────────────────────────────────────────────────┤
│ Window  Edit  Options                                               Help │
├─────────────────────────────────────────────────────────────────────────┤
│ al/http                                                                  │
│     http  5624   197  0   Mar 10 ?         0:00 /usr/local/http/httpd -d /usr/loc │
│ al/http                                                                  │
│     http  5121   197  0   Mar 09 ?         0:00 /usr/local/http/httpd -d /usr/loc │
│ al/http                                                                  │
│     http  5109   197  0   Mar 09 ?         0:00 /usr/local/http/httpd -d /usr/loc │
│ al/http                                                                  │
│     root  1078   106  0   Mar 02 ?         0:00 in.rlogind               │
│   dwight  3756   323  0   Mar 07 ?         0:00 /bin/sh -c dtfile -noview │
│     http  5110   197  0   Mar 09 ?         0:00 /usr/local/http/httpd -d /usr/loc │
│ al/http                                                                  │
│     http  5108   197  0   Mar 09 ?         0:00 /usr/local/http/httpd -d /usr/loc │
│ al/http                                                                  │
│   dwight  3757  3756  0   Mar 07 ?         0:05 dtfile -noview           │
│   dwight  5830   334  0   Mar 10 pts/6     0:00 /bin/csh                 │
│     root  7225   106  0 06:41:25 ?         0:00 rpc.rstatd               │
│     root  5101     1  0   Mar 09 ?         0:00 /usr/sbin/aspppd -d 1    │
│   dwight  5836  5835  0   Mar 10 ?         0:18 /usr/openwin/bin/snapshot│
│   dwight  5842  5830  0   Mar 10 pts/6     0:12 imagetool                │
│   dwight  5814  5771  0   Mar 10 ?         3:10 /opt/SUNWwabi/bin/wabiprog -c -s │
│ d:\msoffice\winword\winword.exe /export/home/d                          │
│   dwight  5771  5770  0   Mar 10 ?         0:00 /bin/sh /opt/SUNWwabi/bin/wabi -s │
│ d:\msoffice\winword\winword.exe /export/home/                           │
│ % █                                                                      │
└─────────────────────────────────────────────────────────────────────────┘
```

Oops. The list of processes is so long it scrolls off the screen.

This process makes column headings at the top of the ps listing and the first few processes easy to read. By pressing the space bar, the more program will display the next page of output. However, to locate a particular process, you may have to view several pages of output. At this point, you may wish to further customize the list of processes. The tool to create such a list is **grep**.

```
 ┌─────────────────────────────────────────────────────────────────────┐
 │ ─                            Terminal                            · ░  │
 ├─────────────────────────────────────────────────────────────────────┤
 │ Window  Edit  Options                                           Help │
 ├─────────────────────────────────────────────────────────────────────┤
 │ % ps -ef | more                                                      │
 │     UID    PID  PPID  C    STIME TTY      TIME CMD                    │
 │    root      0     0  0    Mar 01 ?       0:01 sched                  │
 │    root      1     0  0    Mar 01 ?       0:10 /etc/init -            │
 │    root      2     0  0    Mar 01 ?       0:06 pageout                │
 │    root      3     0  0    Mar 01 ?      11:49 fsflush                │
 │    root    259     1  0    Mar 01 console 0:00 /usr/lib/saf/ttymon -g -h -p tigg │
 │ er.net-kitchen.com console login:  -T sun -d /                       │
 │    root    160     1  0    Mar 01 ?       0:01 /usr/lib/lpsched       │
 │    root    258     1  0    Mar 01 ?       0:00 /usr/lib/saf/sac -t 300│
 │    root    100     1  0    Mar 01 ?       0:17 /usr/sbin/in.named     │
 │    root     80     1  0    Mar 01 ?       0:01 /usr/sbin/rpcbind      │
 │    root     82     1  0    Mar 01 ?       0:00 /usr/sbin/keyserv      │
 │    root    197     1  0    Mar 01 ?       0:01 /usr/local/http/httpd -d /usr/loc │
 │ al/http                                                              │
 │    root     88     1  0    Mar 01 ?       0:00 /usr/sbin/kerbd        │
 │    root    106     1  0    Mar 01 ?       0:03 /usr/sbin/inetd -s     │
 │    root    109     1  0    Mar 01 ?       0:00 /usr/lib/nfs/statd     │
 │    root    111     1  0    Mar 01 ?       0:00 /usr/lib/nfs/lockd     │
 │    root    130     1  0    Mar 01 ?       0:00 /usr/lib/autofs/automountd │
 │    root    186     1  0    Mar 01 ?       0:01 /usr/sbin/vold         │
 │    root    134     1  0    Mar 01 ?       0:09 /usr/sbin/syslogd      │
 │    root    150     1  0    Mar 01 ?       0:33 /usr/sbin/nscd         │
 │ --More--                                                             │
 └─────────────────────────────────────────────────────────────────────┘
```

Output of ps -ef filtered by more.

Filtering with grep

The **grep** family of commands (**grep**, **egrep**, **fgrep**) is typically used by beginning UNIX users to find text strings in files. A common usage follows.

```
# grep mercury /etc/inet/hosts
```

The above command would look for the string *mercury* in the host table found in *etc/inet/hosts*. This command might be used when seeking a host's IP address. Similar to **more**, grep can be used to search for any lines with matching text strings in data sent to it via a pipe.

```
┌─────────────────────────────── Terminal ──────────────────────────┐
│ Window  Edit  Options                                        Help  │
├────────────────────────────────────────────────────────────────────┤
│ % ps -ef | grep httpd                                              │
│     root   197     1   0   Mar 01 ?         0:01 /usr/local/http/httpd -d /usr/loc│
│ al/http                                                            │
│     http  5115   197   0   Mar 09 ?         0:00 /usr/local/http/httpd -d /usr/loc│
│ al/http                                                            │
│   dwight  7237  5830   0 06:56:40 pts/6     0:00 grep httpd        │
│     http  5111   197   0   Mar 09 ?         0:00 /usr/local/http/httpd -d /usr/loc│
│ al/http                                                            │
│     http  5112   197   0   Mar 09 ?         0:00 /usr/local/http/httpd -d /usr/loc│
│ al/http                                                            │
│     http  5623   197   0   Mar 10 ?         0:00 /usr/local/http/httpd -d /usr/loc│
│ al/http                                                            │
│     http  5624   197   0   Mar 10 ?         0:00 /usr/local/http/httpd -d /usr/loc│
│ al/http                                                            │
│     http  5121   197   0   Mar 09 ?         0:00 /usr/local/http/httpd -d /usr/loc│
│ al/http                                                            │
│     http  5109   197   0   Mar 09 ?         0:00 /usr/local/http/httpd -d /usr/loc│
│ al/http                                                            │
│     http  5110   197   0   Mar 09 ?         0:00 /usr/local/http/httpd -d /usr/loc│
│ al/http                                                            │
│     http  5108   197   0   Mar 09 ?         0:00 /usr/local/http/httpd -d /usr/loc│
│ al/http                                                            │
│ % █                                                                │
└────────────────────────────────────────────────────────────────────┘
```

Output of ps -ef filtered to capture a specific process by name.

The output lines of **ps -ef**, which contained the name of the process, *httpd,* have been filtered out of the longer listing. The only item missing is the column header that **ps** prints identifying each column of output. To execute this action, another process comes into play.

The **egrep** variant of grep accepts regular expressions in its search string, similar to those used in **vi**. A regular expression that consists of a string, a vertical bar, and another string will cause egrep to match either the first or the second string.

```
 ┌─────────────────────────────────────────────────────────────────────┐
 │ ─                          Terminal                            ·  □  │
 ├─────────────────────────────────────────────────────────────────────┤
 │  Window  Edit  Options                                         Help  │
 ├─────────────────────────────────────────────────────────────────────┤
 │ % ps -ef | egrep "UID|httpd"                                         │
 │     UID   PID  PPID  C    STIME TTY       TIME CMD                    │
 │     root  197     1  0  Mar 01 ?         0:01 /usr/local/http/httpd -d /usr/loc│
 │ al/http                                                              │
 │     http 5115   197  0  Mar 09 ?         0:00 /usr/local/http/httpd -d /usr/loc│
 │ al/http                                                              │
 │     http 5111   197  0  Mar 09 ?         0:00 /usr/local/http/httpd -d /usr/loc│
 │ al/http                                                              │
 │     http 5112   197  0  Mar 09 ?         0:00 /usr/local/http/httpd -d /usr/loc│
 │ al/http                                                              │
 │     http 5623   197  0  Mar 10 ?         0:00 /usr/local/http/httpd -d /usr/loc│
 │ al/http                                                              │
 │     http 5624   197  0  Mar 10 ?         0:00 /usr/local/http/httpd -d /usr/loc│
 │ al/http                                                              │
 │     http 5121   197  0  Mar 09 ?         0:00 /usr/local/http/httpd -d /usr/loc│
 │ al/http                                                              │
 │     http 5109   197  0  Mar 09 ?         0:00 /usr/local/http/httpd -d /usr/loc│
 │ al/http                                                              │
 │     http 5110   197  0  Mar 09 ?         0:00 /usr/local/http/httpd -d /usr/loc│
 │ al/http                                                              │
 │     http 5108   197  0  Mar 09 ?         0:00 /usr/local/http/httpd -d /usr/loc│
 │ al/http                                                              │
 │    dwight 7241  5830  0 06:57:48 pts/6    0:00 egrep UID|httpd         │
 │ % █                                                                  │
 └─────────────────────────────────────────────────────────────────────┘
```

Output of ps -ef filtered to capture the head line and a specific process.

⊷ NOTE: *Take time to note how the pipe character is being used. Without quotation marks, the shell would attempt to interpret the vertical bar as another pipe. Like the dollar sign character ($) in* vi, *the pipe means different things in different contexts. In the previous illustration, the pipe between the* ps *command and the* egep *command indicates a connection of the output of* ps *to the input of* egrep. *However, the vertical bar found inside the quoted search string after the* egrep *command is part of a regular expression which directs the user to "match this or that." The meaning of the vertical bar is controlled by the quotes placed around the search string. The quotes tell the shell to pass the string contained within them to* egrep.

tar Tricks

Another common system management problem is moving directories while preserving the ownership and permissions assigned to the files and subdirectories. This problem often arises when there is a need to move a user's home directory from one disk to another, or even to another machine. The usual solution to this problem is to use the **tar** command to copy the directory onto an archive file on tape, and then extract the copy of the directory from the archive file on tape to the new location.

A faster method, however, is to combine the two tar commands with a pipe. With this procedure, the user avoids storing the archive file created by tar, and instead sends it directly from the tar which creates the archive to the tar which will extract the archive. The next illustration shows a simple example of moving a directory between file systems on a single system.

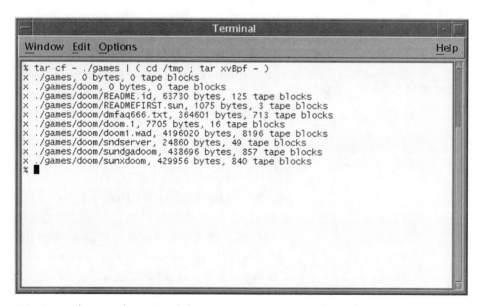

```
% tar cf - ./games | ( cd /tmp ; tar xvBpf - )
x ./games, 0 bytes, 0 tape blocks
x ./games/doom, 0 bytes, 0 tape blocks
x ./games/doom/README.id, 63730 bytes, 125 tape blocks
x ./games/doom/READMEFIRST.sun, 1075 bytes, 3 tape blocks
x ./games/doom/dmfaq666.txt, 364601 bytes, 713 tape blocks
x ./games/doom/doom.1, 7705 bytes, 16 tape blocks
x ./games/doom/doom1.wad, 4196020 bytes, 8196 tape blocks
x ./games/doom/sndserver, 24860 bytes, 49 tape blocks
x ./games/doom/sundgadoom, 438696 bytes, 857 tape blocks
x ./games/doom/sunxdoom, 429956 bytes, 840 tape blocks
%
```

Moving a directory between disks using two tar commands and a pipe.

Reading the command from left to right, **tar** shows the **cf** option flags which mean "create an archive on the specified device or file." The hyphen (-) indicates that the file is standard output. The *./games* element indicates the directory to be archived. The output of this tar command flows through the pipe to the second tar command. The commands on the right side of the pipe within the parentheses constitute a single command to the pipe.

The parentheses create a new shell which interprets the commands within. The first command is the familiar **cd** which indicates where to place the moved directory. Next comes the second tar command, which reads the archive fed to it by the pipe from the first tar command. The second tar uses the **xvBpf** option flags. This is interpreted as "extract the archive, print verbose messages, keep reading even if a full block is not sent, preserve the file modes and ownership, and read from the specified device or file." In this case, the hyphen (-) indicates standard input. The **B** option flag avoids potential problems from a less than full block of data being sent through the pipe. If *root* is running the tar command, the **p** option flag preserves the file modes and ownership. This is an important option, especially when moving a user's home directory. The modes and ownership of a moved directory must exactly match those of the source directory.

⊸ **NOTE:** *When creating archives using* tar, *be aware that the path given for the directory or files to be archived is saved in the archive. If a full path is given (e.g.,* /export/home/stuff*), it will not be possible to extract the archive contents in another location. Instead, use a relative path, one which starts from the current directory (e.g.,* ./stuff*). This can be extracted in any location.*

The same technique can be extended to move directories between machines. Once again, use two **tar** commands and add the **rsh** command. The rsh command runs shell commands on a remote machine with access to the command's standard input and output. This procedure is similar to the single machine example. The first tar command creates an archive on its standard output using the **cf** option flags, the parentheses, and the relative path of the directory to be moved.

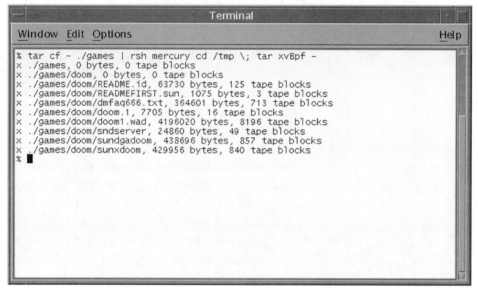

Moving a directory between machines with rsh, two tar commands, and a pipe.

On the receiving end of the pipe is a rsh command. This command connects to the indicated remote system, and then runs the shell commands listed after the machine's name on the remote system. As before, **cd** positions the shell in the proper location for the directory being moved. The backslash in \; prevents the shell from applying a special interpretation to the semicolon character (;). In brief, the semicolon is not processed by the shell on the sending machine, but is passed along via rsh to the remote machine so that two commands, **cd/tmp** and **tar xvBpf -**, can be run there. As before, the second tar command uses the **xvBpf** option flags and a hyphen (-).

rsh, login, telnet, rcp and ftp

The **rsh** command is one of several used to access machines over a network. As shown in the preceding section, rsh can be used to run a command on a remote machine. The **rlogin** and **telnet** commands are used to begin a terminal session on a remote machine. Because they use different protocols when connecting to a remote machine, it is important to note which protocol the remote machine or networked device supports. Some network devices, such as network hubs, terminal servers, and routers can be managed over a network, but only accept telnet connections.

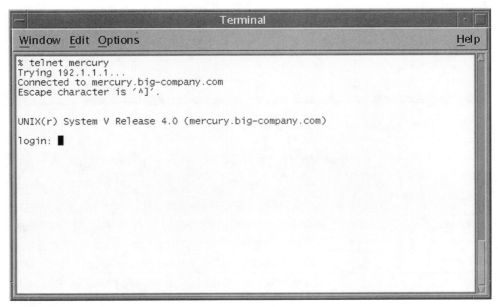

A telnet connection to a remote machine.

Files can be moved between machines on a network by using the **rcp** and **ftp** commands. The rcp command acts like a **cp** command but adds syntax to the file name description to include a machine name. The machine name is added prior to the file's path name and separated from it by a colon. An example of copying a file named *motd* is shown in the next illustration.

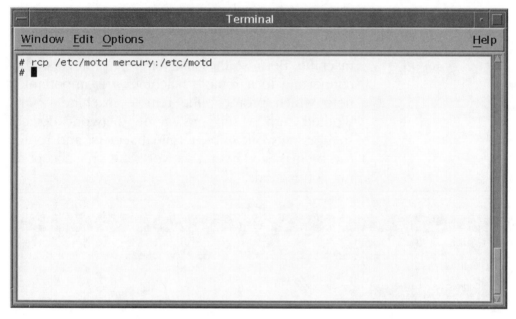

Copying /etc/motd between machines using rcp.

↝ **NOTE:** *The* rcp, rsh *and* rlogin *commands make use of the* .rhosts *file to determine access. Access is granted if the machine and user being contacted have a* .rhosts *file, and if that file contains a line listing both the machine the connection is coming from and the user name being used. Exercise great care before allowing access to the root account via entries in the* / .rhosts *file. See Chapter 18, "Network*

Security," for more information on root account security.

The **ftp** command provides more extensive file manipulation between machines. This command has an exclusive command interpreter allowing mode setting, directory listing, and file copying. The ftp command is a streaming protocol, and is thus efficient for moving files over wide area networks and the Internet. FTP access is accomplished via a user name and password dialog similar to that of logging in for a terminal session.

✓ **TIP:** *By default,* ftp *assumes the files being copied are text files containing seven-bit ASCII characters. To copy a binary file, such as a program, be sure to set* ftp *to binary mode by typing* bin.

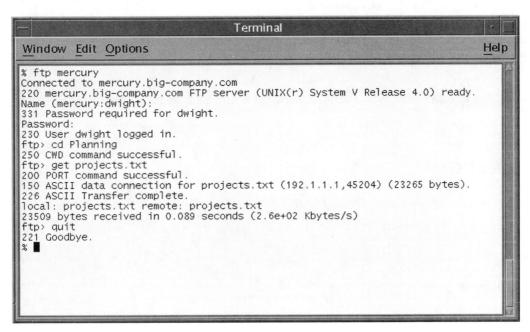

```
% ftp mercury
Connected to mercury.big-company.com
220 mercury.big-company.com FTP server (UNIX(r) System V Release 4.0) ready.
Name (mercury:dwight):
331 Password required for dwight.
Password:
230 User dwight logged in.
ftp> cd Planning
250 CWD command successful.
ftp> get projects.txt
200 PORT command successful.
150 ASCII data connection for projects.txt (192.1.1.1,45204) (23265 bytes).
226 ASCII Transfer complete.
local: projects.txt remote: projects.txt
23509 bytes received in 0.089 seconds (2.6e+02 Kbytes/s)
ftp> quit
221 Goodbye.
%
```

Copying a file via ftp.

Anonymous ftp

Some machines on the Internet offer anonymous FTP access, often for purposes of distributing software. In this case, input *anonymous* for the user name, and provide the user's e-mail address as the password.

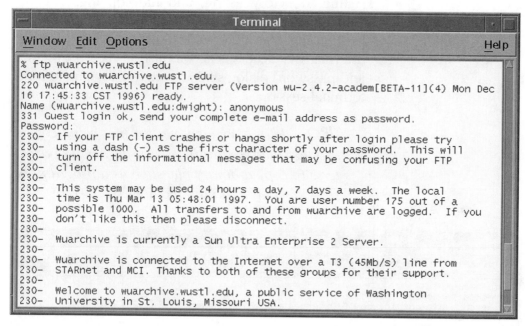

```
% ftp wuarchive.wustl.edu
Connected to wuarchive.wustl.edu.
220 wuarchive.wustl.edu FTP server (Version wu-2.4.2-academ[BETA-11](4) Mon Dec
16 17:45:33 CST 1996) ready.
Name (wuarchive.wustl.edu:dwight): anonymous
331 Guest login ok, send your complete e-mail address as password.
Password:
230-   If your FTP client crashes or hangs shortly after login please try
230-   using a dash (-) as the first character of your password.  This will
230-   turn off the informational messages that may be confusing your FTP
230-   client.
230-
230-   This system may be used 24 hours a day, 7 days a week.  The local
230-   time is Thu Mar 13 05:48:01 1997.  You are user number 175 out of a
230-   possible 1000.  All transfers to and from wuarchive are logged.  If you
230-   don't like this then please disconnect.
230-
230-   Wuarchive is currently a Sun Ultra Enterprise 2 Server.
230-
230-   Wuarchive is connected to the Internet over a T3 (45Mb/s) line from
230-   STARnet and MCI. Thanks to both of these groups for their support.
230-
230-   Welcome to wuarchive.wustl.edu, a public service of Washington
230-   University in St. Louis, Missouri USA.
```

Connecting to an anonymous ftp server.

Summary

This chapter covered a number of basic UNIX tools found on Solaris systems that provide general purpose aids to system management. Consider taking the time to practice using these basic tools, and to study the options not covered here as well as the basic commands in the on-line manual pages. System management involves a commitment to continually learning about the capabilities and features of available tools.

Part II

System Management

Refining System Boot and Shutdown Procedures

Unlike starting up a personal computer, the boot sequence of a Solaris system involves many steps. The Solaris boot sequence can be modified in several places to suit various needs. This chapter explores the terminology used in the Solaris boot sequence, and changes that can be made in the sequence. Several methods of altering the system run level and respective impacts on boot and shutdown sequences are also explored.

Boot Sequence

The term *boot* is derived from the expression, "pick yourself up by the bootstraps." To boot a computer is to start it running from a stopped state, usually right after turning the power on. The following diagram illustrates the steps in the Solaris boot sequence.

Boot sequence.

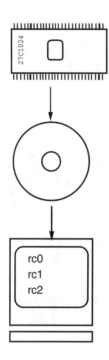

Step 1: PROM monitor executes power-on self test (POST).

Step 2: PROM monitor loads boot block from boot device.

Step 3: Boot block loads /ufsboot from boot device.

Step 4: /ufsboot loads SunOS kernel from boot device.

Step 5: SunOS kernel configures devices, initializes memory, and starts init.

Step 6: init reads /etc/inittab, moves to default run level, executes rc files.

Step 7: rc files execute scripts from /etc/rc{run_level}.d directories.

The first step of the boot process relies on the physical hardware of the workstation to initialize itself and load a small program. This program is usually stored in a ROM (read only memory) or PROM (programmable read only memory) chip. The program which is loaded at power-on is called a *PROM monitor.*

↦ **NOTE:** *Personal computer users may be more familiar with the term basic input output system (BIOS). The BIOS of an x86 system is functionally equivalent to the SparcStation PROM monitor.*

Step 1: PROM Monitor

The PROM monitor has several functions. It can be used to modify basic hardware parameters such as serial port configurations or the amount of memory which should be tested upon system power-up. Another PROM configurable parameter is the system *boot device* specification. This parameter tells the PROM monitor where it should look for the next stage of the boot process. Most importantly, the PROM monitor has routines to load the next stage into memory and start it running.

The next stage of the boot process may be loaded from a disk attached to the system, a floppy diskette, a CD-ROM drive, a tape drive, or over the network from a boot server (in the case of a diskless workstation). Load source depends on the setting of the PROM monitor's boot device parameter.

On a SparcStation, the PROM monitor initially starts in a restricted state. This state is sometimes referred to as the "old" command mode, because this mode was the only one available on some early model SPARC workstations. The old command mode restricts the operations that can be taken to accomplish the following tasks:

❒ Start the boot process.

❒ Continue the running program if the system has accidentally entered the PROM monitor.

❒ Switch to the new command mode. The boot process is invoked by the **b** command. The new command mode is invoked by the **n** command. If the operation of the system has been interrupted, it may be returned to the running state by invoking the **c** (continue) command.

b (boot) Option

The **b** command accepts an optional device argument which indicates the device to use when booting, and an optional file name that dictates the program to be loaded and started. The b command also allows an option flag which controls the boot process. A common usage of the option flag is to specify the **-s** option to cause the system to enter run level 1 (single user mode) so that system maintenance activities may be performed. The format for the b command is *b [device] [file] [options]*. A summary of the b command appears in the following table.

Boot options

Parameter	Allowed values	Default value
[device]	Cdrom, disk, floppy, net, tape	Determined by the PROM monitor boot device parameter. Typically set to SCSI disk 3.
[file]	Any bootable file name	Determined by the PROM monitor boot file and boot mode parameters. Typically set to *vmunix* if the system is in normal mode, and to */stand/diag* if the system is booted in diagnostic mode.
[option]	-a — Prompt for further information	Not applicable.
	-r — Rebuild the kernel device tables	
	-s — Enter run state 1 (single user)	

For example, booting a system from the restricted mode of the PROM monitor is done by typing **b** at the command prompt. The command line below contains the **-r** option which forces the system to perform a reconfiguration boot.

```
b -r
```

The flag indicates that the boot process needs to check for changes in the available system devices and take actions such as loading device drivers, and creating or deleting entries in the */dev* directory to match any changes which have occurred since the system was last booted. See Chapter 8, "Working with Solaris 2 Device Names," for a detailed discussion on device reconfiguration.

n (new) Option

The new command mode, also known as the *forth monitor*, allows an extra level of control and customization. The new command mode is identified by the *ok* prompt instead of the > prompt used in the old command mode. From new command mode two parameters control the system boot state. One of the parameters is a security mode setting which prevents a user from specifying an alternate boot. The other parameter is called *auto-boot*, and if it is set the workstation will bypass the restricted mode and continue on directly with the boot process. The following table presents a few of the most common commands available in new command mode.

Forth monitor options

Command	Description
boot	Similar to the b command in restricted mode.
probe-scsi	Probes the internal SCSI bus and lists the devices found. The probe-scsi-all variant will probe all SCSI bus adapters and report on all devices to any SCSI bus on the system.
reset	Causes the PROM monitor to initialize the system as if power had been cycled.

Command	Description
printenv	Lists the current value of PROM monitor parameters.
setenv	Sets the value of a PROM monitor parameter.

➡ **NOTE:** *Use caution when setting PROM monitor parameters, especially* boot-device *and* boot-file. *The PROM monitor does not check to determine whether parameters are valid; a mistake could prevent the system from automatically booting.*

An example of starting the boot process from the new command mode is shown below.

ok boot

If the *auto-boot* parameter is set to true, or if the **b** or **boot** commands are issued, the PROM monitor proceeds to the next stage of the boot process. The next stage consists of loading and starting a program known as the *boot block*.

Step 2: Boot Block

Similar to the PROM monitor, the boot block gets its name from the location in which it is stored. Typically, the boot block is stored in the first few blocks (sectors 1 to 15) on the hard disk attached to the workstation. The boot block's job is to initialize some of the system's peripherals and memory, and to load the program which will in turn load the SunOS kernel. A boot block is placed on the disk as part of the Solaris installation process, or in some circumstances, by the system administrator using the **installboot** program.

Although a diskless machine does not read its boot block from a disk, the term is still used to describe the small program that the PROM monitor loads into the workstation's memory from a boot server over the network. A boot server is another system on the network which provides network boot services via the trivial file transfer protocol (**tftp**) daemon. When the diskless client's PROM monitor issues a network request for its boot block, the server replies by sending the boot block through the network to the PROM monitor on the client workstation. The PROM monitor on the client workstation loads the program into memory and starts its execution.

Step 3: Boot Program

Depending on the location of the boot block, its next action is to load a boot program such as **/ufsboot** into memory and execute it. The boot program includes a device driver as required for the device (such as a disk drive or network adapter) which contains the SunOS kernel. Once started, the boot program loads the SunOS kernel into memory, and then starts it running. On a diskless workstation the boot program handles setting up NFS access over the network to the workstation's root (/) file system and *swap* area.

Step 4: SunOS Kernel

The *kernel* is the foundation of both SunOS and Solaris. Once loaded into memory by the boot program, the kernel has several tasks to perform before the final stages of the boot process can proceed. First, the kernel initializes memory and the hardware associated with memory management. Next, the kernel performs a series of device probes. These routines check for the presence of various devices such as graphics displays, Ethernet controllers, disk controllers, disk drives, tape devices, and so on. This search for memory and devices is sometimes referred to as *auto-configuration*.

✓ **TIP:** *On certain systems the status lights may blink on devices such as disk drives while the kernel probes for and locates the devices.*

With memory and devices identified and configured, the kernel finishes its start-up routine by creating *init*, the first system process. The init process is given process ID number 1 and is the parent of all processes on the system. Process 1 or init is also responsible for the remainder of the boot process.

Step 5: init and /etc/inittab

The **init** process, the files it reads and the shell scripts it executes are the most configurable part of the boot process. Management of the processes which offer the *login* prompt to terminals, the start-up for daemons, network configuration, disk checking, and more occur during this stage of the boot sequence.

Upon invocation, the init process reads a configuration file named */etc/inittab*. The */etc/inittab* file contains a series of init directives. The format for these directives is *id:run-level:action:process*. The init directives are described below.

❒ id—A one- to four-character label which uniquely identifies the entry.

❒ run-level—The system run level at which the current process will be invoked.

❏ action—This parameter indicates one of a variety of actions to be taken by init.

❏ process—The name of the program which will be invoked depending on the values of the previous two fields (run level and action).

Solaris Run Levels

The run level is a number which specifies a particular system state. Actions to be taken by *init* are keyed to various run levels. The run levels used in Solaris are described in the following table.

Solaris run levels

Run level	Description
0	Power-down or shutdown state: init stops all system activity and causes the kernel to exit, returning control to the PROM monitor.
1	Single user: init executes actions which unmount all file systems except root (/) and /usr, and allows root access only on the system console.
2	Multi-user: resources are not exported to other systems.
3	Multi-user: resources are exported to other systems.
4	Alternate multi-user. Not currently used.
5	"Green" or power saving mode for Sun4m and Sun4u systems.
6	Reboot: shut the system down and reboot.
s,S	Single user mode with all file systems available: terminal logins are disabled, but network logins are enabled.

More than one run level can be listed for */etc/inittab* entries, in which case the process is invoked when init enters any of the run levels listed. An entry with no associated run level is assumed to be valid at all run levels.

init Actions

When init moves to a given run level, each line in the */etc/inittab* file with the current run level is processed. The processes invoked by some inittab entries are run once, while other processes must be restarted whenever they stop running. The actions listed in the following table describe the things init can do with a process.

init Actions

Action code	Description
respawn	Start the process upon entering this run level. Do not wait for process to finish before moving on to next entry in *inittab*. Restart process if it should die.
wait	Process is started and init will wait for process to complete before moving on to next entry in *inittab*.
once	Process is started but init does not wait for process to complete. No init action is taken if process dies.
boot	Process is executed only during the first reading of the *inittab* file.
bootwait	Process is executed when init moves from single user to multi-user, and init waits for process to complete before moving on to next entry in *inittab*. Process is not restarted if it dies.
powerfail	Process is executed when init receives a power fail signal.

Action code	Description
powerwait	Similar to powerfail except that init waits for process to finish before proceeding to next entry in *inittab*.
off	If process is running, init will stop it when entering this run level.
initdefault	Special entry to denote default run level. The run level argument of the line with the *initdefault* action is interpreted as the default run level.
sysinit	Process listed in this entry is to be run before init starts processes associated with allowing logins.

A Closer Look at /etc/inittab

Action codes and run levels are easier to understand when the *inittab* file is examined line by line. Study the following sample *inittab*.

```
ap::sysinit:/sbin/autopush -f /etc/iu.ap
fs::sysinit:/sbin/rcS                  >/dev/console 2>&1 </dev/console
is:3:initdefault:
p3:s1234:powerfail:/sbin/shutdown -y -i5 -g0 >/dev/console 2>&1
s0:0:wait:/sbin/rc0                    >/dev/console 2>&1 </dev/console
s1:1:wait:/sbin/shutdown -y -iS -g0    >/dev/console 2>&1 </dev/console
s2:23:wait:/sbin/rc2                   >/dev/console 2>&1 </dev/console
s3:3:wait:/sbin/rc3                    >/dev/console 2>&1 </dev/console
s5:5:wait:/sbin/rc5                    >/dev/console 2>&1 </dev/console
s6:6:wait:/sbin/rc6                    >/dev/console 2>&1 </dev/console
of:0:wait:/sbin/uadmin 2 0             >/dev/console 2>&1 </dev/console
fw:5:wait:/sbin/uadmin 2 6             >/dev/console 2>&1 </dev/console
RB:6:wait:/sbin/sh -c 'echo "\nThe system is being restarted." ' >/dev/console 2>&1
rb:6:wait:/sbin/uadmin 2 1             >/dev/console 2>&1 </dev/console
sc:234:respawn:/usr/lib/saf/sac -t 300
co:234:respawn:/usr/lib/saf/ttymon -g -h -p " `uname -n` console login: " -T T "uT
sun -d /dev/console -l console -m ldterm,ttcompat
```

In the following discussion, the file is dissected line by line in much the same way that the init process reads the file. A line's two-character *id* field is used below to identify which line is being discussed.

```
ap::sysinit:/sbin/autopush -f /etc/iu.ap
fs::sysinit:/sbin/rcS          >/dev/console 2>&1 </dev/console
```

Because the first two lines, *ap* and *fs* have no associated run levels, and both lines use the action code *sysinit*, init executes each process, and then waits for the process to finish before moving on to process more lines. The *ap* line initializes the keyboard and mouse drivers for use with the console and other terminal lines. The process on the *fs* line is an **rc** (run command) shell script which will be discussed in detail later. The *rcS* script checks the /(root) and */usr* file systems for consistency.

```
is:3:initdefault:
```

The *is* line sets the default system run level to 3 (multiuser) via the *initdefault* action code. The *is* line does not change the run level, but rather specifies which run level init will move to by default when it is started. By specifying run level 3, init will move from level 0 to 1, 1 to 2, and 2 to 3 as the system progresses through the boot process.

```
p3:s1234:powerfail:/sbin/shutdown -y -i0 -g5 >/dev/console 2>&1
```

Upon booting, init moves from run level 0 to the default run level 3 in the example *inittab* file. Because the *p3* line lists several run levels, including run level 3, init performs the *powerfail* action. The powerfail action causes init to "remember" the process listed on the *p3* line as the process to be run if a power failure signal (**sigpwr**) is sent to init. This action is used at sites where a backup power supply and special monitoring software are installed to automatically shut the system down in the event of a power failure.

```
s0:0:wait:/sbin/rc0 off                    >/dev/console 2>&1 </dev/console
s1:1:wait:/sbin/shutdown -y -iS -g0        >/dev/console 2>&1 </dev/console
```

The *s0* line is skipped because it is not valid for run level 3. The *s1* line is skipped for the same reason.

```
s2:2:wait:/sbin/rc2                        >/dev/console 2>&1 </dev/console
```

The *s2* line invokes an **rc** (run control) shell script. The **rc2** script in turn invokes a set of scripts to set the system time zone, and starts several system daemons.

```
s3:3:wait:/sbin/rc3                        >/dev/console 2>&1 </dev/console
```

The *s3* line invokes the **/sbin/rc3** script which in turn invokes a series of other scripts to configure the network, and starts additional service daemons.

```
s5:5:wait:/sbin/rc5 ask                    >/dev/console 2>&1 </dev/console
s6:6:wait:/sbin/rc6 reboot                 >/dev/console 2>&1 </dev/console
of:0:wait:/sbin/uadmin 2 0                 >/dev/console 2>&1 </dev/console
fw:5:wait:/sbin/uadmin 2 2                 >/dev/console 2>&1 </dev/console
RB:6:wait:/sbin/sh -c 'echo "\nThe system is being restarted." ' >/dev/console 2>&1
rb:6:wait:/sbin/uadmin 2 1                 >/dev/console 2>&1 </dev/console
```

The *s5, s6, of, fw,* and *RB* lines are skipped because they are not valid for run level 3.

```
sc:234: respawn:/usr/lib/saf/sac -t 300
co:234:>respawn<D>:/usr/lib/saf/ttymon -g -h -p "`uname -n` console login: " -T
sun -d /dev/console -l console -m ldterm,ttcompat
```

The last two lines, *sc* and *co*, are scanned and the processes listed are invoked by init. These processes are automatically restarted should they die as a result of the *respawn* action specified in the inittab entries. The *sc* line starts the process which monitors the terminal lines and provides a login prompt for users. The *co* line starts the process which monitors the system console and provides a login prompt on that device.

When init finishes processing the entries in */etc/inittab*, the system boot process is complete.

Modifying /etc/inittab

A common boot process modification is to add entries to the **/etc/init.d** script files such that new processes may be started at specific run levels. Verify that these new entries are placed in the correct sequence with respect to other */etc/inittab* entries. For example, to add a line for a network license manager daemon, it would be wise to ensure that the daemon is started after things it depends on (such as network configuration or the NIS+ daemons). Most third party software packages that require modification provide explicit directions on where respective entries must be placed in */etc/inittab*, and which run level to use.

Although modifications to */etc/inittab* can be used to change the final stages of the boot sequence, the most significant drawbacks to this method are listed below.

❏ Difficulty in testing inittab entries
❏ Single-line limit for entries
❏ Difficulty in making entries conditional upon factors other than the run level.

For the above reasons, the use of run control (**rc**) files are the preferred method of implementing boot process modifications.

Run Control Files

The run control (rc) files and associated scripts are the most commonly modified component of the boot sequence. The local administrator can add or modify many system configuration processes in much the same way as for an *inittab* entry, but these files are easily tested and allow the programming features of the Bourne shell. Run control files allow greater flexibility in configuring the system during the boot process.

Each system run level has an associated rc file. Most rc files contain a directory bearing the name of the rc file with a *.d* extension. The rc file is a short Bourne shell which scans its associated directory and invokes the scripts found there in a certain sequence. The rc files are located in the */sbin* directory, and respective script directories are located in the */etc* directory.

The *s3* line in the sample */etc/inittab* file invokes the **/sbin/rc3** script when the system enters run level 3.

```
s3:3:wait:/sbin/rc3                        >/dev/console 2>&1 </dev/console
```

When init encounters the above line upon entering run level 3, it will invoke the /sbin/rc3 script and wait until the script finishes. The rc3 script will in turn scan the */etc/rc3.d* directory in search of scripts to be run. Each script in the */etc/rc3.d* directory handles a specific system function, such as the management of the system print service or NFS services.

For each run level, there is an rc file and (usually) a directory of scripts to handle the start-up or shutdown of the services associated with that run level. The table below shows the rc file, script directory, and a brief description of the functions that the rc file and scripts perform for each run level.

Run control script file relationships

Run level	RC script	RC directory	Common functions
0	/sbin/rc0	/etc/rc0.d	Kills all running processes except init and unmounts all file systems.
1	/sbin/rc1	/etc/rc1.d	Kills all non-essential system daemons.
2	/sbin/rc2	/etc/rc2.d	Kills the NFS, volume management and lp services; mounts file systems; and starts most system services except NFS server on start-up.
3	/sbin/rc3	/etc/rc3.d	Starts the NFS server process.
4	None	None	Not currently used.
5	/sbin/rc5	None	Same as rc0.
6	/sbin/rc6	None	Same as rc0.
S	/sbin/rcS	/etc/rcS.d	Starts basic network services; mounts /usr; and executes reconfiguration commands required for a reconfiguration boot.

The scripts in these directories are run in their ASCII sorting order within two divisions. Files which start with *K* contain commands to kill off processes and are run first when the corresponding run level is entered. The K scripts are invoked with the **stop** argument (e.g., **K20lp stop**). Files which start with *S* contain commands to start processes and are invoked after all *K* scripts for a particular run level have been invoked. The *S* scripts are invoked with the **start** argument (e.g., **S20lp start**). The flow of control from init through an rc file onto a script in an rc directory is illustrated in the following diagram.

*Control flow for
rc files.*

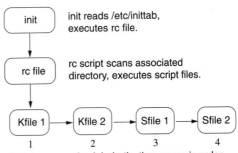

init reads /etc/inittab,
executes rc file.

rc script scans associated
directory, executes script files.

Scripts are run in alphabetic, then numeric order.

While it is sometimes necessary to directly modify an rc file, the more common approach is to add, delete, or modify a script located in the appropriate */etc/rc?.d* directories. By convention, the same script is used to start and stop a given service. The script contains a simple *case* statement and takes different actions depending on the argument it is passed at invocation (*start* or *stop*). The same script may be used at several run levels. To avoid multiple copies of a script and to ease maintenance, symbolic links are used.

All scripts used by the rc files are located in the **/etc/init.d** directory. A symbolic link is made from the script in **/etc/init.d** to a file in the appropriate */etc/rc?.d* directory.

Symbolic Links

A symbolic link, or simply *link*, is a term used in UNIX for another directory entry which points to the same file. A link may be considered an alias because it is another name by which the same information is known. Links are created using the **ln** command shown in the following illustration.

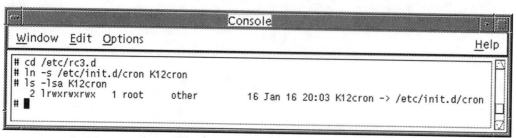

```
# cd /etc/rc3.d
# ln -s /etc/init.d/cron K12cron
# ls -lsa K12cron
   2 lrwxrwxrwx   1 root      other           16 Jan 16 20:03 K12cron -> /etc/init.d/cron
#
```

Creating and examining a file link.

Note that once the link is made, using either the file name or the link name, the contents of the same file are shown. This property allows links to be placed in locations where files would normally be found. Because a link is just another name for the file, all changes made to the file are seen through all respective links. In the case of the scripts found in the rc directories, this makes maintenance easier because only the original script needs to be modified in order to resolve problems.

✓ **TIP:** *It is usually preferable to use symbolic links instead of hard links when linking* rc *scripts to the* /etc/init.d *entries. Symbolic links are pointers to the original file.*

By convention, the files which contain all scripts used by the **rc** files are located in the */etc/init.d* directory. Their names are provided by the function they perform or the system service that they start and stop (e.g., */etc/init.d/nfs.server*).

Each *rc* script file reads and processes links whose names begin with a *K* to stop processes which should not be running at this run level. Next, the *rc* file reads and processes the links which begin with an *S*, to start processes which should be running at this run level.

For example, the script which starts the **cron** program, which runs other programs at scheduled times, is found in */etc/init.d/cron*. Because the cron program should not be running at run levels 0 or 1, links are made from */etc/init.d/cron* to *K* files in the */etc/rc0.d* and */etc/rc1.d* directories (i.e., */etc/rc0.d/K70cron* and */etc/rc1.d/K70cron*).

✓ **TIP:** *It is always good practice to verify that any editor used to modify files in the* /etc/rc?.d *directories does not break the symbolic link and create a new file as part of the editing process. Text Edit and* vi *preserve links.*

The cron script and links in the rc directories.

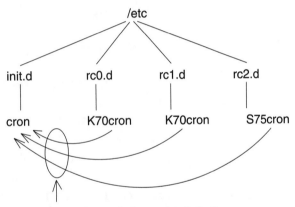

Symbolic links pointing to /etc/init.d/cron

The *70* in the name of the *K70cron* link controls the order in which the links are processed by the **rc** scripts. For example, when the **rc1** script is run by *init*, it runs the scripts in the */etc/rc1.d* directory in the order listed below.

1. K00ANNOUNCE

2. K42audit

3. K47asppp

4. K55syslog

5. K57sendmail

6. K65nfs.server

7. K67rpc

8. K68autofs

9. K70cron

10. K80nfs.client

11. S01MOUNTFSYS

The *K* links are read and processed first in numerical order. The **rc** script runs each *K* script with the single argument, **stop**. For example, the cron link in */etc/rc1.d* is run as if the **/etc/rc1.d/K70cron stop** command had been invoked. Because of the link to */etc/init.d/cron*, the command is equivalent to typing **/etc/init.d/cron stop**.

The **cron** process should be started at run level 2; consequently, a link is made in the */etc/rc2.d* directory for */etc/rc2.d/S75cron*. The *S* scripts are read and processed after the *K* scripts in numerical order based on the two-digit number following the *S*. Like the *K* scripts, the *S* scripts are run as if the **/etc/rc2.d/S75cron start** command had been invoked, which in turn is equivalent to **/etc/init.d/cron start**.

Modifying Run Control Files

Modifications to the run control files are generally made by adding, changing, or deleting one of the scripts in */etc/init.d*, and adding or deleting the corresponding links in the *rc* directories as needed. To add a new system daemon, such as a license manager, a new script is created in */etc/init.d*. At the very least, the script must be able to handle the two arguments below.

❒ start—This argument is given to the script when it is called as an *S* script from one of the rc files. When the script is invoked with the start argument it should take whatever start-up action is required, such as starting processes, clearing out directories, testing for the availability of configuration files, and so on.

❒ stop—This argument is given to the script when it is called as a *K* script from one of the rc files. When the script is invoked with the stop argument, it should take whatever shutdown and clean-up action is required, such as killing off processes, clearing out temporary files, saving state information, and so on.

As an example, the script used to start up a license manager daemon appears below. The daemon controls how many copies of certain programs are running. The programs contact this daemon over the network for permission to run. This script is placed in **/etc/init.d/lmf**. It is written using Bourne shell syntax because the rc files will run these scripts using the Bourne shell.

```
# Start and stop the license manager daemon.
# Save the first argument in the variable state.
```

```
state=$1
# Handle each possible value for state.
```

(1)

```
case $state in
```

(2)

```
'start')
    # Check for the lmf configuration file and if present, start lmfd.
```

(4)

```
[ -f /usr/lib/lmf.conf ] && /usr/lib/lmfd
```

(2)

```
;;
```

(3)

```
'stop')
    # Find lmfd in the output of ps.
```

(5)

```
id=`ps -e | grep lmfd | awk '{print $1}'`
# If a process id was found, then kill it.
if test -n "$id"
then
    kill $id
fi
```

(3)

```
;;
```

(1)

```
esac
```

The heart of the script is the block of lines starting at the *case* and ending at *esac* (see 1). This block is further divided into two sections: the first runs from *'start')* to the double semicolon (see 2), and the second runs from *'stop')* to the double semicolon (see 3).

The *start* block is executed if the first argument given to the script and saved in the *state* variable is the word "start." The *[-f /usr/lib/lmf.conf] &&* phrase (see 4) is Bourne shell syntax for "check to see if the file */usr/lib/lmf.conf* exists, and if so, do the following." The phrase that follows, */usr/lib/lmfd*, invokes the license manager daemon.

Simply stated, the entire block checks for the *lmf* configuration file. If the file exists, the block invokes the **lmf** daemon program.

The *stop* block is a bit more involved. The line which begins with *id=* (see 5) executes three commands joined by pipes, and places the result in the *id* variable. The first command is **ps -e** which lists all processes that are currently running. The output of this command is passed to **grep lmfd** which filters out the line listing the license manager daemon. That line is then passed along to *awk '{print $1}'*.

The *awk* term is a programmable filter. The program is the *{print $1}* phrase. Using this program, *awk* prints the first word, or group of characters separated by spaces in the line passed to it. In this case the output of *awk* is the process ID number. Thus, the entire series of commands finds the process ID number for the license manager daemon and puts that number in the *id* variable.

But what if the license manager daemon was not running? In this case, nothing would be found, and the value of *id* would be empty. The *if* line tests for this condition to ensure that the **kill** command is invoked only if *id* contains a value. The kill command sends the *SIGKILL* signal to the license manager daemon, and the daemon stops running.

Benefits of Using rc Directory Scripts

One of the benefits of using *rc* directory scripts is that they are easily tested. The scripts may be manually invoked with the **stop** and **start** arguments as a check to determine whether they function correctly before creating the symbolic links and trying them under actual system boot conditions. This procedure is recommended because it can help you catch mistakes that might interrupt the boot process and leave the system unusable.

The final step in adding a script is to add links in the appropriate places in the *rc* directories. Because the daemon in this example needs to communicate over the network, it should be started *after* the network utilities are started. One possible choice is to make the following links:

```
ln -s /etc/init.d/lmf /etc/rc0.d/K21lmf

ln -s /etc/init.d/lmf /etc/rc1.d/K21lmf

ln -s /etc/init.d/lmf /etc/rc3.d/S95lmf
```

The above set of links ensures that the license manager is not running when the system is in run level 1 or single user mode. Next, when you use these links the license manager is started after the network services have been started when the system is in run level 3 (multi-user mode).

Changing Run Levels

Up to this point, system run levels have been discussed only with respect to the normal boot sequence. In addition to the boot process several other methods may be employed to change the system run.

↝ **NOTE:** *The following commands are typically reserved for system maintenance activities.*

shutdown Command

The **shutdown** command is the preferred method for shutting the system down. This command will bring the system down to a particular run level in an orderly manner. The shutdown command sends messages to all users warning them of the impending system shutdown. The command will also notify the system syslog daemon of the shutdown. When the specified time arrives, shutdown will cause the system to sync the memory buffers to disk, and then change the system run level to the desired state.

Three of the most common uses of shutdown are to (1) bring the system to a halt; (2) bring the system down to single user mode; and (3) reboot the system.

↝ **NOTE:** *The* /usr/sbin/*shutdown command can only be executed with root access.*

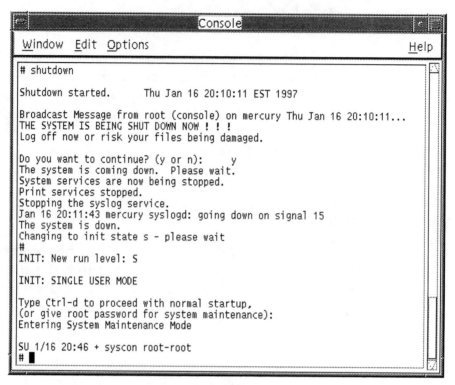

```
# shutdown

Shutdown started.        Thu Jan 16 20:10:11 EST 1997

Broadcast Message from root (console) on mercury Thu Jan 16 20:10:11...
THE SYSTEM IS BEING SHUT DOWN NOW ! ! !
Log off now or risk your files being damaged.

Do you want to continue? (y or n):      y
The system is coming down.  Please wait.
System services are now being stopped.
Print services stopped.
Stopping the syslog service.
Jan 16 20:11:43 mercury syslogd: going down on signal 15
The system is down.
Changing to init state s - please wait
#
INIT: New run level: S

INIT: SINGLE USER MODE

Type Ctrl-d to proceed with normal startup,
(or give root password for system maintenance):
Entering System Maintenance Mode

SU 1/16 20:46 + syscon root-root
#
```

Using shutdown with no command line options.

➥ **NOTE:** *In order to ensure proper operation of the* shutdown *command, the operator should first exit any windowing environment and change directory to the / (root) directory.*

The shutdown command recognizes the following flags:

❑   *-y* tells the system to pre-answer all confirmations with a yes.

❑   *-g N* tells the system to execute the shutdown in *N* seconds. Default is 60 seconds.

❑   *-i N* tells the system to change to the init state specified by *N*.

✓ **TIP:** *It is always a good idea to warn users in advance of a shutdown. This allows them time to save files and log off, avoiding the risk of file damage or loss. If a grace period of 900 seconds is used, the users will be notified with a 15-minute warning of the impending shutdown.*

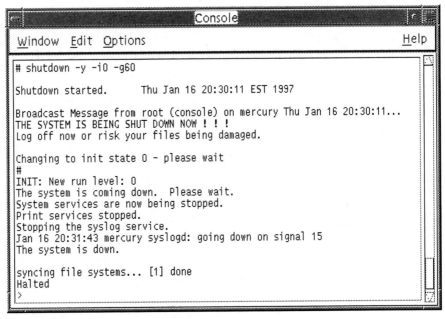

```
# shutdown -y -i0 -g60

Shutdown started.      Thu Jan 16 20:30:11 EST 1997

Broadcast Message from root (console) on mercury Thu Jan 16 20:30:11...
THE SYSTEM IS BEING SHUT DOWN NOW ! ! !
Log off now or risk your files being damaged.

Changing to init state 0 - please wait
#
INIT: New run level: 0
The system is coming down.  Please wait.
System services are now being stopped.
Print services stopped.
Stopping the syslog service.
Jan 16 20:31:43 mercury syslogd: going down on signal 15
The system is down.

syncing file systems... [1] done
Halted
>
```

Using shutdown with command line modifiers.

init 0 Command

Another method for bringing the system to a different run level is through the **init** command. As with shutdown, root access is required to execute this command. The init command does not send warning messages to users. Using the init command to change the system run level is best reserved for instances where nothing else has worked.

✓ **TIP:** *Some system administrators prefer* init 0 *over the* halt *command because* init 0 *will execute the scripts in* /etc/rc?.d. *These scripts cleanly shut down databases and other systems before the system is halted.*

The **telinit** program is a link to the **/sbin/init** program, and allows a single character argument. This argument is the alphanumeric identifier of one of the init states. In addition to the init state identifiers understood by shutdown, telinit understands the following arguments.

❒ a,b,c—The a,b,c init states are pseudo run levels. Using one of these states in the */etc/inittab* file or with the telinit command tells the system to run commands that are tagged as level a,b,c but do not change the system run state.

❒ q,Q—Causes the system to immediately re-examine the */etc/inittab* file.

❒ s,S—Causes the system to change state to the single user mode. The terminal which issues the telinit command becomes the system console.

Using telinit to bring the system down.

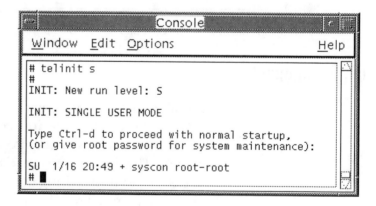

halt Command

Sometimes it is not feasible or necessary to warn users of an impending shutdown. The **/usr/sbin/halt** command initiates an orderly shutdown of the operating system, and does not send warning messages to the users. The halt command will not execute scripts in */etc/rc?.d*, nor does it leave the system in *init* state *s* (single user).

Using halt without flags.

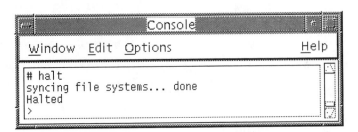

As with shutdown and init, root access is required to execute the halt command. The halt command recognizes the following four flags which control how the system is shut down.

❐ -l—Forces halt not to inform *syslog* of the system halt.

❐ -n—Forces halt not to write (sync) the memory buffers to disk before halting.

❐ -q—Forces halt to perform a rapid halt. This flag will also cause the system not to sync the memory buffers to disk, and may cause some processes to exit less than gracefully.

❐ -y—Forces halt to halt the system, even if the terminal issuing the halt command is a dial-up line.

Using halt with command line flags.

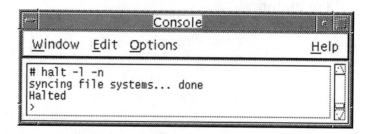

☞ **WARNING:** *Halting a system without writing the memory buffers to disk can be dangerous to the file system! The* halt -n *method is generally used only if memory contents are known to be corrupt. Consequently, executing a sync before halting is not desirable.*

reboot Command

The **reboot** command is another method for bringing the system down. This command causes the system to shut down and reboot to the default state specified in the */etc/inittab* file. The reboot command does not execute the shutdown scripts in */etc/ rc?.d.* The following four flags are recognized by the reboot command.

❑ -d—Causes the reboot process to generate a core dump. This option is provided for compatibility, but is not supported.

❑ -l—Causes the suppression of notification to the syslog daemon.

❑ -n—Causes the system to halt and then reboot without syncing the memory buffers to disk.

❏ -q—Causes the system to ungracefully halt and reboot without syncing the memory buffers to disk.

☞ *WARNING: Halting a system without writing the memory buffers to disk can be dangerous to the file system! The* reboot -n *method is generally used only if memory contents are known to be corrupt. Consequently, executing a sync before halting is not desirable.*

Results of using the reboot command to bring the system down and back up appear below.

```
Booting from: sd(0,0,0)vmunix
Sun05 Release 5.6 Version Beta [UNIX(R) System V Release 4.0]
Copyright (c) 1983-1996, Sun Microsystems, Inc.
configuring network interfaces: le0.
Hostname: redstone
The system is coming up.  Please wait.
checking ufs filesystems
/dev/rdsk/c0t2d0s6: is clean.
/dev/rdsk/c0t3d0s1: is clean.
/dev/rdsk/c0t2d0s3: is clean.
starting routing daemon.
starting rpc services: rpcbind keyserv done.
Setting netmask of le0 to 255.255.0.0
Setting default interface for multicast: add net 224.0.0.0: gateway redstone sys-
log service starting.
Print services started.
volume management starting.


Tue Jan 28 14:20:49 1997


The system is ready.


redstone console login:
```

Stop-A Command

Sometimes a system will lock up, making the use of the above commands impossible. When all else fails, the system can usually be stopped by holding down the <STOP> key (upper left corner of the keyboard) while simultaneously pressing the <A> key. This sequence is known as *Stop-A* (or L1-A). The key sequence stops the system processor, and puts the system under control of the PROM monitor.

```
# STOP-A pressed
Abort at PC 0xF0031870.

>
```

Once the system has been stopped in this manner, it is best to type the **sync** command to the PROM monitor. This will cause the monitor to sync the memory buffers to disk, and then crash the system. This action will force a system crash dump to be saved for post-mortem inspection. It may be possible to determine the reason the system was locked up by examining the crash dump information. Consult the manual page on *crash(1M)* for more information on crash dumps.

Using the PROM monitor to cause a crash dump is demonstrated below.

```
>sync
panic: zero
syncing file systems  ...2 2 2 2 2 2 2 done.
  586 static and sysmap kernel pages
   34 dynamic kernel data pages
  179 kernet-pageable pages
    0 segkmap kernel pages
    0 segvn kernel pages
  799 total pages (799 chunks)
```

```
dumping to vp ff1b9004, offset 181288
117 pages left        109 pages left    101 pages left
5 pages left
799 total pages, dump succeeded
rebooting...
```

Summary

Modifying the system boot process is an important means of tailoring Solaris to a unique environment. Certain software products and procedures may be started or performed at system boot by adding them to the boot sequence. Through the run control files and their associated directories, new boot procedures can easily be added. By using Bourne shell scripts, a new boot procedure can conditionally handle a variety of situations, and is easily developed and tested without the need to repeatedly boot a system. The preferred method of changing the system run level is to use the **shutdown** command. When this is not possible, there are other options available which allow the operator to change the run level of the system.

Creating, Deleting, and Managing User Accounts

Solaris is a multi-user operating environment. Not only can multiple windows and processes be running on a Solaris system, but multiple people can use such a system simultaneously. Unlike a single-user system, such as a PC or Macintosh, Solaris needs to keep track of each user's preferences and the location of each user's files and electronic mail. Each user must also have some way of identifying and authenticating herself to Solaris.

A user account is the name given to the collection of information that defines a user on a Solaris system. This information is employed when a user logs into the system, sends and receives electronic mail, and starts up OpenWindows. The maintenance of user account information is an ongoing task for the system manager. Solaris provides numerous tools to make user account maintenance easier.

User Account

A user account is a collection of information which defines a user on a Solaris system. The definition starts with the user's log-in name, or user name, which is used when logging in and sending or receiving electronic mail. This definition contains various pieces of information used to control access to files as well as a password used for authenticating the user. The component parts of a user account are listed below.

❏ A name by which the user is known to the system.

❏ A set of credentials consisting of a user identification number, group identification(s), and access lists which control ownership and access to directories, files, and devices such as printers and tape drives.

❏ The location of a home directory, where basic user files such as *.login* and *.cshrc* are stored.

❏ The name of a command interpreter, called a shell in UNIX terminology. This is the program run when the user logs in. Different users may have different shells.

❏ An optional disk space usage limit or quota.

❏ Optional information such as a password "age" to force the account's owner to periodically change his/her password.

An account can be considered as analogous to the collections of items in a wallet or purse. The driver's license number is like a user name. The picture on the driver's license allows other people to verify who the license holder is in much the same way the password allows Solaris to verify the user's log-in name. A club membership card grants the card holder certain access to shared resources in the same way that a UNIX group allows a user shared access. Spending limits on charge cards are similar to the usage quotas some systems place on disk usage.

License and credit card analogs to user account items.

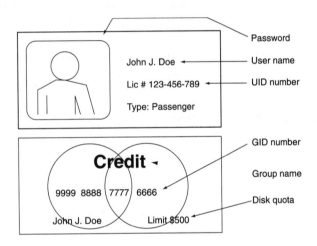

Admintool

Managing the collection of user account information on earlier UNIX systems required editing several files and using a couple of different commands. Under Solaris, account management is brought together under a single tool, **admintool**.

In large networks, consider installing and using Solstice AdminSuite by Sun Microsystems. This product has an interface similar to that provided by admintool, but adds the ability to manage accounts across multiple systems from a single location.

Overview of admintool

The **admintool** is a jumping-off station for accessing a collection of administrative functions. These functions include user account management; editing and updating various system databases; setting up printers, modems, terminals and other devices; and managing host names and software packages. The admintool is discussed throughout this book as each of these functions is addressed. This section focuses on editing the group database file and using the user account management functions.

Using admintool to Manage User Accounts

To begin working with the admintool, log in as *root* and start OpenWindows, or use the **su** command to gain root privileges from a terminal window if Open-Windows is already running. Next, type *admintool &* at the root prompt (#) to start the admintool and put the process in the background. The base admintool window is shown in the next illustration. This window presents a list of user accounts that are ready to be added to, deleted from, or edited. Other admintool functions are accessed via the Browse menu.

```
┌─────────────────────────────────────────────────────────────────┐
│                        Admintool: Users                           │
├───────────────────────────────────────────────────────────────────┤
│  File   Edit   Browse                                      Help   │
├───────────────────────────────────────────────────────────────────┤
│   User Name          User ID    Comment                           │
│  ┌──────────────────────────────────────────────────────────────┐ │
│  │ adm                    4     Admin                             │ │
│    bin                    2                                        │
│    daemon                 1                                        │
│    listen                37     Network Admin                      │
│    lp                    71     Line Printer Admin                 │
│    noaccess           60002     No Access User                     │
│    nobody             60001     Nobody                             │
│    nobody4            65534     SunOS 4.x Nobody                    │
│    nuucp                  9     uucp Admin                         │
│    root                   0     Super-User                        │
│    smtp                   0     Mail Daemon User                   │
│    sys                    3                                        │
│    uucp                   5     uucp Admin                         │
│                                                                    │
│                                    Host: glenn.net-kitchen.com     │
└───────────────────────────────────────────────────────────────────┘
```

The admintool opening screen.

Before diving into setting up an account, there are a number of housekeeping chores that should be taken care of and policy decisions to be made. Accomplishing these tasks now will speed up account creation and lay the foundation for easier management of accounts in the future.

Names, UIDs, and GIDs

Each user account must be assigned a name. This name is used to identify the account for purposes such as electronic mail and logging into a machine. While just about any string of two to eight characters can be used for an account name, many sites prefer to use a systematic naming convention, such as the initials of the person who owns the account (e.g., *ddm*), the owner's last name (e.g., *mckay*) or first name (e.g., *dwight*), or a combination of initials and names (e.g., *ddmckay,* or *dwightm*). A number can be added if there is more than one user with the same name or initials (e.g., *ddm2*). This sort of naming scheme makes it easy for people to recognize a user's account name when sending electronic mail, transferring files, and so on. For example, for a person whose name is John Doe, the system administrator can construct account names such as *doej, johnd, jd, jdoe, doe, john,* and *jd1.*

User IDs

Along with a name, each account is assigned a user identification number, or UID. Solaris, like other versions of UNIX, uses the UID number to tag files owned by the account and to control access permissions to files and devices. The first 100 UID numbers (1 through 100) are reserved by the Solaris system to be assigned to system accounts such as *root* and *uucp*. Likewise, UIDs above 60000 are assigned to special purposes. However, assigning a UID is a bit more complicated than simply picking a number between 100 and 60000.

❑ UID numbers *must* be unique for each user account. File ownership is recorded and controlled by the UID number. Accounts with the same UID number are effectively the same in regard to file ownership and other owned items such as processes and accounting records.

❑ In order for NFS file sharing to work correctly, UID numbers *must* be consistent between machines on a network. NFS uses the UID number when determining if a given user has access to a given file.

➥ *NOTE: Avoid reusing UID numbers whenever possible. The UID number is kept in several data files, all of which must be carefully checked and cleaned up before a UID can be reused. Reuse occurs when a UID number from a deleted account is assigned to a new account.*

Group IDs

A second number, called a *group identification number* (GID), is also assigned to each account. Unlike UID numbers, multiple users can share GID numbers. Groups and respective GID numbers are used to provide shared file access. Like UID numbers, Solaris also tags files with a GID number to allow access to other users who share the same GID number. In addition, the first 100 GID numbers (1 through 100) are reserved by the Solaris system to be assigned to special uses. GIDs must also be consistent between machines which share file systems via NFS. GIDs are an important and powerful mechanism in Solaris for controlling file sharing.

Upon examining a long file listing produced by **ls -l**, you will see that each file listed has both an owner and a group associated with it. There are separate permission bits for reading, writing, and execution associated with the file's owner, group, and everyone else. These bits can be set to make files readable or writeable by members of a group, while remaining inaccessible to other user accounts. The GID given to a user's account specifies the *primary* group a user is in. A user can also be made a member of up to 15 secondary groups (to make a total of 16 groups) by using the admintool to add that user's account name to a group listed in the group file.

Devising a scheme for grouping accounts together before creating new accounts is highly recommended. Consider the ways in which the various user accounts need to be able to work together. Do groups exist for each project? What about functional groupings such as accounting, engineering, or clerical? Creating groups first and then assigning accounts to groups makes file sharing and resource management easier later on.

Consider the following guidelines, which are similar to those for UID numbers.

❑   GID numbers *must* be unique for each group.

❑   GID numbers *must* be consistent between machines on a network in order for NFS file sharing to work correctly.

↝ **NOTE:** *Avoid reuse of GID numbers as well. While the GID number is kept in fewer data files than the UID number, the system administrator should nevertheless carefully check for potential problems before reusing a GID number.*

✓ **TIP:** *Consistency in UID and GID numbering can be easily maintained by using NIS+ to manage the various account database files. See Chapter 21, "Network Name Services," for details.*

Creating a Group

To create a group, select Groups from the Browse menu in the Admintool window.

Group Database window.

As an example, assume that a group of users will be using an accounting software package. They need permission to access the package's data, but that data should be protected from access by other users. To meet this need, create a group called *account* by selecting Add from the Edit menu. A small window containing a form appears. Fill in the form to specify the group name and identification number.

If additional groups are required, the process can be repeated. The admintool has the very helpful ability to add groups and user accounts in any order at any time, making it unnecessary to define the entire grouping and account naming and numbering scheme at the outset. Accounts and groups can be added as needed.

Add Group form.

```
 ┌─────────────────────────────────────────────────┐
 │ ─              Admintool: Add Group              │
 ├─────────────────────────────────────────────────┤
 │                                                  │
 │    Group Name: │account│                         │
 │      Group ID: ││101        │                    │
 │   Members List:│I                            │   │
 │                                                  │
 ├─────────────────────────────────────────────────┤
 │   ┌────┐  ┌──────┐  ┌──────┐  ┌──────┐  ┌──────┐ │
 │   │ OK │  │Apply │  │Reset │  │Cancel│  │ Help │ │
 │   └────┘  └──────┘  └──────┘  └──────┘  └──────┘ │
 └─────────────────────────────────────────────────┘
```

Preconfiguring with Skeleton Files

When a user logs into her account, a command interpreter or shell is started. The setup of the shell, how it
behaves, where it looks for programs, and so forth
can be modified to fit a particular environment, and
automatically configured for each user account when
it is created. As with account naming and UID and
GID numbering, planning and preparation prior to
setting up new accounts will save time later on.

Solaris provides two levels of shell customization: a
system-wide set of files which are read by every shell
started by every user, and a set of files which can be
copied into the user's home directory when an
account is created. The global files are */etc/.login* for
the C shell, and */etc/profile* for the Bourne and Korn
shells. The global files should contain only the settings needed for *all* shells to access when they start.
These global initialization files have the advantage of
not being modifiable by individual users. However,
any initialization files the user creates will override
the shell settings given by the global files.

In */etc/skel*, the system administrator can place the initialization files which will be copied into a user
account's home directory when the user account is

created. Admintool copies the files */etc/skel/local.cshrc* and */etc/skel/local.login* into a user's home directory as *.login* and *.cshrc*, respectively, if the user is assigned */bin/csh* as his shell. If the user is assigned */bin/sh*, then */etc/skel/local.profile* is copied to his home directory as *.profile*.

The advantage of skeleton files over global initialization files is that skeleton files are given to the user and can be modified by the user if and when necessary. The advantage to the system manager is that each user is given a predefined set of shell initialization files to ensure that items such as user search paths are set correctly, without requiring users to create initialization files from scratch.

Sun provides sample initialization files for Bourne, Korn and C shells in */etc/skel* which serve as a basis for customized files for a particular site. In version 2.3 of Solaris, these initialization files include useful start-up procedures for OpenWindows as well as search path settings for commonly used Solaris programs and tools.

Creating an Account

With the preparatory work in place, select Add User from the Edit menu in **admintool** to begin adding a new user account.

The Add User form that appears is divided into three sections, each dealing with a different aspect of the account to be created. Each section is discussed in subsequent sections as the form is filled in to create an account.

Add User form.

```
┌─────────────────────────────────────────────┐
│ ─                Admintool: Add User          │
├─────────────────────────────────────────────┤
│  USER IDENTITY                                │
│        User Name: [            ]              │
│          User ID: [1001]                      │
│     Primary Group: [10                ]       │
│   Secondary Groups: [                       ] │
│          Comment: [                         ] │
│       Login Shell: Bourne ⌐   /bin/sh         │
│  ACCOUNT SECURITY                             │
│         Password:   Cleared until first login ⌐│
│       Min Change: [        ] days             │
│       Max Change: [        ] days             │
│      Max Inactive: [       ] days             │
│   Expiration Date: None ⌐    None ⌐   None ⌐  │
│       (dd/mm/yy)                              │
│          Warning: [        ] days             │
│  HOME DIRECTORY                               │
│    Create Home Dir: ☑                         │
│             Path: [                         ] │
│                                               │
│  ┌────┐  ┌──────┐  ┌──────┐  ┌──────┐ ┌─────┐│
│  │ OK │  │Apply │  │Reset │  │Cancel│ │Help ││
│  └────┘  └──────┘  └──────┘  └──────┘ └─────┘│
└─────────────────────────────────────────────┘
```

User Identity

The User Identity section contains the basic name and
number information for the account. The User Name
field is limited to eight characters. The User ID is the
number discussed earlier that identifies this user. The
primary group is the group ownership given by
default to files created by this user. In the example, the
account group—created earlier using admintool—
will be used. The Secondary Groups are other groups

in which this account is a member. Under Solaris, a user can be a member of as many as 16 groups.

The comment field is typically used to contain additional identification information, such as the real name of the account user. Additional information may be placed in this field, such as the user's office phone extension number. Be aware that all data in this field will show up when anyone uses the **finger** command to inquire about the user account. The administrator can hide information from this command by inserting a comma between the name and any additional information. The finger command will only display the information up to the first comma. However, the information will still be visible in the user's entry in */etc/passwd*.

The Login Shell menu selects the command interpreter or shell that will be started by the log-in process when the user logs in using this account. Every shell has advantages and disadvantages; some users prefer one shell over another. In the example, the account will be given the default Bourne shell.

☛ **WARNING:** *Be very careful when choosing whether to change the shell of the root user. The setting provided with Solaris is chosen to allow for system operation even if problems such as disk failures occur which may make other shells inaccessible. It is possible to render a system unusable by changing the root shell without first ensuring that the shell is available under all possible conditions.*

✓ **TIP:** *It is often useful to have a root window available while experimenting with changes to the root shell and/or system security. If something goes* *wrong, root access to the system is still available to carry out repairs.*

Account Security

The Account Security section of the Add User form deals with the password used by the account user to authenticate herself to the system. The password itself can be set in the following four different states via the Add User form.

❏   *Cleared until first log-in.* This is the default setting. An account created in this manner will prompt the user to input a password the first time she/he logs in to the account. While this is handy, consider the implications of setting an account's password to this state, especially in an environment where the account may not be used and is likely to be discovered by someone other than its owner. See the discussions of system security in Chaper 7, "Managing System Security," and Chapter 18, "Network Security."

❏   *Account is locked.* In this state, the account exists but cannot be logged into until and unless a password is set for the account. An account can be put in this state automatically as described below.

❏   *No passwd -- set uid only.* This setting creates an account which cannot be logged into. It is used when an account is needed for a daemon or system program to use. An example of this is the *uucp* account which allows the **uucp** program to own files and directories required for that program to function.

❏   *Normal Password.* This selection brings up a small form in which a password can be assigned to the account.

The three date fields that follow the Password menu control a function called *password aging*. The aging process allows the system manager to enforce the practice of changing account passwords on a regular basis. Reusable passwords, such as those found in Solaris, are vulnerable to being discovered if they remain unchanged for long periods of time. Changing the password periodically helps to reduce this risk.

The downside to password aging is the psychological factor. Some users dislike changing passwords. Being asked to change with no warning may contribute to a user choosing a simpler, easily guessed password, or the user may simply enter a new password and then change back to the old password immediately afterward. Password aging is most effective when the account user understands the reasons for periodically changing a password and the definition of a good password, and is given a chance to choose a good password.

The Solaris password aging system gives the system administrator control over three aspects of the process. The Min Change field specifies the minimum time in days between password changes. Setting this to a value other than zero prevents a user from setting a new password and then immediately switching back to the old password until the time specified has passed. The Max Change field specifies the age the password must reach before the password aging system requests that the user change the password. The Warning field, at the bottom of this section, sets the number of days that a user will be notified prior to being required to change the password. This period gives users time to think of a new password before they are required to make the change.

For further protection, Solaris adds an inactivity timer and an expiration date. These features help to prevent inactive accounts or improper use of accounts owned by people who may no longer use them. The Max Inactive field sets the maximum time in days that an account can be left idle before it is locked. The expiration date sets a date at which the account is considered expired and is locked. These mechanisms are especially handy for accounts which are transient in nature, such as student accounts in an academic environment and accounts created for temporary employees.

Home Directory

The Home Directory section allows the option of having the User Account Manager automatically create and populate the user's home directory as part of account creation. This option provides considerable time savings and is worth examining in detail.

The Path field contains the UNIX directory path to the user's home directory. Typically the path is something like */export/home/smithb*, as it would be in the case of the user account being set up in this example. The Create Home Dir check box tells the admintool to create the user's home directory. Uncheck the box if the user's home directory is being provided by another machine via NFS or has otherwise already been created.

*Completed
Add User form.*

```
┌─────────────────────────────────────────────────┐
│ ─               Admintool: Add User              │
├─────────────────────────────────────────────────┤
│ USER IDENTITY                                     │
│       User Name: │smithb│                         │
│          User ID: │1001│                          │
│     Primary Group: │101│                          │
│   Secondary Groups: │10│                          │
│         Comment: │Bob Smith│                      │
│       Login Shell:  │ C  ⌐│ /bin/csh              │
│                                                   │
│ ACCOUNT SECURITY                                  │
│        Password:   │ Cleared until first login ⌐│ │
│                                                   │
│      Min Change: │1│      days                    │
│      Max Change: │180│    days                    │
│     Max Inactive: │30│    days                    │
│   Expiration Date: │None ⌐│ │None ⌐│ │None ⌐│     │
│     (dd/mm/yy)                                    │
│         Warning: │7│      days                    │
│ HOME DIRECTORY                                    │
│    Create Home Dir: ☑                             │
│                                                   │
│           Path: │/export/home/smithb│             │
│                                                   │
├─────────────────────────────────────────────────┤
│  │  OK  │   │ Apply │   │ Reset │  │ Cancel │  │ Help │ │
└─────────────────────────────────────────────────┘
```

Modifying and Deleting Accounts

With the foundation work required to set up accounts completed, using the **admintool** to modify and delete accounts becomes very easy. Modifying an account is simply a matter of opening the admintool as before and double-clicking on a user name in the list or selecting Modify from the Edit menu.

Selecting an account to be modified.

```
┌─────────────────────────────────────────────────────────┐
│─                    Admintool: Modify User               │
├─────────────────────────────────────────────────────────┤
│  USER IDENTITY                                            │
│           User Name: │smithb                      │       │
│             User ID: 1001                                 │
│        Primary Group: │101                        │       │
│      Secondary Groups: │staff                     │       │
│             Comment: │Bob Smith                   │       │
│          Login Shell:    C   ▭  /bin/csh                  │
│  ACCOUNT SECURITY                                         │
│            Password:    Cleared until first login  ▭      │
│          Min Change: │1          │ days                   │
│          Max Change: │180        │ days                   │
│         Max Inactive: │30        │ days                   │
│      Expiration Date: None ▭    None ▭    None ▭          │
│         (dd/mm/yy)                                        │
│             Warning: │7          │ days                   │
│  HOME DIRECTORY                                           │
│                Path: │/export/home/smithb          │      │
│                                                           │
│   ┌────┐   ┌───────┐   ┌───────┐   ┌────────┐   ┌──────┐ │
│   │ OK │   │ Apply │   │ Reset │   │ Cancel │   │ Help │ │
│   └────┘   └───────┘   └───────┘   └────────┘   └──────┘ │
└─────────────────────────────────────────────────────────┘
```

All fields filled in during account creation can now be edited and changed. Common activities include locking and unlocking accounts or changing an account's password using the Password menu. Selecting Normal Password will access a small form in which a new password can be entered.

However, the Modify User form does not completely handle all changes the administrator may wish to make. Watch out for changes to fields listed in the next table.

Password fields to watch out for

Field name	Description
User ID	While the Modify User form will change the UID number for the user in the /etc/passwd file or in the NIS+ database, it will not change the ownership of the files in the account's home directory or elsewhere (e.g., /tmp, /var/mail). This must be done manually as described below.
Primary Group	The same rule applies for the primary group. The /etc/passwd or NIS+ database files will be changed, but not the group ownership assigned to files in the account's home directory or elsewhere (e.g., /tmp, /var/mail).
Path	The user's home directory path will be changed only in the /etc/passwd and NIS+ database files. The directory itself must be moved manually *before* changing the account form, or the account form will report an error because it will not be able to find the new path.

Changing File Ownership

If the UID or GID numbers assigned to an account are modified, the ownership of the files and directories in the account's home directory must be modified manually so that the account will continue to have access to those files and directories. This is easy to accomplish with the **chown** and **chgrp** commands. These two commands work in a similar manner: both require the name or number of the UID or GID to change and the name of a file. Common usage when changing the UID or GID number for an account is to use the **-R** option to **chown** and **chgrp** to recursively change the ownership of all files in a sub-tree.

❐ The **chown** command sets the file ownership by changing the UID number the file is tagged with. The UID can be specified as a number or as a user name. If the user name is specified, chown will search for the number in the password file. The **-R** option will recursively descend through any directories it finds in the list of files given it, and change the ownership of the directory and the files under the directory. Examples of typical chown commands appear below; no output is produced unless there is an error.

```
chown dwight file1 file2 file3
chown -R dwight big-directory
```

❏ The **chgrp** command sets the file group ownership by changing the GID number the file is tagged with. As with chown, the GID can be specified as a number or as a group name; if the group name is used, chgrp will search for the number. This command also uses the **-R** option to recursively descend through directories and change the group ownership of all files and directories found. Typical chgrp commands also produce no output unless there is an error. Examples appear below.

```
chgrp 105 file
chgrp system file4 file9
chgrp -R accounting secret-directory
```

✓ **TIP:** *A common mistake is to specify .* as the list of files to be changed when using* chown -R *or* chgrp -R, *as in* chown -R dwight .*. *Unfortunately, the .* expression will match not only the files beginning with a period (.) in the current directory, such as* .cshrc *or* .login, *but also the directory above the current directory as referenced by its nickname, two periods (..). To avoid this, use a more restrictive expression such as .??*, which will match any file that begins with a period (.) and has at least two characters after the period. As always, be very* *careful using wildcard expressions such as the asterisk (*) while working as the root user.*

For example, if the UID and GID numbers were changed to 1002 and 1000 for the *smithb* account, you might use the sequence of commands shown in the next illustration to change the user and group ownerships of the files in the *smithb* account home directory.

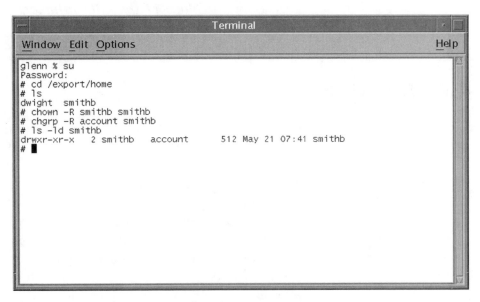

```
Terminal
Window  Edit  Options                                                      Help
glenn % su
Password:
# cd /export/home
# ls
dwight   smithb
# chown -R smithb smithb
# chgrp -R account smithb
# ls -ld smithb
drwxr-xr-x   2 smithb    account        512 May 21 07:41 smithb
#
```

Changing user and group ownerships.

The sequence of commands used above follows: (1) The **su** command is used to become *root*. The root account has permission to change the ownership of any file. (2) The current directory is changed to the directory *above* the account's home directory. This positions the shell to simplify the next two commands. (3) The **chown** and **chgrp** commands are used to change the UID and then the GID numbers assigned to the files in the *smithb* directory. Directories contained inside the *smithb* directory can be changed by using the **-R** option flag with the same command.

↝ **NOTE:** *If the* smithb *account's UID and GID have already been changed using the User Account Manager, you can use the account name,* smithb, *in place of the UID number, and the group name,* account, *in place of the GID number. Both the* chown *and* chgrp *commands will search for the numbers from the group and password databases.*

Moving Account Paths

Moving an account's home directory from one location to another on the same server or between servers requires a bit more work than setting UID and GID numbers. With the use of the **tar** command, certain shell magic directories can be moved without resorting to tapes or temporary files. The basic sequence is to create a tar archive of the files and directories in the current location, and by means of a pipe (|) or connection between processes, send that archive to a second invocation of tar which extracts the archive in the new location. The procedure appeared in Chapter 4, "Fundamental System Administrator Tools," but bears repeating and expanding upon here. In the following illustration, the *smithb* directory is moved to another file system on the same server.

✓ **TIP:** *While you could use the* cp *command to move user accounts, using* tar *provides a major advantage. The* tar *command preserves symbolic links and file ownerships, while* cp *does not.*

```
┌─────────────────────────────────────────────────────────────────┐
│ ─                          Terminal                         · □ │
├─────────────────────────────────────────────────────────────────┤
│ Window  Edit  Options                                      Help │
├─────────────────────────────────────────────────────────────────┤
│ # cd /export/home                                            ▲  │
│ # ls                                                            │
│ dwight   smithb                                                 │
│ # tar cf - ./smithb | ( cd /export/home2 ; tar xvBpf - )       │
│ tar: blocksize = 8                                              │
│ x ./smithb, 0 bytes, 0 tape blocks                             │
│ x ./smithb/.login, 575 bytes, 2 tape blocks                   │
│ x ./smithb/.cshrc, 124 bytes, 1 tape blocks                   │
│ # cd /export/home2                                             │
│ # ls                                                           │
│ smithb                                                         │
│ # █                                                            │
│                                                                │
│                                                                │
│                                                             ▼  │
└─────────────────────────────────────────────────────────────────┘
```

Moving an account's home directory.

After moving to one of the home file systems on this machine, a long command is issued consisting of several parts. The first part has the **tar** command creating an archive file of the *smithb* directory and putting it on the standard output. The *./smithb* path is used because it is a relative path which will allow the archive created by tar to be extracted in any location. Along the line to the right is a pipe (|), or the shell pipe character. The pipe takes the standard output of the tar command and routes it into the next command to the right in the command line.

The two commands inside the parentheses are executed in a new shell created for them. They act as a single command and will read from the pipe. The two commands move to the new location for this path (cd) and then carefully extract the archive file coming in from the pipe (tar). The options used with the second tar command cause it to extract the archive (x), maintain the same file ownership and permission bits (p), continue to read even if the archive file is slow in coming through the pipe (B), and produce a verbose list of the files as they are extracted from the archive (v).

Once the above operation is executed, modify the account's path using the User Account Manager and then delete the old directory. However, prior to removing the original path, checking to verify that the account can be logged into and that no mistakes occurred is good practice. A few minutes of checking can save hours of work recovering files and directories from backup tapes.

Moving a Home Directory Between Servers

Moving an account's home directory between servers uses the same general principle of creating a tar archive file and using a pipe to send it to another **tar** command to extract the contents. Appearing in the following illustration is an example of moving the *smitha* account from the server named *glenn* to the server named *grissom*.

The change of directory and the first tar command used are the same as in the previous example. On the right side of the pipe, the **rsh** command is used to execute commands on the *grissom* server instead of creating a second shell on the local server. The \; is used to prevent the semicolon from being processed on the local machine. This allows two commands to be passed to the remote machine via **rsh**: one to change to the appropriate directory, and the other to extract the archive file coming through the pipe using tar.

```
─                           Terminal                          ▪ □
 Window  Edit  Options                                      Help
 # cd /export/home
 # ls
 dwight  smithb
 # tar cf - ./smithb | rsh grissom cd /export/home2 \; tar xvBpf -
 tar: blocksize = 8
 x ./smithb, 0 bytes, 0 tape blocks
 x ./smithb/.login, 575 bytes, 2 tape blocks
 x ./smithb/.cshrc, 124 bytes, 1 tape blocks
 # rsh grissom ls /export/home2
 smithb
 # ▮
```

Moving a home directory between servers.

Compared to creating or modifying an account, deleting an account is simple. Just click the account to be deleted from the User Account Manager window and select Delete User from the Edit menu. The Delete User form which appears presents the option of having the account manager delete the account's home directory, or it can be manually deleted by the system administrator. Depending on the particular environment, it is often desirable to make a backup copy of the contents of any account before deleting it, just in case.

Managing Accounts Without the admintool

On systems lacking access to OpenWindows, tools like the **admintool** are not available. In these cases, such as file servers without any graphics displays, you must resort to the **useradd**, **usermod**, and **userdel** commands for account management; **groupadd**, **group-**

mod and **groupdel** for group management; and **passwd** or the **vipw** command provided in the Berkeley UNIX compatibility package to control passwords.

The basic strategy for account management is the same as that presented above for use with the admintool. The system administrator should carry out preparatory work to decide on an account naming scheme, UID numbers, required groups, and GID numbers. Then, local modifications to the shell initialization files must be made. With these foundations in place, groups and accounts can be created.

groupadd, groupmod, and groupdel

This trio of commands works on the group database. Unlike the admintool, the command line group management commands *do not* update the NIS+ databases. These databases must be updated manually as described later in this book.

Adding a group is accomplished by specifying the group name and GID number as options to the **groupadd** command. The following command line creates the *account* group.

```
/usr/sbin/groupadd -g 1000 account
```

Group modification is executed by using **groupmod**. Changing the *account* group GID to 1234 could be accomplished as seen below.

```
/usr/sbin/groupmod -g 1234 account
```

⟿ **NOTE:** *The* groupmod *command changes the GID number in the* /etc/group *file only. File ownership must be changed manually using the* chgrp *command, as shown in the examples in the preceding section.*

Similarly, groups are deleted using the **groupdel** command. Deleting the *account* group would be accomplished as follows:

```
/usr/sbin/groupdel account
```

useradd, usermod, and userdel

This trio of commands performs the same functions as the User Account Manager via a series of command line options similar to the **groupadd** family. In the following string, the *smithb* account is created with the **useradd** command in the same way an account was created with the User Account Manager.

```
/usr/sbin/useradd -u 1000 -g account -c Bob W. Smith -s /bin/sh -m -d /export/
home/smithb smithb
```

The above command is lengthy, but easy to follow. As described in the table below, the individual option flags parallel the four sections of the Add User form.

Option flags for the useradd command

Option	Description
-u 1000	Sets the UID number for account.
-g account	Sets the primary group.
-c Bob W. Smith	Sets the comment field.
-s /bin/sh	Sets the account shell to the Bourne shell.
-m	Creates the home directory.
-d /export/home/smithb	Sets the path to the account's home directory.
smithb	Account user name.

• **NOTE:** *The useradd command copies all files from the skeleton directory* /etc/skel, *but it does not* *rename the files to their proper names (e.g., local.cshrc to .cshrc) as the admintool does. One workaround option is to manually rename each file. Another workaround would be to set up a separate directory for each shell containing the dot file (e.g., /etc/skel/csh/.cshrc) and use the -k option of useradd to tell it to copy the dot files from the specified directory.*

The **useradd** command also has options to set the account's expiration date (**-e**) and the maximum number of days the account can be idle before being locked (**-f**). This command does not have options for setting the password aging time period. In addition, the useradd command does not handle setting the user's password.

The **usermod** command uses the same set of option flags as useradd. An example of changing the comment field for *smitha* via usermod appears below.

```
/usr/sbin/usermod -c William W. Smith smitha
```

Finally, the **userdel** command simply takes the name of the account to be deleted.

```
/usr/sbin/userdel smitha
```

passwd

To set a user's password from the shell, use the **passwd** command from the root account and give the user account's name as the first argument.

```
passwd smitha
```

The passwd command is also used to control the password aging time limits. The **-n** option sets the minimum number of days between password changes. The

-x option sets the maximum number of days between password changes. The **-w** sets the number of days before the expiration date that the user will be warned about password expiration. The command line to set these values for the example user account follows:

```
passwd -n 2 -x 180 -w 7 smitha
```

As demonstrated below, the password aging feature can be turned off by using **-1** for the maximum age.

```
passwd -x -1 smitha
```

The passwd command also has the ability to give the root user password status information. The **-s** option prints a single line of information on the account listed. The line contains the account name, status (NP = no password, LK = locked, PS = normal), the date of the last password change, and the minimum, maximum, and warning password aging values.

```
# passwd -a smitha
smitha PS 7/1/94 2 180 7
```

You can obtain a list of the password status for all accounts on a system by using the **-a** option along with the **-s** option.

```
# passwd -s -a
root PS
daemon LK
dwight PS
smitha PS 7/1/94 180 7
 (...)
```

/etc/shadow

The password for an account is not stored in the */etc/passwd* file because the file is typically world readable. A simple method of breaking system security is to obtain a copy of the encrypted passwords, and then run a program to crack the encryption by guessing passwords, encrypting them, and comparing the results.

To provide more secure passwords, Solaris stores the encrypted passwords in the */etc/shadow* file. This file is readable by the *root* account only. All log-in functions use password retrieval routines which know to check the */etc/shadow* file for the password. By keeping the passwords in a locked file, system crackers cannot easily obtain a copy of encrypted passwords, and system security is maintained.

For Experts Only: vipw

If the Berkeley UNIX compatibility package is installed, the **vipw** command can be used to add, modify, and delete account records from the */etc/passwd* file. This command simply allows the administrator to directly edit this data file using an editor, typically the **vi** editor from which the command gets its name.

Direct editing of the password file is prone to errors. One must meticulously adhere to the file format and be careful not to leave incomplete or malformed entries in the file. The vipw command performs no checking other than to verify that the *root* account is sufficiently correct to allow root to log in. Incorrect entries in the password file can prevent users from logging in to their accounts. Unless you are accustomed to this sort of file editing from working on other UNIX systems, stay away from **vipw** and use the other commands described above. These commands are designed to prevent errors.

Summary

An account is a collection of identification, authentication, and permission information stored in data files or an NIS+ database. By using OpenWindows, the admintool provides an easy-to-use graphical series of forms for adding, modifying, and deleting user accounts. However, attention must be devoted to preparing for new accounts and handling situations not covered by the admintool. If OpenWindows is not installed on the system being managed, you can use a series of command line tools which cover nearly all functions found in the admintool.

Managing System Security

System security is probably the most talked about topic in the computer industry today. Collections of workstations and servers running Solaris present a variety of security challenges for the system manager. It is essential to understand the level of security required by a particular site and what must be done to establish that security. Maintaining the appropriate level of security for a particular environment requires a balance between technical considerations and the needs of the organization.

For Additional Information

In this chapter, the discussion is limited to file and directory security, user account security, and the **aset** tool. Network security is covered in Chapter 17. Readers whose Solaris systems are currently or soon to be connected to the Internet, or whose environment requires a more thorough approach, are strongly encouraged to read further on this topic. References which specifically address UNIX system security should be considered required reading for managers of systems with advanced security requirments. A short list of suggested readings appears at the end of Chapter 18, "Network Security."

How Much Security?

When considering system security, the contents of users' files constitute the most basic item being protected. The system administrator needs to ask several questions. Who must be able to read or modify the file contents? Would knowing the contents of a certain file allow someone to gain access to other files? Would knowing the contents of the file release information that should not be disseminated?

Think about an ordinary paper document (file) for a moment. If the file is very sensitive one might lock it in a file cabinet or even a safe. The same holds true for the computer equivalent of such files. If the contents of a file on the system are highly sensitive, should the file be available on-line at all? Consider the hazards. Workstations, disk drives, and especially notebook computers are easy to carry off. Networked computers allow files to be copied quickly and invisibly.

How can system administrators protect the most sensitive data? Files which demand the highest security yet still need to be manipulated on a computer should be stored on removable media such as a floppy, removable disk, or tape which can be physically locked up in the same way as a paper file of similar importance.

Encrypting Sensitive Data

Another approach for securing sensitive files is to encrypt them using the Solaris **crypt** command. The command uses a mathematical function to scramble the file contents in such a way as to make it unreadable until the file is decrypted.

Encryption's Achilles Heel

While encryption may seem to be a less bothersome method of protecting a sensitive file than removing the file from the system, be warned that encryption only protects a file for a limited period of time. Given time, the key or password used to encrypt the file will be discovered. Automated methods to decipher such passwords are available. The length of time required to identify an encryption key is a function of the key length, the encryption method used, other information known about the contents of the encrypted file, and the computer resources available to the person committed to discovering the key. Strong encryption techniques may require years to circumvent. When something is known about the key or file contents, or when weak methods with short keys are used, encryption can be circumvented in hours or less.

Owners, Groups, and Permissions

The contents of most files typically do not require high-level protection. It is sufficient to ensure that they are readable or modifiable only by certain users. Access to the contents of a file on a Solaris system is controlled by a combination of the file's user and group ownership and the protection bits associated with the file. The output of **ls -l** in a typical directory appears in the following illustration.

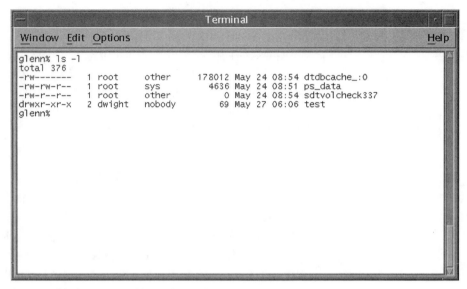

```
glenn% ls -l
total 376
-rw-------    1 root     other        178012 May 24 08:54 dtdbcache_:0
-rw-rw-r--    1 root     sys            4636 May 24 08:51 ps_data
-rw-r--r--    1 root     other             0 May 24 08:54 sdtvolcheck337
drwxr-xr-x    2 dwight   nobody           69 May 27 06:06 test
glenn%
```

Output of ls -l.

The long listing option (**-l**) of the **ls** command produces several columns which contain various pieces of information about the files in the current directory. The columns from left to right are listed below.

1. Symbolic representation of the permission bits, sometimes called the *permission* mode or simply the *mode*.

2. Number of file links made to the file.

3. File owner.

4. File group ownership.

5. File size in bytes.

6. Date and time the file was last modified.

7. Name of the file.

The first, third, and fourth columns from the left contain the security information.

File Protection Bits

Every file has three levels of security associated with it which match the three types of users that may access that file. File protection according to the three categories of users is reviewed below.

❏ *owner*—The assigned owner of the file.

❏ *group*—A logical group the file and user belong to (like a department within a company).

❏ *other*—All other users who either do not own the file or are not a member of the file's group.

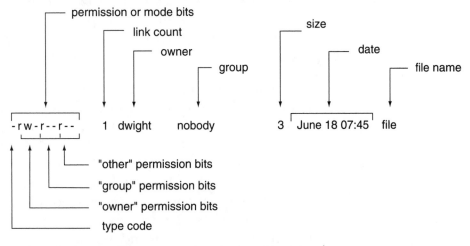

A closer look at the file protection bits displayed by ls-l.

The previous illustration shows the grouping of protection bits as represented by **ls -l**. Each character in the security field represents a significant feature of the file. The first character indicates type information. Type codes are listed in the following table.

Protection type codes

Type code	Meaning
-	File.
d	Directory.
b	Block special device entry. A file system entry which SunOS maps to a specific device. Block specials are usually buffered devices such as disks.
c	Character special device entry, or a file system entry which SunOS maps to a specific device. Character specials are "raw" or unbuffered devices such as terminal ports.
p	Named pipe. A special entry used for interprocess communication.
l	Symbolic link. Unlike a file link, a symbolic link contains the path to another file instead of a direct pointer to it. Symbolic links can span file systems and even point to files available via NFS from other machines.

⊷ **NOTE:** *In the case of a symbolic link, the permissions of the referenced file apply. Permissions pertaining to the link are not used to determine file access.*

The nine remaining characters are treated as three groups of three. As shown in the diagram, the first group indicates permissions for the file owner, the second for group ownership, and the third for other users. Each character triplet denotes various permissions for each entity. The presence of a character such as *r* indicates that permission is granted, while a dash (-) in place of the character indicates that permission is denied. Each triplet is independent; all three permission bits can be set individually for the owner, group members, and others. However, the letter spaces in the triplet are "overloaded"; that is, several less commonly used bits show up as letters other than the common *rwx* to signify special permissions. The permissions and their meanings for files are listed in the following table.

File permissions

Permissions	Meaning
r	Read. Programs can open and read the contents of a file.
w	Write. Programs can store new contents in the file.
x	Execute. Depending on the "magic number" stored in the first two bytes of the file, one of two things happens. If the number indicates that this is a program binary, it is loaded into memory and starts running. If the magic number is not present or indicates that the file is a shell script, a shell is started and the file is read by the shell.
s	Setuid or setgid.
t	The "sticky" bit found in the third position of the third triple. This bit indicates that the file is to be left on the swap device after it finishes execution. It was used to improve performance for frequently executed commands on older versions of UNIX, but has since become obsolete as a result of improved forms of mangement.
l	Mandatory file locking. This letter indicates that the setgid is on and the group execution bit is off. If a program locks this file, the kernel will prevent other programs from accessing it until the lock is cleared.

➡ **NOTE:** *The "magic number" is a special value placed at the beginning of a file. This number tells the system how to load and run the contents of the file. It is consulted if the file's execution bit(s) are set and the user enters the name of the file as a command. The two most common magic numbers are used for program binaries and shell scripts. By default,* /bin/sh *is used to run files which do not appear to have a magic number. To use another shell, place the string #! and the path to the shell as the first line of the file. The string #! is itself a magic number and causes the loader to use the remainder of the first line as the path to the command to be run. For example,* #!/bin/csh *as the first line of a file would cause* /bin/csh *to be started and the remainder of the file to be read by* /bin/csh.

☞ **WARNING:** *The* setuid *and* setgid *bits are very powerful and easily misused, and can result in security problems. The* setuid *bit is most often used to allow unprivileged users to execute a command as if they had root privileges. Do not use this facility lightly. Read the detailed description of* SetUID *problems below.*

As shown in the next table, the protection bits have similar meanings for directories.

Directory permissions

Function	Meaning
r	Read permission. Files in the directory can be listed.
w	Write permission. Files or links can be added or removed from the directory.
x	Execute permission. Files can be opened in the directory. The shell can change to this directory (cd).
s	Setgid. In a directory, setting the setgid bit causes all files created in the directory to be in the same group as the directory itself.
t	The sticky bit. Setting the sticky bit on a directory means that only the owner of a file in the directory can remove it. The sticky bit is most often used for directories like /tmp.

Note the difference between *read* and *execute* permissions for a directory. In the following example, pay special attention to how a directory with only the *read* permission allows files to be listed but not examined, even though the files have the required permissions.

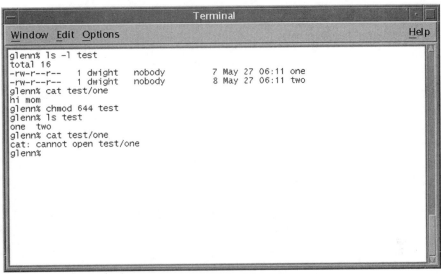

Directory with read permission.

➝ **NOTE:** *The accessibility of a given file is a combination of the permissions and ownerships of all directories along the path to the file as well as the permissions and ownerships of the file itself. When planning file protection schemes, consider which files need to be read and which files should be hidden. Grouping files in directories for security is an approach worth considering.*

chown and chgrp Commands

Changing file ownerships and groups is easily executed with the use of the **chown** and **chgrp** commands.

```
# chown username item
# chgrp group item
```

The *item* listed above is a file or a directory. The *username* and *group* are the user account name and group name, respectively, as found in */etc/passwd* and */etc/group*, or in the NIS or NIS+ databases, or the numerical UID or GID numbers. To change an entire directory and all its subdirectories, both chown and chgrp accept the **-R** option.

Setting Protection Using Symbolic Names

Specifying changes to the protection bits is more involved. Make the changes with the **chmod** command, which has an **-R** option like **chown** for handling entire directories. Two methods of specifying the permission bit changes are possible. The first method is symbolic. A set of symbols is used to specify the entity (user, group, other) and the mode (read, write, execute) to be changed. The symbols are listed in the next table.

Symbols for chmod command

Entity	Symbol
User	u
Group	g
Everyone (other)	o
All entities	a

Mode	Symbol
Read	r
Write	w
Execute	x
Mandatory locking	l
SetUID or SetGID	s
Sticky bit	t

The symbols are combined with the plus sign (+) operator to add the permission, the dash (-) operator to remove the permission, and the equals sign (=) operator to assign the permission. For example, you could remove the *read* permission for other from the file named *bob-stuff* with the following command line:

```
chmod o-r bob-stuff
```

Multiple permissions can be changed at the same time by specifying lists of symbols and operators. An example of setting the execute bits for all and restricting the read access on a file named *test99* follows:

```
chmod a+x,g-r,o-r test99
```

Developing a habit of checking modes and owners on files after setting them is recommended. Use **ls -l** and check the symbolic permission bit column to see who can read and execute the file. Check the owner and group. It is very easy to make a typographical error and set the modes to unintended values.

Using Octal Numbers

The second form of specifying the permission bits is to provide them as a base eight or octal number as shown in the previous directory example. This looks confusing at first, but it is not difficult to learn and has the advantage of compactly representing all protection bits in a single easy-to-type form. A simple way to derive this octal number is to add up the numbers assigned to each permission bit from a table like the following one.

Octal number permissions

Number	Bit
4000	SetUID
2000	SetGID
400	Read permission for owner

Number	Bit
200	Write permission for owner
100	Execute permission for owner
40	Read permission for group
20	Write permission for group
10	Execute permission for group
4	Read permission for other
2	Write permission for other
1	Execute permission for other

Using this table, it is possible to specify permissions for a file that has read, write, and execute permissions for the owner, and read and execute permissions for group and other as follows: 400 + 200 + 100 + 40 + 10 + 4 + 1 = 755.

Use the total number with **chmod** as follows:

```
chmod 755 demofile
```

Setting the Default File Mode

Created files are given a default set of permission bits that is easily customized. The default is set in one of the shell initialization files with the **umask** command. This command takes a single argument which is the complement of the mode the newly created files will receive. The argument can easily be calculated by subtracting the desired file mode in octal from the number *777* for directories and *666* for files. The numbering is such that by default directories will have the execute bit set on, and files will not.

For example, to create a very secure *600* file mode which would give read and write permission to the file owner only, calculate the umask as follows: 666 - 600 = 066.

A more friendly or less secure mode of *644* would be set with a value of *022*. To set the value, simply use it as the first argument to umask as follows:

```
umask 022
```

Many sites set their local default value for permission bits by placing a line including the umask command in the system-side shell initialization files. On a Solaris system, the administrator can also set umask in the */etc/default/login* file. This sets it for all users who log into a system. See Chapter 6, "Creating, Deleting, and Managing User Accounts," for additional information on these files and other initialization files where umask might be placed.

In general, you should consider the contents of a file and who should have access to it when specifying the owner, group, and permission bits. System files that do not need to be read by user programs should be made unreadable.

Implementing File Security Using Groups and File Protection Modes

Effective file security using UNIX groups and file protection modes requires forethought and organization. First, consider the structure of information within the organization. Next, consider which users need to share information and which users should not share nor participate in the exchange of information (i.e., guest accounts, remote log-ins, etc.). Users who need to share files should be put in a UNIX group. For example, you might put the members of the accounting department into a group called *account*. See Chapter 6, "Creating, Deleting, and Managing User Accounts" for instructions on how to accomplish this.

Continuing with the example, files such as an accounting database or other records can be given the group ownership *account.* File modes can be set to allow members of the *account* group access to the accounting database. By selecting a mode which gives read and write permissions to the owner and group, but not to others, files can be effectively shared within a selected group but protected from other users. Using the combination of a group and the group file protection bits is an often overlooked method of obtaining secure but shared files.

Be Careful with chmod

Like the **rm** command, **chmod** does not check the option flags provided to determine whether they make sense; it simply makes the change. Sometimes the changes to the protection bits lead to unexpected results. Be careful not to fall into the common traps described below.

❑ You can inadvertently remove permissions for the file or directory owner. If you suddenly learn of a file which someone owns but cannot read or write, check that the owner still has permission to read or write the file.

❑ Directories require execute permission to access contents. Removing the execute permission on a directory leaves it in the confusing state of being readable but inaccessible.

❑ Directories require read permission for their contents to be seen by **ls** and other commands. If you wish to create a directory that others can read and execute items from, but do not wish them to see the contents of the directory, remove the read permission while leaving the execute permission.

Be warned that this does *not* prevent others from guessing at the names of other files in the directory, nor does it protect them from being read. This procedure only hides them from being listed by commands such as ls.

Dangers of SetUID and SetGID

The following is a scenario that could easily endanger system security. A system administrator logs in to a workstation as *root.* The administrator needs to leave the workstation for a while and forgets to log off or lock the screen. Has the administrator left the system in peril? Could someone use the unsecured session to gain root access later on without the administator's knowledge? Sadly, the answer is yes. One approach available to this "someone" is to use the *setuid* bit. For example, assume that an unscrupulous person typed the following into an unsecured *root* session:

```
# cp /bin/sh /tmp/gotcha
# chmod 4755 /tmp/gotcha
```

The above commands create a copy of the Bourne shell that when run will execute with full *root* privileges. The person who typed in the above two statements will not require the *root* password, nor will she or he be required to log in as *root.* The person need only run the copy of the shell binary which contains the *setuid* bit set.

There are many other methods that a determined person could use to exploit the setuid bit. Great care should be exercised when considering which programs are given this ability. For example, suppose a setuid *root* copy of the **vi** editor is made. Be aware that a user can start up a shell from within vi. What privileges would that shell have?

Taking Precautions

General guidelines to avoid *setuid/setgid* problems are summarized below.

❑ Avoid using setuid/setgid whenever possible. If there is another way to implement the function which does not require setuid/setgid, strongly consider using it.

❑ Make the program setuid/setgid to the least privileged user account possible. Avoid setuid root programs unless the program *must* have that privilege to function.

❑ *Never* make a shell script setuid or setgid. Shell scripts are too easily subverted and used to create setuid shells.

❑ When developing new setuid/setgid programs, consider releasing the privilege as soon as the privileged functions are completed. There are specific system library functions for setting and removing privileges in this manner.

❑ Consider having all programs which are setuid or setgid write to a log file when they are run. This provides an audit trail allowing the system administrator to monitor privileged usage.

Rooting Out Potential Problems

Another precaution worth taking is to scan the files on all systems for setuid and setgid programs. One method of accomplishing this is with the **ncheck** command. The ncheck command scans on a file system by file system basis. An example of ncheck usage follows:

```
# ncheck -F ufs -s /dev/dsk/c0t3d0s6
```

The last argument is the device entry for the file system to be checked. To identify the device entry for a given file system, type **df**. The list produced by ncheck shows the *inode* or index node number for each file it found, and the file name relative to the mount point of the file system. In other words, if the device entry of the */usr* file system were entered, the list of file names produced would start at */usr.* A name listed as */lib/ acct/accton* would have a full path of */usr/lib/acct/ accton.*

Once a list of files has been generated, use **ls** to examine each file to identify its owner, group, and permission bits. To simultaneously accomplish these tasks, process the output of ncheck and feed it directly to ls. An example of this processing as it would be used to check the */usr* file system on a typical system follows:

```
# ls -l `ncheck -F ufs -s /dev/dsk/c0t3d0s6 | cut -f2 | sed 's:^:/usr:'`
```

While appearing complex, this command line elegantly handles the task. The **ls -l** command will produce listings of the owner, group, and permissions for each file name given to it as the output of the phrase between the back quotes. That phrase runs ncheck to produce the list of setuid or setgid files. The second column of that list, the file names, is clipped out by using the **cut** command. Then the list of files is passed through **sed**, which uses a regular expression to prepend */usr* to each file name to make them full path names. The result is a list of the setuid and setgid files on the */usr* file system complete with respective owners, groups, and permissions.

Account Security

Chapter 6 discussed how user accounts were created and managed. One of the features Solaris offers to aid account management is a variety of controls over the account password. The password is the mechanism by which the system authenticates the user of a given account. The file security mechanisms described above depend on this authentication to ensure that the person who has logged on to a particular account is really the file's owner.

Solaris stores account passwords and most other account information in separate places. On a system which is not using NIS or NIS+, these places are the */etc/passwd* and */etc/shadow* files. Examine the owners and permissions for these files. The */etc/passwd* file is readable by all because many programs make use of the information in this file. The */etc/shadow* file is protected, and contains the encrypted version of every user account password along with password aging information. Because the encrypted passwords, like all encrypted text, are susceptible to being discovered, the contents of the */etc/shadow* file must be kept unreadable (except by the root account).

Educate Users About Safe Passwords

It is also important to educate users with regard to password security. Writing the password on paper should be discouraged, as should sharing of passwords among users. How important this is depends on the environment and policies of the system owner organization. The more secure the environment must be, the more thorough user education and password protection should be.

The next line of defense is to ensure that the password chosen by each user is not easily guessed. Personal information should not be used as a password. Items such as the user's real name, phone number, license plate number, address, spouse's or children's names, office number, and so on should be avoided because these items are too easy for a determined individual to guess.

The use of common words or names is not a good approach either. A determined individual can make use of high-speed password guessing software in combination with large dictionaries of words. While this approach is time-consuming, it is getting less so as password guessing tools improve and faster computer hardware becomes readily available. Be especially careful to avoid famous names such as Superman because these will most certainly be in even the shortest dictionary used to guess passwords.

Better choices for passwords include the following properties: punctuation characters, numbers, and modified words. Some suggestions are word combinations such as *best1boy*; the first letter or number of words in a phrase (e.g., *towamfn* from "There once was a man from Nantucket"); or weird combinations of numbers and letter phrases (e.g., *10%grade*, or *any1410s* from "Anyone for tennis?"). Do not use anything suggested here or in other books, as you can be sure a determined individual trying to guess at an account password will know of these examples.

Using the Solaris Password Aging Feature

As discussed previously, encrypted text only remains protected for the time period required to discover the encryption key or password. For this reason, user account passwords should be changed on a regular basis. This is the purpose of the Solaris password aging mechanism. The frequency of password changing depends on the security requirements of the environment. More frequent changes are better for security, but more annoying for the account user. Frequent changes may even prompt an account user to select a poor password just to get back to work. Be sure to spend the time to educate and explain the reasons behind good passwords and changing passwords to all users. The security of the system as a whole depends in part on each user's choice of a password.

For the system administrator, no password is more important to protect than the password for the *root* account. The root password gives access to *all* data stored on the system and may provide access to other systems on the network. The importance of preceding comments about user account passwords is doubled for the root account. Carefully choose a root password and change it frequently. Make a habit of frequently checking the log files in */var/adm*, especially */var/adm/sulog*, which logs valid and failed uses of the **su** command. The su command is often used to gain root privileges from an ordinary user account.

Tracking Security Using aset

Checking for *setuid* files and monitoring log files can be quite a chore. Fortunately, Solaris provides **aset**, a security tool that will help with these chores and act on problems automatically.

The aset tool consists of a set of tasks or separate small programs which check specific areas of system security. The available tasks are **tune**, **cklist**, **usrgrp**, **sysconf**, **env**, **eeprom**, and **firewall**. The aset tool is run at a specific security level: low, medium or high. The tasks report and take actions to aid in securing the system based on the selected security level. The following table describes the tasks and respective actions at each security level.

List of aset tasks by security level

Task	Function	Low	Medium	High
tune	Sets system file permissions.	Sets permissions to release values.	Tightens permissions to improve security.	Sets permissions to be highly restrictive.
cklist	Checks system files against master list.	Compares some files against previously saved permissions settings and reports changes.	Increases the number of areas checked.	Checks the most files.
usrgrp	User/group checks.	Checks for consistency and accuracy of password and group files. Also checks NIS+ password if the *ypcheck* environment variable is set.	Same.	Same.
sysconf	System configuration file checks.	Checks numerous files in */etc*.	Makes changes to restrict access.	Makes the most restrictive set of changes.
env	Root environment check.	Checks "dot" files for path and umask settings; checks *root* and other user accounts as listed.	Same.	Same.

Task	Function	Low	Medium	High
eeprom	EEPROM security check.	Checks the eeprom security setting and password. Does not make changes, but reports recommendations.	Same.	Same.
firewall	Sets system to act as network firewall.	Reports changes required to convert system to a network firewall.	Same.	Takes actions required to convert system to a network firewall.

Taking a walk through a typical **aset** run will further explain the items in the previous table and will also show how aset is used and can be modified to fit a given environment. Starting aset is a matter of becoming *root* and typing the command.

```
# /usr/aset/aset
```

To start **aset** at the medium or high security levels, use the **-l** option. An example follows:

```
# /usr/aset/aset -l medium
```

For the */bin/csh* shell, use the following to set ASET-SECLEVEL:

```
# setenv ASETSECLEVEL medium
```

For the */bin/sh* shell, use the following to set ASETSE-CLEVEL:

```
# ASETSECLEVEL=medium ; export ASETSECLEVEL
```

While aset is running, it is possible to check on its progress by using the **taskstat** utility.

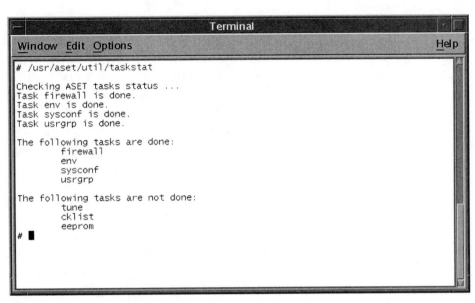

```
┌─────────────────────────────────────────────────────────────┐
│ ─                        Terminal                       ▪ □  │
├─────────────────────────────────────────────────────────────┤
│  Window  Edit  Options                                 Help  │
├─────────────────────────────────────────────────────────────┤
│ # /usr/aset/util/taskstat                                    │
│                                                              │
│ Checking ASET tasks status ...                               │
│ Task firewall is done.                                       │
│ Task env is done.                                            │
│ Task sysconf is done.                                        │
│ Task usrgrp is done.                                         │
│                                                              │
│ The following tasks are done:                                │
│         firewall                                             │
│         env                                                  │
│         sysconf                                              │
│         usrgrp                                               │
│                                                              │
│ The following tasks are not done:                            │
│         tune                                                 │
│         cklist                                               │
│         eeprom                                               │
│ # ▪                                                          │
│                                                              │
│                                                              │
└─────────────────────────────────────────────────────────────┘
```

Checking aset progress with taskstat.

As aset runs it will generate a large amount of disk activity. Consequently, running aset when system usage is at a low level is recommended. It is also a good idea to run aset periodically, especially in a networked environment, to check for any changes that may have occurred. As shown below, both of these conditions are met by using the **-p** option to aset.

```
# /usr/aset/aset -p
```

The previous statement will put an entry for aset in the *root crontab* file. The **cron** daemon reads each user's *crontab* files and runs programs at the times listed in the crontab file. The **aset -p** command inserts a scheduled run of aset to be performed nightly at midnight. This can be checked by typing **crontab -1** and looking for the line which contains aset. To remove this regularly scheduled aset run from the root crontab, type **crontab -e** as root, and delete the line containing aset.

Each aset run creates a set of log files which detail any problems found by aset and all corrective actions taken. The log files are found in directories in */usr/aset/reports*. The directory names correspond to the date and time of the aset run which created the reports. For instance, the directory *0716_07:47* contains the reports for the 7:47 am aset run on July 16. The symbolic link */usr/aset/reports/latest* is set to point to the most recent report directory.

The report directory contains files with the *.rpt* extension. These are the reports generated by various aset tasks. At the low security level used in the above example, very little is written into most reports. However, even at this level, the system has a few minor problems worth correcting. The contents of *env.rpt*, the root environment report, for the aset in the previous example appear below.

```
# more env.rpt
*** Begin Environment Check ***
Warning! umask set to umask 022 in /etc/profile - not recommended.
*** End Environment Check ***
#
```

The above report indicates a questionable setting for **umask** in */etc/profile*. This setting would create files that are globally readable. If global read access is not desirable in a given environment, a change to the umask line in */etc/profile* is in order. Another error was spotted by the **sysconf** (system file configuration task), and reported in the *sysconf.rpt* as seen here.

```
# more sysconf.rpt
*** Begin System Scripts Check ***
Warning! The use of /.rhosts file is not recommended for system security.
*** End System Scripts Check ***
#
```

Although a /.*rhosts* file is a convenience, allowing remote access to *root* across a network requires that the remote machine be every bit as secure as the local machine. This warning should be carefully considered and the /.*rhosts* file removed if it is not needed in a given environment.

Reading through all the reports the first time you run them is worthwhile to get a sense of what aset checks for and the problems that may need correction. For periodic monitoring, however, reading reports is time-consuming. Another approach is to use the **diff** command to compare reports and see if any significant changes have occurred since the last aset run. This command helps to highlight problems which have recently emerged and reduces the repetitive nature of reading the entire report series. A quick shell script which compares the reports in a named directory against the latest reports is reproduced below.

```
#!/bin/sh -f
#
# Compare files in $1 with latest
cd /usr/aset/reports/$1
for i
in *.rpt
do
          echo == $i
          diff $i ../latest/$i
done
```

Running the above script provides the following results on the example system.

```
# /tmp/sample 0716_07:47
== cklist.rpt
```

```
4,6c4,7
<< No checklist master - comparison not performed.
<< ... Checklist master is being created now. Wait ...
<< ... Checklist master created.
---
>> ... Checklist snapshot is being created. Wait ...
>> ... Checklist snapshot created.
>>
>> No differences in the checklist.
== eeprom.rpt
== env.rpt
== firewall.rpt
== sysconf.rpt
== tune.rpt
== usrgrp.rpt
#
```

Because the July 16 run was the first time **aset** was run on this system, the **cklist** task had not yet created a master checklist. The **diff** command caught the change and reported the changed lines. No changes were found in the other reports. A short display of differences like this one can be a real time saver, especially if you are managing numerous systems. With a little work, this simple shell script can be extended to mail any changes to a specific user account. Mail would be sent only if changes were found, further reducing the effort required for security monitoring of large collections of workstations.

aset Customization

Of course, the items checked by **aset** may not be appropriate for a given site. There are two ways to control what aset does and which files and directories it checks.

The first method is to use the *aset* environment variables. ASETSECLEVEL was already mentioned; setting it controls the security level at which aset runs. In addition, the ASETDIR variable specifies an alternative directory for aset to use when looking for files and writing reports.

↝ NOTE: *The environment variables are typically used for interactive runs of* aset. *When* aset *is run by the cron daemon, it uses command line options to select the security level (e.g.,* -l medium*) and alternative directory (e.g.,* -d /home/security*).*

The second method of customization is to modify the various configuration files for aset. The first of these is **asetenv**. This file is usually found in */usr/aset*. It contains a set of variables near the top which may be customized to suit the local environment. The variables are listed in the following table.

List of asetenv configuration variables

Variable	Description
TASKS	Sets the list of tasks to be performed by aset.
PERIODIC_SCHEDULE	Sets the schedule to be entered into the root crontab file when aset -p is run.
UID_ALIASES	Sets the name of the file to consult for a list of UIDs which can be shared by different account names.
YPCHECK	Specifies whether a check of the INS or NIS+ database will be made. This check will only report problems; it does not take action to fix problems.
CKLISTPATH_LOW	Sets the list of paths used at the low security level by the cklist task.
CKLISTPATH_MED	Sets the list of paths used at the medium security level by the cklist task.
CKLISTPATH_HIGH	Sets the list of paths used at the high security level by the cklist task.

The CKLISTPATH variables contain a set of directories separated by colons. For example, to set CHKLISTPATH_LOW to check */usr/bin, /sbin,* and */etc,* enter the following command line:

```
setenv CHKLISTPATH_LOW /usr/bin:/sbin:/etc
```

The second set of customization files is found in */usr/aset/master* and consists of the **tune** family of files. These files contain the lists of files the tune task checks for integrity. The format is described in the top few lines of the file and consists of the path name to the file, its permission bits in octal notation, owner, group, and type. You can add path names of files to be checked by following the format shown. Add files in the tune file for the security level at which aset will run: *tune.low, tune.med,* or *tune.high.*

Restoring Permissions to Pre-aset Values

To conclude the discussion of **aset**, one more handy aset tool exists: **/usr/aset/aset.restore**. This program restores system file permissions to the state they were in before aset was first run. You might wish to use this tool to restore settings before removing aset, or if file permission settings were found to be too restrictive after an initial aset run.

Summary

This chapter discussed the basics of the UNIX file permission scheme, and how to view and modify file permissions. The dangers of the setuid and setgid bits were discussed; these should not be overlooked when creating programs which have these attributes. By using the password aging and aset tools found in Solaris, many aspects of system security can be controlled. These tools and techniques are sufficient for many environments. However, system administrators whose sites demand more strict security would be well advised to read further on this topic and consider employing more advanced tools and techniques.

Working with Solaris 2 Device Names

One of the major functions performed as a system administrator is the installation of and communication with devices on the system. This chapter presents an overview of Solaris devices, followed by discussion of device types and methods which may be used to access these devices. Finally, the chapter covers naming conventions used for system devices under Solaris, how to determine which devices are connected to systems, and how to inform Solaris that those devices are available.

Device Terminology

Understanding device terminology is the first step toward allowing an administrator to troubleshoot problems with devices on a system. Most devices are named with an acronym derived from the device type. For instance, a SCSI disk drive is an *sd* device. But a SCSI disk is also known as a *dsk* device. A SCSI tape drive is an *st* device, but this same tape drive also has an *rmt* device name associated with it.

In brief, under Solaris a device can have several names. The name of the device is also known as the device identifier. The following table lists a few of the common devices on SPARC-based systems.

Device Names

Device name	Description	Device name	Description
cn	System console	st	SCSI tape
esp	System SCSI adapter	zs	On-board serial ports
/dev/fd0a	Solaris 1 compatibility links to Solaris floppy device	fd0/	Standard input file descriptor
/dev/fd0b		/fd1	Standard output file descriptor
/dev/fd0		/fd2	Standard error file descriptor
/dev/diskette	Solaris 2 native mode floppy diskette device	hme0	100 Mbit Ethernet (UltraSPARC E series)
/dev/ rdiskette		sd	SCSI disk drive
le	Lance Ethernet devices	/dev/ttya or /dev/term/a	Serial port a
mbus	System mbus adapter	qe	Quad Ethernet devices
Sbus	System Sbus adapters		

Hardware Level Device Access

SPARC based systems allow the administrator to access system devices from two levels: hardware level device access, and operating system device access. Hardware level access involves the use of OpenBoot monitor commands. These commands allow the administrator to view and manipulate devices without booting the operating system, and can be very useful when new hardware is being added to the system.

The operating system device access scheme employs utilities available once the operating system is executing. These commands allow the operator to add, query, and troubleshoot device problems while the operating system is executing. Both access schemes are grounded on an understanding of how these devices are connected to the system, and the concept of a "device information tree."

Device Information Tree

On SPARC-based systems, the input/output devices are attached to a set of interconnected buses. The OpenBoot monitor recognizes these interconnected buses and respective devices as a tree of nodes. This representation of the device interconnection is called the device information tree.

It is possible to become familiar with the device information tree structure by using the OpenBoot monitor **show-devs** command. At the root of the tree is a node which describes the entire system. The leaf (child) nodes are indented (with respect to their parent node) to show the hierarchical relationships between nodes.

Typical device information tree as illustrated by the OpenBoot monitor show-devs command.

```
ok show-devs
/options
/fd@1,f7200000
/virtual-memory@0,0
/memory@0,0
/sbus@1,f8000000
/auxiliarry-io@1,f7400003
/interrupt-enable@1,f5000000
/memory-error@1,f4000000
/counter-timer@1,f3000000
/eeprom@1,f2000000
/audio@1,f7201000
/zs@1,f0000000
/zs@1,f1000000
/openprom
/packages
/sbus@1,f8000000/cgsix@2,0
/sbus@1,f8000000/SUNW,bpp@1,200000
/sbus@1,f8000000/SUNW,lpvi@1,300000
/sbus@1,f8000000/le0@0,c00000
/sbus@1,f8000000/esp@0,800000
/sbus@1,f8000000/dma@0,400000
/sbus@1,f8000000/esp@0.800000/st
/sbus@1,f8000000/esp@0.800000/sd
/packages/obp-tftp
/packages/deblocker
/packages/disk-label
ok █
```

Nodes on the device information tree which have children generally represent system buses and the associated controllers on the buses. Each parent node is allocated a physical address space and a major device number that distinguishes the devices on this node from one another. Each child of a parent node is allocated a minor device number, and a physical address within the parent node's address space.

The physical address of the nodes is generally assigned in relation to the device's physical characteristics, or the bus slot in which the controller is installed. This is accomplished in an attempt to keep device addresses from changing as systems are upgraded, and new devices are installed on the system. Each device node on the device information tree can have the attributes listed below.

❏ Properties or data structures which describe the node and associated devices.

❏ Methods or software procedures used to access the devices.

❏ Children (devices or other nodes) that are attached to the node and lie directly below it in the tree.

❏ A parent, or the node that lies directly above the current node in the device tree.

Accessing System Devices with the OpenBoot Monitor

The OpenBoot monitor refers to devices by their points of connection to the system. The operating system refers to devices by mnemonic names and device special files. Most of the device names that system administrators deal with are actually aliases which point to some physical device on the system.

For example, the root file system may reside on a disk which is referred to as either */dev/dsk/c0t3d0s0* or *sd3a* by the administrator. This same disk may be known as *disk3* or */sbus@1,f8000000/esp@0,4000/ sd@3,0:a* to the OpenBoot monitor. On a SPARCStation 2 system these names are all references to the device that the operating system refers to as */devices/ sbus@1,f8000000/esp@0,4000/sd@3,0:a*. Examination of the OpenBoot monitor addressing scheme will shed some light on this (at first glance, archaic) device naming system.

➣ **NOTE:** *On the UltraSPARC systems the root file system is located on* /dev/dsk/c0t0d0s0 *(SCSI disk zero) instead of* /dev/dsk/c0t3d0s0 *(SCSI disk 3).*

OpenBoot Access Methods

The OpenBoot monitor allows two methods of accessing devices: by use of an easy to remember alias, and by physical address references. Until the operating system is up and running, these are the only two methods to access the system devices.

The device physical addresses are determined as part of the system Power On Self Test (POST). As the system runs the power-on diagnostics, the firmware probes all system buses, and builds a table of devices. The device aliases are hardcoded into the OpenBoot monitor firmware.

EEPROM Device Mappings

The OpenBoot monitor employs an electronically erasable programmable read only memory (EEPROM) chip to store certain vital system information. Among the many parameters stored in this chip are the mappings between the device aliases and the physical device addresses. For example, the alias *disk3* refers to the SCSI disk drive connected to the on-board SCSI controller at target address three. This mapping is controlled by the value of the *scsi-probe-list* variable within the OpenBoot monitor.

The eeprom device mappings may be viewed using the OpenBoot monitor **printenv** command. The alias to physical address mappings can be altered by the OpenBoot monitor **setenv** command.

⊷ **NOTE:** *Because the* eeprom *mappings differ between SPARC systems, it is best to consult the system hardware reference manuals for more information on the contents of the OpenBoot monitor* eeprom.

OpenBoot Monitor Physical Names

The OpenBoot monitor's physical name for a device is constructed by concatenating the physical names of the nodes above the device in the device information tree with a forward slash character (/). This concatenation of nodes yields the generic form of a device physical name: *name@address:arguments*. Fields in the device name are described below.

❐ name—A name is a text string that generally has a mnemonic value. For example, **sd** represents SCSI disk. Many controller names use some portion of the manufacturer's name or model number.

❐ @—The at sign must precede the address parameter.

❐ address - The address is a text string representing the physical address of the device on the system. The address is usually in the following form: *hex_number,hex_number.*

❐ :—A colon must precede the arguments parameter.

❐ arguments—The arguments comprise a text string whose format depends on the device. The arguments may be used to pass additional information to the device's driver software.

OpenBoot Monitor Aliases

The generic OpenBoot monitor address for *disk3* may be used to illustrate how the *disk3* device alias is mapped to the physical name for this device.

1. By default, the *sbus-probe-list* variable contains the value, *0123*.

2. The eeprom defines four disk aliases: *disk0*, mapped to the first address in the probe list; *disk1*, mapped to the second address in the probe list; *disk2*, mapped to the third address in the probe list; and *disk3*, mapped to the last address in the probe list.

3. The eeprom maps the *disk3* alias to the SCSI disk connected to the on-board SCSI controller as target 3. This device is mapped to the following physical address: */sbus@1,f8000000/esp@0,4000/sd@3,0:a.*

•◦ **NOTE:** *The above mappings may be altered by changing the value of the* sbus-probe-list *variable.*

Determining Device Connectivity

Examining the physical address of a device allows the system administrator to determine the point of connection for that device. To determine the point of connection for the *disk3* device, simply break down the physical address */sbus@1,f8000000/esp@0,4000/sd@3,0:a* into the components listed below.

❑ /sbus@—Device connected to an SBus node.

❑ 1,f80000000—An address on the main system bus. Therefore, the SBus in question is an on-board device.

❑ esp@—The esp device is the on-board SCSI adapter.

❏ 0,4000—An SBus slot number and an address off-set within that slot. This means that the esp SCSI device is in SBus slot 0 with a hexadecimal address offset of 4000 from the parent node (main system bus) address.

❏ sd@—Device in question is a SCSI disk.

❏ 3,0:a—Denotes SCSI target 3, logical unit number 0, partition a of the disk.

Operating System Device Access

The operating system provides the administrator with access to the contents of the OpenBoot eeprom via the **eeprom** utility. The eeprom variables may be examined or changed with this utility. Next, the operating system provides the administrator with device access methods similar to the methods available under the OpenBoot monitor. Before examining operating system device access the administrator should have an understanding of how the UNIX operating system services input/output requests.

Device Special Files

The UNIX operating system treats input/output devices as "special" files. When the operating system references one of these device special files, it is actually invoking a device driver routine to handle the input/output request.

```
┌────────────────────────────────── Console ──────────────────────────────────┐
│ Window  Edit  Options                                                   Help │
├──────────────────────────────────────────────────────────────────────────────┤
│ # ls -lsa /devices                                                            │
│ total 16                                                                      │
│     2 drwxrwxr-x   4 root     sys          512 Dec 18 15:49 .                  │
│     2 drwxr-xr-x  25 root     root        1024 Feb  6 19:31 ..                 │
│     0 crw-------   1 root     other     28,  0 Dec 18 14:30 audio@1,f7201000:sound,audio │
│     0 crw-------   1 root     other     28,128 Dec 18 14:30 audio@1,f7201000:sound,audioctl │
│     0 crw-------   1 root     sys       68, 11 Dec 18 14:30 eeprom@1,f2000000:eeprom │
│     0 brw-rw-rw-   1 root     sys       36,  0 Dec 18 14:30 fd@1,f7200000:a     │
│     0 crw-rw-rw-   1 root     sys       36,  0 Dec 18 14:30 fd@1,f7200000:a,raw │
│     0 brw-rw-rw-   1 root     sys       36,  1 Dec 18 14:30 fd@1,f7200000:b     │
│     0 crw-rw-rw-   1 root     sys       36,  1 Dec 18 14:30 fd@1,f7200000:b,raw │
│     0 brw-rw-rw-   1 root     sys       36,  2 Dec 18 14:30 fd@1,f7200000:c     │
│     0 crw-rw-rw-   1 root     sys       36,  2 Jan 22 20:23 fd@1,f7200000:c,raw │
│     0 crw-------   1 root     sys       74,  0 Dec 18 14:30 profile:profile     │
│    10 drwxr-xr-x   2 root     sys         4608 Dec 18 15:52 pseudo             │
│     2 drwxr-xr-x   3 root     sys          512 Dec 18 14:30 sbus@1,f8000000    │
│     0 crw-rw-rw-   1 root     sys       29,  0 Dec 18 14:30 zs@1,f1000000:a     │
│     0 crw-------   1 uucp     uucp      29,131072 Dec 18 14:30 zs@1,f1000000:a,cu │
│     0 crw-rw-rw-   1 root     sys       29,  1 Dec 18 14:30 zs@1,f1000000:b     │
│     0 crw-------   1 uucp     uucp      29,131073 Dec 18 14:30 zs@1,f1000000:b,cu │
│ #                                                                             │
└──────────────────────────────────────────────────────────────────────────────┘
```

Typical device special file entries.

The device special files have two unique characteristics not found in typical data files. First, the device special files are associated with "major" and "minor" device numbers. The major device number identifies the "device class," and the minor device number identifies a specific device within a class. These major and minor device numbers are used as entry points into the device driver routines.

The operating system kernel contains a "jump table" which cross-references each device and the major/minor numbers. When the system needs to communicate with a device, it consults the jump table to locate the appropriate device major/minor numbers for this transaction.

The system invokes the device driver routine by referencing the appropriate device special file. The device driver routine determines which to open by extracting the major/minor device number of the device special file that was referenced by the operating system.

Another unique characteristic of device special files is the concept of "mode bits." The mode bits determine the type of data transfer supported by the device. Types of device special files include character and block mode devices. *Character mode devices*, also known as raw mode devices, transfer data in single byte quantities. *Block mode devices*, also known as cooked mode devices, transfer data in larger quantities (typically disk sectors, or blocks).

Some UNIX variants place all device special files in the */dev* directory. Under Solaris, the device special files are contained in the */devices* directory. Solaris uses the entries in the */dev* directory as device aliases. These alias files are symbolically linked to the device special files in the */devices* directory.

Viewing the Device Information Tree

Under Solaris the physical device interface to the nodes on the device information tree lies in the */devices* directory. This directory contains subdirectories which store device special files for audio devices, floppy disk drives, the eeprom, pseudo devices, SBus adapters, and zs (serial port) devices.

```
|total 20
    2 drwxrwxr-x   4 root    sys        512 May  7 20:43 .
    2 drwxr-xr-x  23 root    root       512 Oct  5 19:40 ..
    0 crw-------   1 curt    nobody  28,   0 Apr  9  1994 audio@1,f7201000:sound,audio
    0 crw-------   1 curt    nobody  28,128 Apr  9  1994 audio@1,f7201000:sound,audioctl
    0 crw-------   1 root    sys     68, 11 Apr  9  1994 eeprom@1,f2000000:eeprom
    0 brw-rw-rw-   1 root    sys     36,   0 Apr  9  1994 fd@1,f7200000:a
    0 crw-rw-rw-   1 root    sys     36,   0 Apr  9  1994 fd@1,f7200000:a,raw
    0 brw-rw-rw-   1 root    sys     36,   1 Apr  9  1994 fd@1,f7200000:b
    0 crw-rw-rw-   1 root    sys     36,   1 Apr  9  1994 fd@1,f7200000:b,raw
    0 brw-rw-rw-   1 root    sys     36,   2 Apr  9  1994 fd@1,f7200000:c
    0 crw-rw-rw-   1 root    sys     36,   2 Apr  9  1994 fd@1,f7200000:c,raw
    0 crw-------   1 root    sys     74,   0 Apr  9  1994 profile:profile
   14 drwxr-xr-x   2 root    sys       6656 May  7 20:34 pseudo
    2 drwxr-xr-x   3 root    sys        512 Apr  9  1994 sbus@1,f8000000
    0 crw-------   1 lp      sys     29,   0 Apr  9  1994 zs@1,f1000000:a
    0 crw-rw-rw-   1 root    sys     29,131072 Apr  9  1994 zs@1,f1000000:a,cu
    0 crw-rw-rw-   1 root    sys     29,   1 Apr  9  1994 zs@1,f1000000:b
    0 crw-rw-rw-   1 root    sys     29,131073 Apr  9  1994 zs@1,f1000000:b,cu
~
```

A typical listing of the /devices directory.

Within each subdirectory of */devices* there are more entries which may be special device files, or directories which contain special device files. The directory hierarchy simplifies the task of locating a specific device by grouping device entries for that type of device in a single directory.

```
┌────────────────────────────────────────────────────────────────────┐
│▓▓▓▓▓▓▓▓▓▓▓▓▓▓▓▓▓▓▓▓▓▓▓▓▓▓▓│Console│▓▓▓▓▓▓▓▓▓▓▓▓▓▓▓▓▓▓▓▓▓▓▓▓▓▓▓│
├────────────────────────────────────────────────────────────────────┤
│  Window   Edit   Options                                      Help  │
├────────────────────────────────────────────────────────────────────┤
│ # ls -lsa /devices/sbus@1,f8000000                                  │
│ total 10                                                            │
│    2 drwxr-xr-x   3 root    sys        512 Dec 18 14:30 .            │
│    2 drwxrwxr-x   4 root    sys        512 Dec 18 15:49 ..           │
│    0 crw-------   1 root    other   39,   0 Dec 18 14:30 cgsix@3,0:cgsix0 │
│    6 drwxr-xr-x   2 root    sys       2560 Feb  4 18:22 esp@0,800000 │
│    0 crw-------   1 root    sys     69,   0 Dec 18 14:30 sbusmem@0,0:slot0 │
│    0 crw-------   1 root    sys     69,   1 Dec 18 14:30 sbusmem@1,0:slot1 │
│    0 crw-------   1 root    sys     69,   2 Dec 18 14:30 sbusmem@2,0:slot2 │
│    0 crw-------   1 root    sys     69,   3 Dec 18 14:30 sbusmem@3,0:slot3 │
│ # ■                                                                 │
└────────────────────────────────────────────────────────────────────┘
```

A typical listing of the / devices/SBus directory.

✔ **TIP:** *A simple way to identify the address of any device on the system is to use the following command:* ls -lsaR/devices/devices/*. *The output of* *this command will list the device special files for all devices on the system.*

Logical Device Names

As seen in the previous section, physical device names can be difficult to decipher, and even more difficult to remember. Fortunately, Solaris provides an easy-to-use, easy-to-remember way to access devices that does not require the memorization of several eight-digit hexadecimal addresses.

Alphabet Soup Device Names

Like the OpenBoot monitor, Solaris allows multiple names for each device on the system. As discussed in the section on device special files, the operating system's link to the physical device is through a file in the */devices* directory. But Solaris also includes aliases for these */devices* entries. Such alias entries reside in the */dev* directory, and are symbolically linked to the entries in the */devices* directory.

For example, the */dev/dsk* directory contains entries with names such as */dev/dsk/c0t3d0s0*. This form of device name is often referred to as the "alphabet soup" designator for the device. These logical names for devices are much easier to remember than names such as */devices/sbus@1,f8000000/esp@ 0,800000/ sd@3,0:a,raw*.

An alphabet soup name for a device consists of four fields: CN, TN, DN, and SN. A breakdown of these fields appears below.

❑   CN represents controller number N. This refers to the logical controller number of the device interface. For instance, a system with a single SCSI interface would have all SCSI devices connected to controller *c0*. A system with two SCSI interfaces may have SCSI devices connected to both the *c0* and the *c1* SCSI interfaces.

❏ TN represents the target number. This is the SCSI target ID (or SCSI address) of a device connected to the controller.

❏ DN represents the drive or unit number of the device connected to target controller TN, which is in turn connected to bus controller CN. Some peripherals allow the connection of several devices to a single target controller on the SCSI bus. The Solaris device naming scheme uses the DN field of the device name to refer to these child devices. One such example is an ESDI disk controller. A typical ESDI disk controller may connect to the SCSI bus as target two. The ESDI controller may in turn allow the connection of two disk drives. These disk drives may be referred to as */dev/dsk/c0t2d0s2* and */dev/dsk/c0t2d1s2*.

❏ SN represents the slice or partition number of the device. See Chapter 9, "Disk Subsystem Hardware," on managing disks for more information on disk slices and partitions.

⟜ **NOTE:** *Under Solaris, disk drives are the only devices which use alphabet soup designators. Other System V Release IV Operating Systems may use similar designators for tape drives and other system devices.*

Legacy Device Names

Solaris also includes mappings for certain legacy device aliases. Many UNIX variants name SCSI disk devices with the following format: *sd#P,* where *#* refers to the SCSI target address, and *P* refers to a file system partition on that drive. For example, */dev/sd3a* on a SunOS 4.1.3 system refers to the "a" partition of the disk connected to the on-board SCSI controller at address 3.

These legacy device aliases are also contained in the */dev* directory hierarchy. Symbolic links connect the aliases to the appropriate device special files under the */devices* directory. For example, the */dev* directory contains an entry for the *sd3a* device. The link for this device points to another alias: */dev/rdsk/c0t3d0s0.* This file is symbolically linked to the following device special file: */devices/sbus@1,f8000000/esp@0,800000/ sd@3,0:a.*

Resolving *sd3a* to the device name */devices/SBus@ 1,f8000000/ esp@0,800000/sd@3,0:a* is shown below.

```
# ls -lsa /dev/sd3a
  2 lrwxrwxrwx  1 root     root      12 Oct  3 08:31 /dev/sd3a -> dsk/c0t0d0s0

# ls -lsa /dev/dsk/c0t0d0s0
  2 lrwxrwxrwx  1 root     root      73 Jul  1 08:48 /dev/dsk/c0t0d0s0 -> ../../devices/io-unit@f,e0200000/sbi@0,0/dma@0,81000/esp@0,80000/sd@0,0:a

# ls -las /devices/io-unit@f,e0200000/sbi@0,0/dma@0,81000/esp@0,80000/sd@0,0:a
  0 brw-r-----  1 root     sys   32,  0 Jul  1 08:15 /devices/io-unit@f,e0200000/sbi@0,0/dma@0,81000/esp@0,80000/sd@0,0:a

# ls -las /devices/io-unit@f,e0200000/sbi@0,0/dma@0,81000/esp@0,80000/sd@0,0:a,raw
  0 crw-r-----  1 root     sys   32,  0 Jul  1 08:16 /devices/io-unit@f,e0200000/sbi@0,0/dma@0,81000/esp@0,80000/sd@0,0:a,raw
```

Disk drives are not the only devices which map to legacy device names. For example, to see how the serial port *ttyb* is connected to the system, use *ls -lsa /dev/ttyb.* The output of the **ls** command shows that */dev/ttyb* is a symbolic link to */dev/term/b.* The */dev/term/b* entry is actually a symbolic link to */devices/zs@1,f1000000:b.* Similarly, */dev/rst0* is a symbolic link to */dev/rmt/0.* This file will point to */devices/sbus@1,f8000000/esp@0,800000/st@4,0:.*

Displaying System Configurations

It is frequently necessary to examine the configuration of the devices connected to a system. Whenever a new peripheral is to be added, the administrator will need to know how to address the device and where to install it. One way of displaying the system configuration is to use the OpenBoot monitor **show-devs** command. This command causes the OpenBoot monitor to display information about all devices in the device information tree.

The Solaris operating system also provides utilities which may be used to examine the system configuration information. Two utilities are described in the following sections.

prtconf

The Solaris **prtconf** utility allows the administrator to print the system configuration to a terminal or line printer. This utility provides a list of devices and drivers loaded in the system. The device information tree data are preserved by indenting child node information with respect to the parent nodes in the prtconf output.

When used with the *-p* flag, prtconf will display the OpenBoot prom information about the devices. The *-v* flag tells prtconf to display the output in verbose format.

Typical output of the prtconf command.

```
System Configuration:  Sun Microsystems  sun4c
Memory size: 28 Megabytes
System Peripherals (Software Nodes):

4_75
    packages (driver not attached)
        disk-label (driver not attached)
        deblocker (driver not attached)
        obp-tftp (driver not attached)
    options, instance #0
    aliases (driver not attached)
    openprom (driver not attached)
    zs, instance #0
    zs, instance #1
    audio (driver not attached)
    eeprom (driver not attached)
    counter-timer (driver not attached)
    memory-error (driver not attached)
    interrupt-enable (driver not attached)
    auxiliary-io (driver not attached)
    sbus, instance #0
        dma, instance #0
        esp, instance #0
            sd (driver not attached)
            st (driver not attached)
            sd, instance #0
            sd, instance #1 (driver not attached)
            sd, instance #2 (driver not attached)
            sd, instance #3
            sd, instance #4 (driver not attached)
            sd, instance #5 (driver not attached)
            sd, instance #6 (driver not attached)
        le (driver not attached)
        cgsix, instance #0
    memory (driver not attached)
    virtual-memory (driver not attached)
    fd, instance #0
    pseudo, instance #0
```

sysdef

The Solaris **sysdef** utility also allows the administrator to examine system configuration information. The sysdef utility provides a list of devices defined on the system and loadable driver modules, as well as the state of many tunable kernel variables. The device information tree information is preserved by indenting child node information with respect to the parent nodes in the sysdef output.

Typical output of the sysdef command.

```
┌─────────────────────────── Console ───────────────────────────┐
 Window  Edit  Options                                      Help
├────────────────────────────────────────────────────────────────┤
 # sysdef -d

 Node 'TWS,Tatung 4_75', unit #-1
         Node 'packages', unit #-1 (no driver)
                 Node 'disk-label', unit #-1 (no driver)
                 Node 'deblocker', unit #-1 (no driver)
                 Node 'obp-tftp', unit #-1 (no driver)
         Node 'options', unit #0
         Node 'aliases', unit #-1 (no driver)
         Node 'openprom', unit #-1 (no driver)
         Node 'zs', unit #0
         Node 'zs', unit #1
         Node 'audio', unit #0
         Node 'eeprom', unit #-1 (no driver)
         Node 'counter-timer', unit #-1 (no driver)
         Node 'memory-error', unit #-1 (no driver)
         Node 'interrupt-enable', unit #-1 (no driver)
         Node 'auxiliary-io', unit #-1 (no driver)
         Node 'sbus', unit #0
                 Node 'dma', unit #0
                 Node 'esp', unit #0
                         Node 'sd', unit #-1 (no driver)
                         Node 'st', unit #-1 (no driver)
                         Node 'sd', unit #3
                         Node 'st', unit #4
                 Node 'le', unit #0
                 Node 'cgsix', unit #0
         Node 'memory', unit #-1 (no driver)
         Node 'virtual-memory', unit #-1 (no driver)
         Node 'fd', unit #0
         Node 'pseudo', unit #0
                 Node 'clone', unit #0
                 Node 'ip', unit #0
                 Node 'tcp', unit #0
                 Node 'udp', unit #0
                 Node 'icmp', unit #0
                 Node 'arp', unit #0
                 Node 'sad', unit #0
                 Node 'consms', unit #0
                 Node 'conskbd', unit #0
                 Node 'wc', unit #0
                 Node 'iwscn', unit #0
                 Node 'ptsl', unit #0
                 Node 'tl', unit #0
                 Node 'cn', unit #0
                 Node 'mm', unit #0
                 Node 'openeepr', unit #0
                 Node 'kstat', unit #0
                 Node 'log', unit #0
                 Node 'sy', unit #0
                 Node 'pm', unit #0
                 Node 'vol', unit #0
                 Node 'ptm', unit #0
                 Node 'pts', unit #0
                 Node 'ksyms', unit #0
                 Node 'winlock', unit #0
 #
```

Device Instance Names

In the sample **prtconf** and **sysdef** output, many of the devices have an instance number associated with them. These instance numbers refer to the device address or ID on the system. For example, the disk with instance number 3 refers to disk */dev/dsk/c0t3d0s0*, or *sd3a*. The esp adapter with an instance number of 0 is the on-board SCSI adapter on this system.

Other adapters and devices follow these conventions, thereby allowing the administrator to determine the structure of the device information tree by looking at the instance numbers of the attached devices.

Reconfiguring Device Information

Sometimes when a new peripheral is installed in a system, it becomes necessary to reconfigure other peripherals. For instance, it may be necessary to move an SBus adapter card from one slot to another to allow the installation of a new SBus adapter card.

Because the device information tree contains information on the slot number of each device on the SBus, the device information tree must be rebuilt whenever devices are moved or new devices are installed on the system. Solaris provides the following three methods to rebuild the device information tree.

❐ boot -r—This command tells the OpenBoot monitor to reconfigure the device information tree before starting the operating system. Whenever a device is added or removed, the system should be restarted with the boot -r command. This forces the operating system to rebuild the device information tree and load the correct drivers.

❐ touch /reconfigure; reboot—The Solaris start-up code looks in the root directory for the */reconfigure* file. If this file exists, Solaris will rebuild the device information tree as part of the boot process.

A more dangerous way to reconfigure the system on the fly is to execute the **drvconfig**, **disks**, **tapes**, and **devlinks** commands. These four commands cause the system to probe for new disk and tape devices, and add them to the device information tree while the system is up and running. While these commands are available and perform the task, it is always best to use the **boot -r** command to let the system find and install the devices at reboot time.

Summary

This chapter examined how devices on Solaris systems are named and how they are addressed. Several examples detail the relationship between device special files, logical device names, device aliases, physical device names, and addresses of the devices. The examples also show how to track down the device physical address information for several typical device aliases.

The example output also shows how to identify the devices and driver modules loaded on a system, how to use this information to determine where to install new devices on the system, and how to force the system to reconfigure the device information when devices are added to or removed from the system.

Disk Subsystem Hardware

Previous chapters explored the naming conventions of devices under Solaris 2. The next few chapters focus on specifics of the devices that system administrators deal with every day: disks, tapes, terminals, modems, and printers.

One of the most important peripherals on any computer system is the mass storage subsystem comprised of disk drives, tape drives, and optical media (e.g., CD-ROM drives). Proper system optimization requires understanding of how the mass storage subsystem is organized. This chapter is focused on the disk subsystem on SPARCStations running the Solaris operating system.

Understanding Disk Geometry

Formatted disk drives are composed of data storage areas called *sectors*. Each Solaris data sector is capable of storing 512 bytes of data. A UNIX file system is comprised of a particular number of sectors that have been bound together by the **newfs** command. But how are these sectors organized on a disk drive? Which sectors should be bound into a file system? Can the size of a file system be changed? To answer these questions, an examination of the physical disk media is the first step.

Typical internal disk layout.

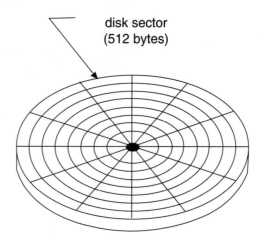

disk sector
(512 bytes)

Physical Disk Media

A disk drive contains several magnetic surfaces, called platters, which are coated with a thin layer of magnetic oxide. A typical SCSI disk may contain nine platters. Depending on the disk drive, there are one or more read/write heads positioned over each platter surface.

Disk sectors are arranged across the surface of the platter into concentric horizontal rings called *cylinders*. On the smaller innermost cylinders, the platter may be divided into a small number of sectors. On the outermost cylinders the platter could be divided into a larger number of sectors.

In order to transfer data to and from these sectors, the read/write head is passed over the magnetic surface of the platter. The read/write head converts electrical impulses from the system into magnetic patterns on the platter surfaces. The magnetic patterns represent the data stored on the platter.

Drive sectors are numbered sequentially beginning with the first sector on the first head of the first cylinder of the drive. On a drive with 55 sectors per track, the first 55 sectors will be numbered zero through 54. The 55th sector will be the first sector on the second platter of the drive. The 110th sector on the drive will be the first sector on the third platter of the drive, and so on. Upon reaching the last sector on the last platter, the heads are moved to the next cylinder, and sector numbering continues at head zero of the new cylinder.

•• **NOTE:** *Due to the incorporation of alternate sectors, sector numbering may not always be as described above. Many factors may affect the sequence of sectors. For instance, a bad sector may be mapped to a spare sector such that there would be an out of order sector in the counting sequence.*

Logical Disks and Partitions

In order to use a disk drive under Solaris, the drive must be partitioned, and the sectors must be bound together with **newfs** to form a file system. But why are disks partitioned in the first place? In order to understand the need to partition the disks, it is necessary to review a page of computer history.

At the time UNIX was first released, most machines that ran UNIX used 16-bit hardware. Consequently, a 16-bit unsigned integer number could address 65,536 sectors on the disk drive. As a result, the disk drive could be no larger than 65,536 sectors * 512 bytes/sector, or roughly 32 Mb.

When large disk drives (300 Mb or more) became available, provisions were made to allow their use on UNIX systems. The solution was to divide such drives into multiple "logical drives," each comprised of 32 MB of data. By creating several logical drives on a single physical drive, the entire capacity of the 300 Mb disk drive could be utilized. These logical drives became known as partitions. Each disk drive may have eight *partitions*. Under Solaris, the partitions are numbered zero (0) through seven (7).

Early system architecture required that large drives be partitioned into smaller logical units.

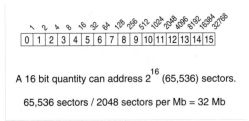

A 16 bit quantity can address 2^{16} (65,536) sectors.

65,536 sectors / 2048 sectors per Mb = 32 Mb

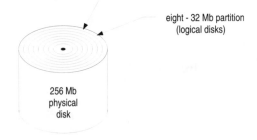

eight - 32 Mb partition
(logical disks)

256 Mb
physical
disk

Because more recent computer hardware employs 32- or 64-bit processors, the system can directly address much larger drives. A 32-bit signed variable can address 2,147,483,648 bytes (2.1 gigabytes or Gb). Under Solaris, the disk driver software is able to address nearly a terabyte of disk space, with files as large as two gigabytes.

Adding a Disk

The general procedure to add a new SCSI disk to a system is very simple. While this chapter provides details on the first six steps of physical installation of the new disk, the entire process is outlined below for completeness. Steps 7 through 12 are covered in Chapter 10.

1. Identify the SCSI target address that the new drive will use. Use **prtconf** to determine the addresses in use.

2. Use **touch /reconfigure** to cause the system to look for new devices upon reboot.

3. Shut the system down using the **shutdown** command.

4. Set the drive address jumpers, and install the new drive on the system.

5. With the **boot -s** command, boot the system to the single-user state.

6. If necessary, use the **format** utility to format and partition the new drive.

7. Use **newfs** to bind sectors into the partitions.

8. Use **fsck** to check file system integrity.

9. Edit the */etc/vfstab* file to add partitions to the system tables.

10. Use **mkdir** to create the mount point directories for the partitions.

11. Use **/etc/halt** to bring the system down.

12. Use boot to restart the system.

Pre-format Considerations

The **format** utility allows the administrator to format, partition, label, and maintain disk drives under Solaris. The command uses the information found in the */etc/format.dat* file to determine the parameters to be applied during the format process.

➤ *NOTE: Solaris versions 2.0 through 2.5.1 support SCSI, IPI, and SMD disk subsystems. Beginning with Solaris 2.6, only SCSI disk subsystems are supported.*

The following sections discuss the installation of drives already known to the format software, as well as how to develop a *format.dat* entry for a drive that is not known to the format software. These methods will work for SCSI and IPI disk drives. The procedure for developing *format.dat* entries for SMD disks is similar to the SCSI and IPI procedure.

Installing Drives Known to format.dat

The **format** utility reads a table of available drive types from the */etc/format.dat* file. If the new drive is described in the file, the drive is easily formatted, partitioned and made available for use.

If an exact description is not available for the drive to be installed, it may be possible to use a Sun-supplied definition. For instance, when attempting to install a 2.4 Gb disk drive, the format utility may be able to use Sun's SUN2.1G entry to format the disk. Although this entry may waste a bit of disk space on the drive, time spent determining an exact *format.dat* entry for the drive will not be necessary.

↝ ***NOTE:*** *When formatting SCSI disks, the Solaris* format *command can assign a default partition table and disk label to unknown drive types. This constitutes a change from the way the* format *utility worked under SunOS.*

✓ ***TIP:*** *Several disk brands and models may be used in Sun systems. Drives of similar capacity bear the same disk geometry label for purposes of convenience. When installing a drive model known to ship with Sun hardware, take advantage of the* format.dat *entry for the drive.*

Understanding format.dat Entries

If a description of the drive is not contained in *format.dat*, the administrator must create an entry that describes the new drive. A brief examination of the fields of a *format.dat* entry will help you understand how to develop such entries. The layout of a typical *format.dat* file appears in the next illustration.

```
# cat /etc/format.dat
#
#ident   "@(#)format.dat 1.17    96/06/13 SMI"
#
# Copyright (c) 1991 by Sun Microsystems, Inc.
#
# Data file for the 'format' program.  This file defines the known
# disks, disk types, and partition maps.
#
#
# This is the list of Sun supported disks for embedded SCSI.
#
disk_type = "SUN1.3G" \
        : ctlr = SCSI : fmt_time = 4 \
        : trks_zone = 17 : asect = 6 : atrks = 17 \
        : ncyl = 1965 : acyl = 2 : pcyl = 3500 : nhead = 17 : nsect = 80 \
        : rpm = 5400 : bpt = 44823

disk_type = "SUN2.1G" \
        : ctlr = SCSI : fmt_time = 4 \
        : ncyl = 2733 : acyl = 2 : pcyl = 3500 : nhead = 19 : nsect = 80 \
        : rpm = 5400 : bpt = 44823

disk_type = "SUN2.9G" \
        : ctlr = SCSI : fmt_time = 4 \
        : ncyl = 2734 : acyl = 2 : pcyl = 3500 : nhead = 21 : nsect = 99 \
        : rpm = 5400
#
# This is the list of partition tables for embedded SCSI controllers.
#
partition = "SUN1.3G" \
        : disk = "SUN1.3G" : ctlr = SCSI \
        : 0 = 0, 34000 : 1 = 25, 133280 : 2 = 0, 2672400 : 6 = 123, 2505120

partition = "SUN2.1G" \
        : disk = "SUN2.1G" : ctlr = SCSI \
        : 0 = 0, 62320 : 1 = 41, 197600 : 2 = 0, 4154160 : 6 = 171, 3894240

partition = "SUN2.9G" \
        : disk = "SUN2.9G" : ctlr = SCSI \
        : 0 = 0, 195426 : 1 = 94, 390852 : 2 = 0, 5683986 : 6 = 282, 5097708
#
```

Layout of a typical format.dat file.

Listed in the first portion of the *format.dat* file are type, format parameters, and controller combination for each drive. The fields listed below are mandatory in every *format.dat* entry, and must contain a value.

❏ ctlr—Type of controller the drive emulates. Solaris SCSI drives use the term *SCSI* in this field.

❐ ncyl—Number of data cylinders on the drive according to the operating system.

❐ acyl—Number of alternate cylinders to be used for bad block remapping.

❐ pcyl—Actual physical number of cylinders on the drive (usually derived from the drive technical manual).

❐ nhead—Number of data heads per drive according to the operating system; sometimes referred to as "tracks."

❐ nsect—Number of sectors per track according to the operating system. Note that most current SCSI drives provide a variable number of sectors per track. The nsect entry is typically an average number derived by dividing the total number of sectors by the total number of tracks.

❐ rpm—Rotational speed of the drive.

❐ bpt—Number of bytes per track on the drive.

The *format.dat* fields listed below are optional.

❐ cache—Value used to disable the drive cache during the format procedure.

❐ fmt_time—Value used to tell the format program of the amount of time necessary to format the drive.

❐ trks_zone—Typically seen on drives that utilize a variable number of sectors per track. The drive is divided into zones and there are *N* tracks per zone.

❐ atrks—Number of alternate tracks available (per zone) for bad block remapping.

❐ asect—Number of alternate sectors available (per zone) for bad block remapping.

The final section of the *format.dat* file contains partition maps for each drive/controller combination. Fields in the partition map section of the *format.dat* file follow.

❒ partition—ASCII label used to identify the partition map.

❒ disk—ASCII label that will be used to identify the type of disk drive.

❒ ctlr—Type of controller the drive is connected to.

❒ 0=0,N;—Patterns that define the slice or partition number {0 through 7}, the starting cylinder of the partition, and the size (in sectors) of the partition.

•• *NOTE: Because slice 2 is always defined as encompassing the entire drive, the partition starts at cylinder zero, and encompasses all sectors between the first sector and the maximum sector on the drive.*

Creating *format.dat* Entries for SCSI Disks

What do the numbers in the *format.dat* entry really mean? To the operating system, a SCSI disk appears to be a series of contiguous sectors. The drive controller manages the sector number to cylinder/head/sector address mapping. Consequently, the numbers in the *format.dat* entry of a SCSI disk can be just about anything, as long as the entry does not exceed the formatted capacity of the drive. To calculate a valid *format.dat* entry, you need to know one of the following bits of information: (1) number of sectors available on the formatted drive, or (2) the drive's formatted capacity (in megabytes).

➥ **NOTE:** *For SMD drives, it is also necessary to know the number of sectors per track and the number of bytes per sector.*

Knowing the number of bytes per track and rotational speed of the drive is also useful. However, these items of information can be derived as explained below.

When the number of sectors on the drive is known, *format.dat* entries can be developed in two steps. First, use */usr/games/factor** to factor the number of sectors to prime numbers. Next, use the prime factors to develop the *format.dat* entry

➥ **NOTE:** *The /user/games/factor program is a "game" that ships with Solaris. The program will factor the numeric input argument into a series of prime numbers, and print those prime numbers in the output.*

For example, assume a drive with 10,000 sectors. The */usr/games/factor* program reports that 10,000 = 2*2*2*2*5*5*5*5. Upon grouping these factors into cylinder/head/sector groupings, you can develop a *format.dat* entry as follows: 5*5*5*2 cylinders, 2*2 heads, 2*5 sectors (250 cylinders * 4 heads * 10 sectors).

Assume that the formatted capacity of a drive (in megabytes) is known (512 Mb). Additional steps in the calculations appear below.

1. Divide the capacity by the sector size (512 bytes/ sector): 512,000,000 / 512 = 1,000,000 sectors.

2. Feed the resulting number to */usr/games/factor* : */usr/games/factor* 1,000,000 = 2*2*2*2*2*2*5*5* 5*5* 5*5.

3. Use the resulting prime factors to develop the *format.dat* entry by organizing them into cylinder/head/sector groupings: 5*5*2*2 sectors (1,250 cylinders * 8 heads * 100 sectors/track)

Because the number of sectors on real drives never factor so easily, the time has come to apply the method on a real drive. One particularly popular drive has 1,658 data cylinders, 15 heads, and a formatted capacity of 1,079.1 megabytes. The drive manual lists the drive capacity as 50,910 bytes per track with a rotational speed of 3,600 RPM. Based on this information, the second method would be used to develop the *format.dat* entry reproduced below.

❑ 1,079,100,000 bytes / 512 bytes = 2,107,617 sectors.

❑ */usr/games/factor* 2107617 = 3 * 702,539. The result does not constitute adequate prime factors to develop a *format.dat* entry. Subtract a few sectors from the total, and try again. Subtract 17 sectors, and try factoring 2,107,600 sectors. The result is */usr/games/factor* 2107600 = 2 * 2 * 2 * 2 * 5 * 5 * 11 * 479.

❑ By organizing the factors into cylinder/head/sector groupings, develop a *format.dat* entry: 479*4 cylinders, 2*2*5 heads, 5*11 sectors (1,916 cylinders, 20 heads, 55 sectors).

If the speed of the drive is not known, use 3,600 RPM for drives that transfer at 5 Mb/second or less. If the drive transfers at 5 Mb/second or more, use 5,400 or 7,200 RPM. If the drive manual does not specify the number of bytes per track, calculate an entry by multiplying the number of bytes/sector by the number of sectors/track.

↝ **NOTE:** *Most manufacturers have bulletin boards or technical support phone numbers that users* *can contact to obtain the manufacturer's for-mat.dat information for most of the popular sys-tems on the market. Because the manufacturers know their drives better than anybody else, fol-lowing their recommendations whenever possible is recommended.*

Using the format Utility

Once a *format.dat* entry exists for a drive, it is time to begin the format process. Invoke the **format** utility by logging in as root, and typing format at the prompt.

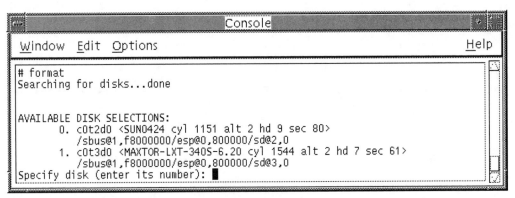

Starting the format utility.

The format utility will respond by listing all drives it can find attached to the system. Select the new drive by entering the number for the desired disk drive and pressing <Enter>. If the drive is already formatted, format will display this information and then display the main format menu. If the drive is not formatted, the format command will determine a default label for SCSI drives and ask if it is okay to use the label.

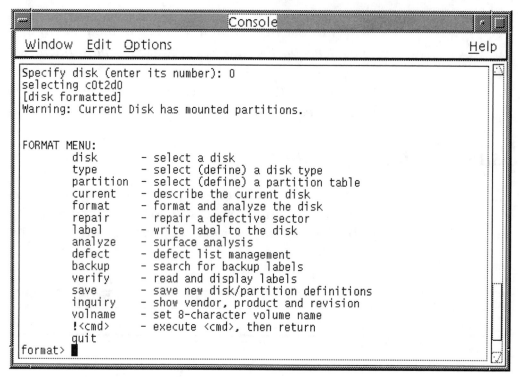

The format utility main menu.

The main menu allows configuration of format operation, defect list extraction and management, drive partitioning, drive labeling, and surface verification. In the following sections, each of the menu items will be examined.

☛ **WARNING:** *When used to format a disk or to perform surface analysis,* format *is a destructive procedure. Verify that file system backups have been performed before formatting a drive that has been in operation.*

disk Command

The **disk** menu is used to select the disk drive to be modified. This menu is a near duplicate of the **format** command menu. When formatting several disks it is possible to simply select a new disk via the disk menu instead of exiting and restarting format for each drive. The disk menu item displays the current controller and drive geometry information for each drive available on the system.

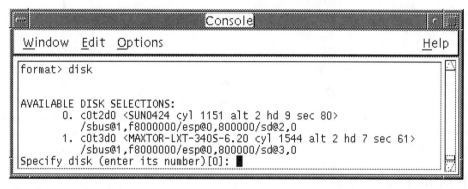

```
                               Console
  Window  Edit  Options                                Help

format> disk


AVAILABLE DISK SELECTIONS:
      0. c0t2d0 <SUN0424 cyl 1151 alt 2 hd 9 sec 80>
         /sbus@1,f8000000/esp@0,800000/sd@2,0
      1. c0t3d0 <MAXTOR-LXT-340S-6.20 cyl 1544 alt 2 hd 7 sec 61>
         /sbus@1,f8000000/esp@0,800000/sd@3,0
Specify disk (enter its number)[0]: █
```

The format utility disk menu.

type Command

The **type** menu allows the selection of the appropriate drive type, and geometry information for the drive to be formatted. The geometry and type information are read in from the entries in the *format.dat* file.

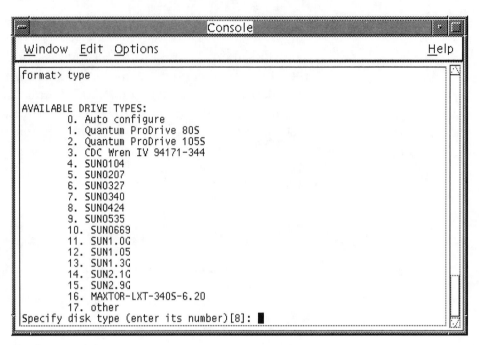

```
                                  Console
  Window  Edit  Options                                        Help

 format> type

 AVAILABLE DRIVE TYPES:
         0. Auto configure
         1. Quantum ProDrive 80S
         2. Quantum ProDrive 105S
         3. CDC Wren IV 94171-344
         4. SUN0104
         5. SUN0207
         6. SUN0327
         7. SUN0340
         8. SUN0424
         9. SUN0535
         10. SUN0669
         11. SUN1.0G
         12. SUN1.05
         13. SUN1.3G
         14. SUN2.1G
         15. SUN2.9G
         16. MAXTOR-LXT-340S-6.20
         17. other
 Specify disk type (enter its number)[8]: █
```

The format utility type menu.

partition Command

The **partition** menu is used to examine and modify the partition information for the drive that is about to be formatted.

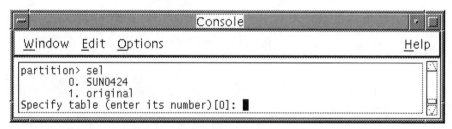

```
┌─────────────────────── Console ───────────────────────┐
│ Window  Edit  Options                            Help  │
│┌──────────────────────────────────────────────────────┐│
││format> part                                          ││
││                                                       ││
││PARTITION MENU:                                        ││
││        0      - change `0´ partition                 ││
││        1      - change `1´ partition                 ││
││        2      - change `2´ partition                 ││
││        3      - change `3´ partition                 ││
││        4      - change `4´ partition                 ││
││        5      - change `5´ partition                 ││
││        6      - change `6´ partition                 ││
││        7      - change `7´ partition                 ││
││        select - select a predefined table            ││
││        modify - modify a predefined partition table  ││
││        name   - name the current table               ││
││        print  - display the current table            ││
││        label  - write partition map and label to the disk ││
││        !<cmd> - execute <cmd>, then return           ││
││        quit                                          ││
││partition> █                                          ││
│└──────────────────────────────────────────────────────┘│
└────────────────────────────────────────────────────────┘
```

The format utility partition menu.

The numbered entries in the partition menu refer to available drive partitions. The partitions are always numbered zero through seven. Remaining entries in the partition menu are submenus that allow customization of the partition tables.

```
┌─────────────────────── Console ───────────────────────┐
│ Window  Edit  Options                            Help  │
│┌──────────────────────────────────────────────────────┐│
││partition> sel                                        ││
││        0. SUN0424                                    ││
││        1. original                                   ││
││Specify table (enter its number)[0]: █                ││
│└──────────────────────────────────────────────────────┘│
└────────────────────────────────────────────────────────┘
```

The partition select submenu.

The **select** menu allows selection of a predefined partition map as supplied in */etc/format.dat*. It is also possible to "copy" the partition map of another drive of the same type on the system.

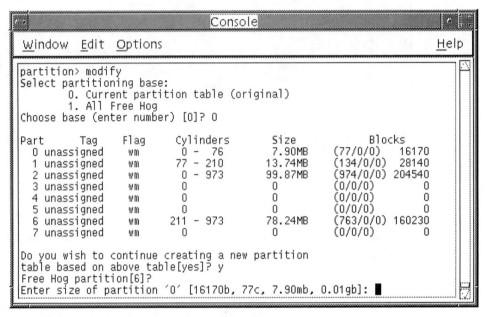

The partition modify submenu.

The **modify** menu allows modification of the current partition map. This menu is used to set the size and name of the file systems on the drive. Within this menu, commands are provided to allow the partition map to be examined, printed, or saved to a file.

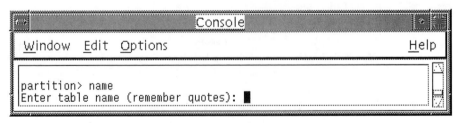

The partition name submenu.

The **name** menu allows selection of a partition name for the current partition map. The name is saved in the */etc/format.dat* file for future reference.

```
┌─────────────────────────────Console────────────────────────────┐
│ Window   Edit  Options                                     Help  │
├──────────────────────────────────────────────────────────────┬──┤
│ partition> print                                              │▲ │
│ Current partition table (test):                              │  │
│ Total disk cylinders available: 1151 + 2 (reserved cylinders)│  │
│                                                               │  │
│ Part      Tag    Flag    Cylinders       Size          Blocks │  │
│    0      root    wm      0 -    45    16.17MB   (46/0/0)    33120│ │
│    1      swap    wu     46 -   136    31.99MB   (91/0/0)    65520│ │
│    2    backup    wu      0 - 1150    404.65MB (1151/0/0) 828720│  │
│    3 unassigned   wm      0              0       (0/0/0)        0│  │
│    4 unassigned   wm      0              0       (0/0/0)        0│  │
│    5 unassigned   wm      0              0       (0/0/0)        0│  │
│    6       usr    wm    137 - 1150    356.48MB (1014/0/0) 730080│  │
│    7 unassigned   wm      0              0       (0/0/0)        0│  │
│ partition> █                                                  │▼ │
└──────────────────────────────────────────────────────────────┴──┘
```

The partition print submenu.

The **print** submenu makes it possible to view the partition map that is being modified. The information provided in the printout allows the determination of the size of the partition(s), starting cylinder for the partition(s), and other information about the partition(s) on the disk drive.

The partition label submenu.

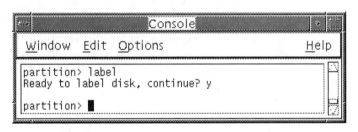

The **label** menu includes the current partition size, and name information to be written into the volume label. This command performs in the same way as the label menu item in the format utility main menu.

current Command

The **current** menu selection displays information about the currently selected drive. Information provided includes the addressing information for the controller for the drive and the drive geometry information.

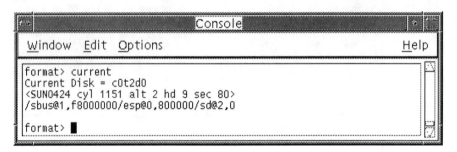

The format utility current menu.

repair Command

The **repair** menu selection provides the ability to add and subtract entries to and from the current drive defect list. Bad spot locations are specified via an absolute block number (for SCSI and IPI disks), or by giving the cylinder/head/sector information for the bad spots (on SMD drives).

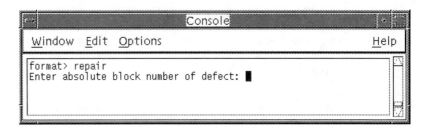

The format utility repair menu.

label Command

The **label** menu selection writes a label on the current drive. This selection operates the same as the label selection in the partition submenu. The disk label contains information on the disk drive geometry and the size of the partitions, including starting cylinder and number of data sectors.

The format utility label menu.

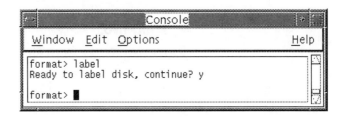

analyze Menu

The **analyze** menu is used to set the surface analysis parameters for the verification portion of the format utility. This menu also allows various surface analysis programs to be executed independent of the format unit operation.

```
┌──────────────────────────────────────────────────┐
│                    Console                         │
├──────────────────────────────────────────────────┤
│ Window  Edit  Options                       Help   │
├──────────────────────────────────────────────────┤
│ format> analyze                                    │
│                                                    │
│                                                    │
│ ANALYZE MENU:                                      │
│         read     - read only test   (doesn't harm SunOS)│
│         refresh  - read then write  (doesn't harm data) │
│         test     - pattern testing  (doesn't harm data) │
│         write    - write then read     (corrupts data)  │
│         compare  - write, read, compare (corrupts data) │
│         purge    - write, read, write  (corrupts data)  │
│         verify   - write entire disk, then verify (corrupts data)│
│         print    - display data buffer             │
│         setup    - set analysis parameters         │
│         config   - show analysis parameters        │
│         !<cmd>   - exeucte <cmd> , then return      │
│         quit                                       │
│ analyze> █                                         │
└──────────────────────────────────────────────────┘
```

The format utility analyze menu.

The **read**, **refresh**, **test**, **write**, **compare**, and **purge** menu items allow control of respective test sequences.

☞ *WARNING: The write, compare, and purge options will destroy data stored on the disk. The read, refresh, and test items do not destroy data stored on the disk drive.*

The **print** item prints out a disk data buffer to the screen, which permits the operator to view the pattern currently being written to the drive. The **setup** option in the analyze menu allows configuration of the surface analysis portion of format.

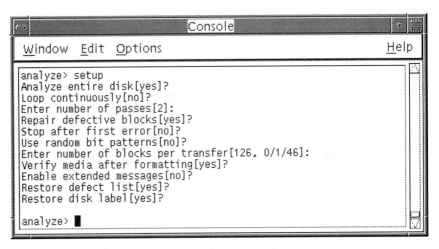

The analyze setup submenu.

The **setup** menu allows selection of the portion of the disk to analyze, as well as the transfer size to use. This menu also provides control over whether bad spots are mapped out or merely reported to the operator. The **analyze config** menu item allows verification of current settings before continuing with the format operation.

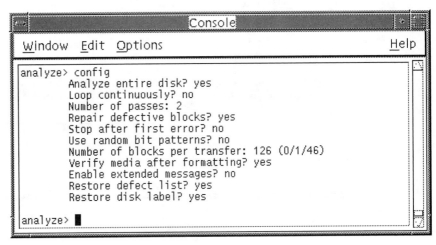

Using the analyze config command to show current setup.

defect Menu

The **defect** menu of format is used to read and manipulate the manufacturer defect list from a drive. It also allows defects to be added or removed from the defect list.

```
┌─────────────────────────── Console ───────────────────────────┐
│  Window   Edit  Options                              Help     │
│┌──────────────────────────────────────────────────────────────┐│
││format> defect                                                 ││
││                                                               ││
││                                                               ││
││DEFECT MENU:                                                   ││
││        primary  - extract manufacturer's defect list         ││
││        grown    - extract manufacturer's and repaired defects lists│
││        both     - extract both primary and grown defects lists││
││        print    - display defect list                        ││
││        dump     - dump defect list to file                   ││
││        !<cmd>   - execute <cmd>, then return                 ││
││        quit                                                   ││
││defect> █                                                      ││
│└──────────────────────────────────────────────────────────────┘│
└────────────────────────────────────────────────────────────────┘
```

The format utility defect submenu.

The **primary** menu item allows the vendor defect list to be read from the disk media. The **grown** and **both** menu items allow the vendor defect list, and the repaired defect list to be extracted from the disk media.

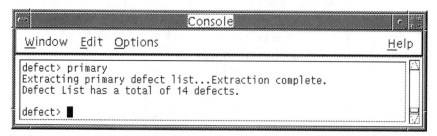

```
┌─────────────────────────── Console ───────────────────────────┐
│  Window   Edit  Options                              Help     │
│┌──────────────────────────────────────────────────────────────┐│
││defect> primary                                                ││
││Extracting primary defect list...Extraction complete.         ││
││Defect List has a total of 14 defects.                        ││
││                                                               ││
││defect> █                                                      ││
│└──────────────────────────────────────────────────────────────┘│
└────────────────────────────────────────────────────────────────┘
```

Using the defect menu primary command to extract the vendor defect list.

Once the defect list(s) have been extracted from the media, this information can be saved in a file with the **dump** menu item. The print menu allows the defect list to be viewed on the screen.

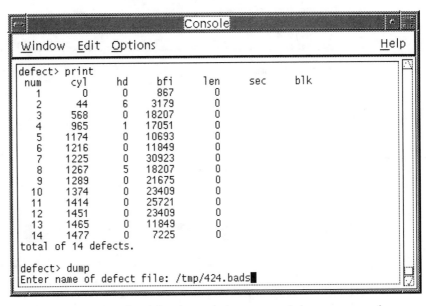

```
                                   Console
  Window   Edit   Options                                          Help

  defect> print
   num    cyl    hd     bfi    len    sec    blk
    1      0      0     867     0
    2     44      6    3179     0
    3    568      0   18207     0
    4    965      1   17051     0
    5   1174      0   10693     0
    6   1216      0   11849     0
    7   1225      0   30923     0
    8   1267      5   18207     0
    9   1289      0   21675     0
   10   1374      0   23409     0
   11   1414      0   25721     0
   12   1451      0   23409     0
   13   1465      0   11849     0
   14   1477      0    7225     0
  total of 14 defects.

  defect> dump
  Enter name of defect file: /tmp/424.bads
```

Viewing and saving the defect list with the print and dump commands.

backup Menu

The **backup** menu selection is used to look for backup copies of the disk label. This can be particularly useful for drives on which the primary label has somehow been damaged. If the primary label is intact, **format** will display this information, and ask if the search for backup labels should continue. Once a valid label is found, the backup command will write this label to the disk.

```
┌─────────────────────────────────────────────────────────────┐
│                         Console                               │
├─────────────────────────────────────────────────────────────┤
│  Window   Edit   Options                              Help    │
├─────────────────────────────────────────────────────────────┤
│ format> backup                                                │
│ Disk has a primary label, still continue? y                   │
│ Searching for backup labels...found.                          │
│ Restoring primary label.                                      │
│                                                               │
│ format> █                                                     │
└─────────────────────────────────────────────────────────────┘
```

Using the format utility backup command to restore the drive label.

verify Menu

The **verify** menu selection searches the disk for a
valid label. Once the label is found, verify displays the
information on the screen.

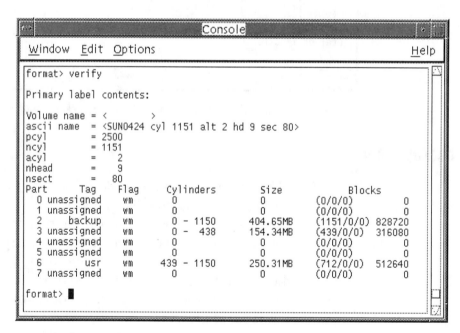

```
┌─────────────────────────────────────────────────────────────┐
│                         Console                               │
├─────────────────────────────────────────────────────────────┤
│  Window   Edit   Options                              Help    │
├─────────────────────────────────────────────────────────────┤
│ format> verify                                                │
│                                                               │
│ Primary label contents:                                       │
│                                                               │
│ Volume name = <         >                                     │
│ ascii name  = <SUN0424 cyl 1151 alt 2 hd 9 sec 80>            │
│ pcyl        = 2500                                            │
│ ncyl        = 1151                                            │
│ acyl        =    2                                            │
│ nhead       =    9                                            │
│ nsect       =   80                                            │
│ Part     Tag    Flag   Cylinders      Size        Blocks      │
│   0 unassigned  wm     0              0        (0/0/0)      0  │
│   1 unassigned  wm     0              0        (0/0/0)      0  │
│   2    backup   wm     0 - 1150    404.65MB  (1151/0/0) 828720 │
│   3 unassigned  wm     0 -  438    154.34MB   (439/0/0) 316080 │
│   4 unassigned  wm     0              0        (0/0/0)      0  │
│   5 unassigned  wm     0              0        (0/0/0)      0  │
│   6      usr    wm   439 - 1150    250.31MB   (712/0/0) 512640 │
│   7 unassigned  wm     0              0        (0/0/0)      0  │
│                                                               │
│ format> █                                                     │
└─────────────────────────────────────────────────────────────┘
```

Using the format utility to verify the disk label.

save Menu

The **save** menu selection saves the current setup and partitioning information to the system */etc/format.dat* file. The user is given the choice to select another file to which the information will be saved.

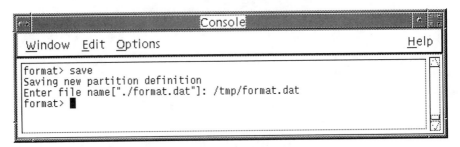

Using the format utility to save the partition information.

inquiry Menu

The **inquiry** menu selection probes the disk drive and displays the vendor name, product, and firmware revision level of the current disk.

Using the format utility inquiry command.

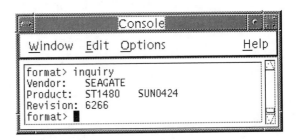

volname Menu

The **volname** menu selection allows the creation of a volume name for the drive. Once the volume name is entered, the volname command writes the volume name to the current drive.

Using the format utility volname command.

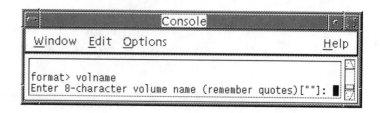

format Command

Once all setup and customization are completed, it is time to format the drive. This is accomplished with the **format** menu selection on the main menu.

The format menu selection is used to start the format process. Staring the process will cause the system to issue the **format-unit** command (for IPI and SCSI drives), or, for SMD disk drives, to begin writing format information to the disk drive via program control. Once the formatting is complete, the format utility will run a few passes of surface analysis on the drive. When the surface analysis is complete, a geometry label is written to the disk and control switches back to the format utility main menu.

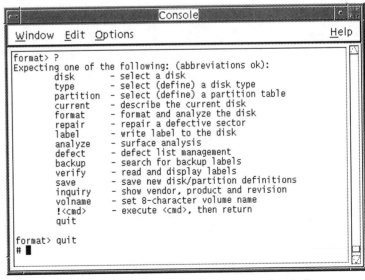

```
format> format
Ready to format.  Formatting cannot be interrupted
and takes 9 minutes (estimated). Continue? y
Beginning format. The current time is Tue Mar 11 14:40:18 1997


Formatting...
done

Verifying media...
        pass 0 - pattern = 0xc6dec6de
   973/4/28

        pass 1 - pattern = 0x6db6db6d
   973/4/28

Total of 0 defective blocks repaired.
format> quit
#
```

The format utility format menu.

quit Menu

The **quit** menu selection exits the **format** utility, and returns you to the root shell prompt.

```
format> ?
Expecting one of the following: (abbreviations ok):
        disk       - select a disk
        type       - select (define) a disk type
        partition  - select (define) a partition table
        current    - describe the current disk
        format     - format and analyze the disk
        repair     - repair a defective sector
        label      - write label to the disk
        analyze    - surface analysis
        defect     - defect list management
        backup     - search for backup labels
        verify     - read and display labels
        save       - save new disk/partition definitions
        inquiry    - show vendor, product and revision
        volname    - set 8-character volume name
        !<cmd>     - execute <cmd>, then return
        quit

format> quit
#
```

Using the format utility quit command to return to the shell.

✓ **TIP:** *One of the simplest ways to breathe new life into a sluggish system is to add disk space and I/O channels. These procedures allow the I/O load of the system to be dispersed across more channels and devices, thereby reducing the load on existing devices.*

Summary

This chapter focused on the mass storage subsystem, one of the most critical Solaris subsystems. Topics included how to install a new disk drive, and formatting and partitioning a disk drive. In the following chapters, selected topics will be revisited and refined as more advanced troubleshooting and storage subsystem management techniques are presented.

Disk Subsystem
Software

Once new disks have been formatted and before they are put into operation, strategic planning questions are in order. For instance, you should determine the size of the file systems on the disk, how the space will be optimized, and how the file systems are to be mounted. This chapter focuses on software issues associated with disk subsystem management, including how to determine partition sizes, mounting file systems, and the use of the volume manager.

Calculating Disk Slice Sizes

What happens if the company purchases a disk drive that is not formatted and partitioned for Solaris systems? How do you change the formatting and partitioning such that the drive is useful for a particular application? As shown in Chapter 9, "Disk Subsystem Hardware," the **format** utility provides the capabilities of formatting the drive, and writing a Solaris partition map on the drive.

The partition maps define how the drive is sliced into logical units. The partitions are sometimes referred to as *disk slices*. You can use a default partition map or develop a partition map which customizes the file system(s) for the local environment. In order to ensure that operations are performed on the correct entity, you need a method of referring to a specific disk slice.

Disk Slice References

Using block numbers to describe the partition start and end points is cumbersome. Most UNIX operating systems define a symbolic system which is used to describe the partitions on a hard disk drive. Under Solaris, different regions of the disk are called *slices*. The drive label contains partition information which details the blocks encompassed by each slice of the drive. Under Solaris, the disk slices are numbered to make things simple for the administrator.

Disk slice	Typical usage
0	Root file system (bootable partition)
1	Swap space (swapfs file system which is not mounted)
2	Entire drive (sometimes called the overlap partition)
3	/var (optional)
4	/usr/openwin (optional)
5	/opt (optional)

Disk slice	Typical usage
6	/usr filesystem (default)
7	/export/home partition (default)

↝ **NOTE:** *The floppy diskette device uses an older convention of naming slices with the letters "a" through "h." This convention held that the third or "c" slice would be sized to encompass the entire disk.*

Every disk drive stores the partition information for the file systems in the disk label on the drive. The disk label contains the volume table of contents (VTOC) and other information about the disk drive. The next section describes the VTOC and the information it stores.

Volume Table of Contents (VTOC)

A label including the VTOC is written on the disk as part of the format process. The VTOC contains information about the drive geometry (number of cylinders, heads, and sectors) and the partition map for the drive, as well as information about how each partition is being used by the system. Much of this information is summarized by the tags and flags fields of the VTOC.

Slice Tags and Flags

The information about mount points, and mount options, is stored in a portion of the VTOC called the "tags and flags" fields. These fields identify how the partitions are used and mounted. The tag field is used to identify how the slice is being used. Valid entries in this field are listed in the following table. Note that the tag field value is not the same as the slice number.

Tag Field Descriptions

Tag field value	Description
0	Unassigned partition
1	Boot partition
2	Root partition
3	Swap partition
4	User partition
5	Backup partition
6	Stand partition
8	Home partition

The flag field is used to determine how the partition is to be mounted. Valid entries are summarized in the next table.

00	Mounted read/write
01	Not mounted
10	Mounted read only

Viewing the VTOC with the prtvtoc Utility

It is possible to view the contents of the VTOC using the **/etc/prtvtoc** utility. The prtvtoc utility reads the disk label, and lists the geometry of the drive, the size of each partition, the mount point for the partition (if known), and other information stored in the VTOC and drive label. The syntax for the **prtvtoc** command follows:

```
# prtvtoc /dev/rdsk/c#t#d#s2
```

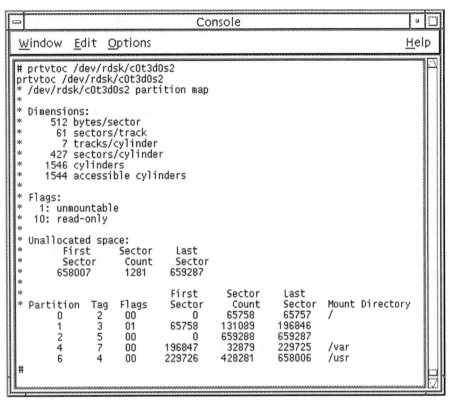

Using prtvtoc to view the volume table of contents (VTOC).

Overallocation Overhead

As mentioned in Chapter 2, "Solaris Concepts and Terminology," a typical rule of thumb when partitioning disks is to allow for 10 percent of the sectors in a file system to be consumed by *overhead*. The simplest way to illustrate overhead is by examining the output of the **/bin/df** command.

In the example below it is obvious that the *used* and *avail* numbers provided by df do not add up to the *kbytes* amount. The difference between these quantities points out the existence of file system overhead.

Part of this difference may be attributed to the way the kernel reserves space on a file system. The UNIX operating system reserves a certain number of sectors for use in short-term overallocation situations. The amount of space reserved for such emergencies is controlled by the parameters used when the file system is created.

```
# df -k
Filesystem              kbytes      used      avail     capacity    Mounted on
/dev/dsk/c0t3d0s0       32431       32262     0         100%        /
/dev/dsk/c0t3d0s6       240463      77640     38783     36%         /usr
/proc                   0           0         0         0%          /proc
fd                      0           0         0         0%          /dev/fd
/dev/dsk/c0t1d0s6       30651       12436     15155     46%         /var
/dev/dsk/c0t1d0s4       239263      79710     135633    38%         /usr/vice/cache
swap                    267916      4         267912    1%          /tmp
```

➼ **NOTE:** *When a file system reaches the 100% full level, the system will generate a message on the system console. The administrator should examine the file system and remove any temporary or unused files. If this operation does not free enough space to allow for normal operation, you will need to take more drastic measures (such as moving files to another file system) to ensure proper system operation.*

Super-block Overhead

Another source of file system overhead is the support structure required to keep track of the files in the file system. One of these support structures is called the *super-block*. The super-block can be considered the index of the file system contents. Any references to a file or directory on the partition must first reference the super-block to determine which sectors are part of that file or directory.

The primary super-block is stored in one of the first few sectors of the partition. Backup or alternate copies of the super-block are stored in several places on the partition in case of failure of the primary super-block. On small file systems there are few copies of the alternate super-blocks. On large file systems, there are many copies of the alternate super-blocks.

General Rules to Determine Partition Sizes

A few general rules which help to simplify the task of determining the appropriate size for a partition are listed below.

1. Determine how much space the application currently occupies.

2. Add 50% to the number from step 1 for growth and expansion of the file system.

3. Add 10% more space for file system overhead.

4. Divide the number from the previous step by the sector size (512).

5. Divide the number from the previous step by (Nheads * Nsects) to determine how many cylinders are required. Round up to the nearest whole number.

For example, an application that requires 100 Mb of storage space, and a disk drive with a geometry of 20 heads and 97 sectors/track would yield the calculations below.

❏ Original package size, N = 100 Mb

❏ Add space for growth, M = N + (N * .5) = 150 Mb

❏ Add 10% overhead space, O = M + (M * .1) = 165 Mb

❏ Divide by 512 bytes per sector, P = 173,015,040 bytes / (512 bytes/sector) = 337,920 sectors required

❏ Ncyl = 337,920 sectors / (20 heads * 97 sectors per head per cylinder) = 175 cylinders required for the partition

✓ *TIP: The following formula is a quick way to estimate disk space requirements: 2,048 sectors * 512 bytes/sector is equivalent to one (1) Mb of disk space.*

Repartitioning Disk Drives

As file systems outgrow current partitions or as file systems become unnecessary, you may have occasion to repartition a disk. It should be apparent from the contents of the VTOC that repartitioning a disk is a destructive process.

☛ *WARNING: Before repartitioning a disk drive, it is wise to save all data stored on the target disk drive. Failure to do so will result in data loss.*

If it becomes necessary to repartition a disk, save all files from the disk drive by dumping them to tape. Once all information from the disk has been saved, you can use the **format** utility to reformat/repartition the drive. Once the drive has been repartitioned, restore the files from the backup tape.

Types of File Systems

Once a disk is partitioned, you need to create a file system on each partition such that the disk space is available to the system. Solaris supports a variety of file system types which can be mounted as part of the overall file tree. File system types must be understood in order for you to create the proper type for use within the local environment.

File system types supported by Solaris

File system type	Hardware medium	Description
ufs	hard disk	Standard UNIX file system using the BSD fast file system.
hsfs	CD-ROM	High Sierra or ISO 9660 file system; can be mounted read-only.
pcfs	floppy disk	MS-DOS file system.
nfs	network file system	Network based file system.
tmpfs	memory/swap area	Special purpose, high performance file system for temporary data.
cachefs	hard disk and nfs file systems	Special purpose, high performance file system to cache CD-ROM and NFS disks.
lofs	loopback	Special purpose file system which allows multiple mount points for a single file system.
procfs	memory	Special purpose memory based file system for use by debuggers and other development tools.
swapfs	hard disk	Special purpose file system used to extend virtual memory on the system.

The first five file system types listed—*ufs, hsfs, pcfs, nfs,* and *tmpfs*—are the most commonly used. In the following diagram showing a sample UNIX file tree and file system mount points, the */tmp* file system is a *tmpfs* file system, while the other file systems are *ufs* file systems. This is a fairly common situation for a standalone workstation. A workstation on a network may employ many types of file systems.

Sample Solaris file tree.

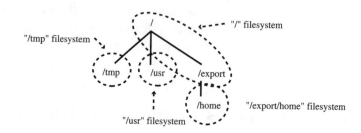

ufs File Systems

The *ufs* file system is the standard UNIX file system. Under Solaris, the ufs file system is based on the Berkeley fast file system. This file system implemented several modifications to early UNIX file systems in order to improve disk subsystem performance and reliability. Many system utilities will work only on *ufs* file systems. The most notable of these utilities are the **ufsdump** and **ufsrestore** commands.

hsfs File Systems

The High Sierra file system (HSFS, also known as the ISO 9660 file system) is a special purpose file system developed for use on CD-ROM media. The geometry and structure of CD-ROM media are different from those of a hard disk drive. The Solaris *hsfs* file system also supports the Rock-Ridge extensions to the HS file system. Consult the manual page for *hsfs* for more information about HS file systems.

pcfs File Systems

The *pcfs* file system is a UNIX implementation of the DOS file attribute table (FAT) file system. This file system allows Solaris systems to mount PC-DOS formatted file systems. The *pcfs* file system allows UNIX users direct read/write access to PC-DOS files using standard UNIX utilities (e.g., **cat**, **ls**, and **rm**). Files in the *pcfs* file system are limited to the standard 8.3 file name conventions (eight characters followed by a dot, followed by three characters). File names are created using all upper-case characters. Consult the manual page for pcfs for more information on pcfs file systems.

nfs File Systems

Network file systems (NFS) allow users to share files between many types of systems on the network. NFS provides a method of making a disk drive on one system (a server) appear as though it was connected to another system (a client). Extra kernel level code is required to ensure that valuable information is not lost from client machines if and when a file server is down. Consult the manual page for *nfs* for more information about nfs file systems.

tmpfs File Systems

The *tmpfs* file systems are a memory-based special purpose file system which is created and destroyed every time a system is rebooted. The memory based file system provides much better performance than a disk based file system. The tmpfs file systems are often used to provide temporary scratch storage which may be used by system utilities, and users, while using the system. Consult the manual pages for tmpfs for more information on tmpfs file systems.

cachefs File Systems

The *cachefs* file systems are an offshoot of the nfs file systems. Modern disk subsystem technology provides much better throughput than exists on many networks. Network bottlenecks typically occurred when an *nfs* client required large quantities of data from a server. The server could get the information from its local disk very quickly, but had to "feed it through a funnel" to get it to the client.

The *cachefs* is a disk based cache on the client machine. The *cachefs* software uses predictive algorithms to try and determine which disk blocks the client machine will request. These blocks are copied to the disk based *cachefs* file system on the client machine. When the client machine needs the data, it can access it from its own disk drive. If the cached data is changed by the server, the *cachefs* is informed that the data are "dirty," which causes the client to request a fresh copy from the server. Any changes to the cached data made by the client machine are written directly to the server's disk via NFS.

lofs File Systems

The *lofs* file systems are virtual file systems which may be used to create new file systems that allow alternate paths to the files in existing file systems. Consult the manual page for *lofs* for more information.

procfs File Systems

The *procfs* file system is similar in concept to the device access methods provided by the file entries in the */dev* directory. The files found in a *procfs* file system are entries which map the address space of a process into a file. Opening one of these files allows you to read directly from the address space of a running process. This file system is used by certain debugging tools as a method for observing and changing the behavior of a process while it is running.

swapfs File Systems

The *swapfs* is a special file system used by the SunOS kernel for paging and swapping. SunOS, like many modern versions of UNIX, manages memory through a system of fixed size units called *pages*. Program instructions and data are broken up into pages and loaded into memory wherever free pages exist. The operating system, by means of a page table and special memory management hardware, presents a linear memory map to each process even though the instructions and data that comprise the process may be scattered throughout memory.

Virtual memory is the process by which physical memory is extended by moving pages which have not been recently referenced out of memory and on to disk. This process of moving program instructions and data in and out of memory is called *paging*. The *swapfs* file system is used by SunOS to store pages that are not currently in use. Because of its special usage, a *swapfs* file system does not appear in the UNIX file tree, and the disk space consumed by a *swapfs* file system does not appear in the output of commands such as **df**.

Using the Swap Command

It is possible to determine the disk space in use by *swapfs* file systems with the **/usr/sbin/swap** command. This command can read and interpret the contents of a *swapfs* file system and report back on its size and current usage. An example appears in the following illustration.

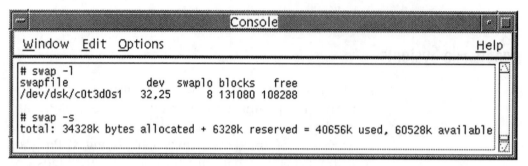

```
# swap -l
swapfile              dev  swaplo blocks   free
/dev/dsk/c0t3d0s1    32,25      8 131080 108288

# swap -s
total: 34328k bytes allocated + 6328k reserved = 40656k used, 60528k available
```

The swap command's two reporting options.

The first form of the swap command, using the **-l** option, gives the size of the *swapfs* file system in 512-byte disk blocks along with its current usage, offset, SunOS device name, and major and minor device numbers. The second form is more frequently used. The **-s** option lists current usage by detailing the amount of space allocated, the space claimed for usage by SunOS but not currently being used, the total of the two previous values, and the amount of unused space. This form of the swap command is often used to help determine how much *swapfs* space is needed by a system when running certain programs.

✓ **TIP:** *If a system requires additional swap space, it is possible to create swap files or swap partitions, and to direct Solaris to add them to the total system virtual memory using the* swap *command with the* -a *option.*

Creating File Systems

Except in special cases, a file system must be created on a partition before the partition can be used under Solaris. The act of creating a file system puts a framework on the partition which enables Solaris to store files there. The file system framework provides index information for the disk blocks which were initialized and mapped when the disk was formatted.

Creation of a new file system can be accomplished by invoking the **/usr/sbin/newfs** utility. The newfs utility allows you to create file systems with default (or custom) file system parameters. In effect, newfs is just a user-friendly front end to the **/etc/mkfs** program. The mkfs command creates the file system super-blocks, divides the file system into several cylinder groups, and takes care of creating all file system sub-structures required by the operating system.

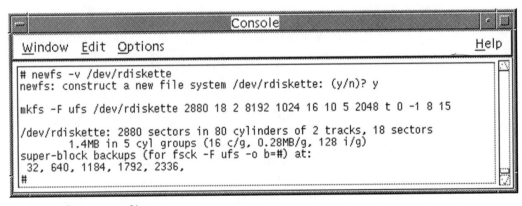

```
# newfs -v /dev/rdiskette
newfs: construct a new file system /dev/rdiskette: (y/n)? y

mkfs -F ufs /dev/rdiskette 2880 18 2 8192 1024 16 10 5 2048 t 0 -1 8 15

/dev/rdiskette: 2880 sectors in 80 cylinders of 2 tracks, 18 sectors
        1.4MB in 5 cyl groups (16 c/g, 0.28MB/g, 128 i/g)
super-block backups (for fsck -F ufs -o b=#) at:
 32, 640, 1184, 1792, 2336,
#
```

Using newfs to create file systems.

Working with the newfs Command

The default type of file system created by **newfs** is a *ufs* file system. The newfs command consults the */etc/*

format.dat file used by **format** and the disk label to identify the specifications for the disk, and build the appropriate file system. The syntax for invoking the newfs command follows:

```
# newfs [mkfs options] /dev/rdsk/c#t#d#s#
```

Invoking newfs with the **-N** option causes newfs to display the basic parameters of the file system without actually creating it.

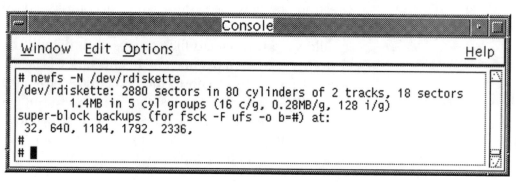

```
# newfs -N /dev/rdiskette
/dev/rdiskette: 2880 sectors in 80 cylinders of 2 tracks, 18 sectors
        1.4MB in 5 cyl groups (16 c/g, 0.28MB/g, 128 i/g)
super-block backups (for fsck -F ufs -o b=#) at:
 32, 640, 1184, 1792, 2336,
#
#
```

Viewing file system parameters.

The newfs command will display information about the new file system as it operates. The first line printed by newfs describes the basic disk geometry. The example shows a disk which is a two-sided (two heads) floppy diskette with 18 sectors and 80 cylinders for a total of 2,880 sectors (or blocks).

The second line of output from newfs describes the *ufs* file system created in this partition. In the example, the file system is 1.4 Mb in size, containing five cylinder groups of 16 cylinders per group. Cylinder groups are derived from the Berkeley UNIX fast file system. Cylinder groups provide a method by which file blocks are allocated close together on the disk to improve performance.

✓ **TIP:** *Varying the size of the cylinder groups is one way to tune the performance of a* ufs *file system.*

The final two lines of newfs output list the locations of the super-block backups. The super-block is the head of the file index node or inode information used by the *ufs* file system. The inodes are "used" by routines in the Solaris kernel to allocate, read, write, and delete files. Backup copies of these super-blocks are created in the event that a system crash, power failure, or other problem occurs which leaves the main copy of the super-block damaged.

How newfs Connects to mkfs

As previously mentioned, the **newfs** command is a front end to **mkfs**. The newfs command invokes mkfs to perform the actual creation of the file system on the target device. The mkfs command requires a long list of parameters. For most applications a detailed knowledge of the mkfs parameters is unnecessary; newfs makes proper choices for the parameters it passes to mkfs. However, it is useful to know the parameters so that modifications may be performed when applications require specialized adjustments to a file system.

For example, it is possible to build a file system tuned to store a large quantity of small, frequently changing files such as those found in the Usenet net-news system. Such a file system would require more index nodes or inodes than are provided by the default options used by newfs. Situations that require manual modification of mkfs parameters are covered in more detail in Chapter 11, "Disk and File System Maintenance."

✓ **TIP:** *Use of the -v option flag with* newfs *will allow the administrator to see which parameters it passes to* mkfs. *The following example uses the -N option as well as to display the parameters without actually making the file system.*

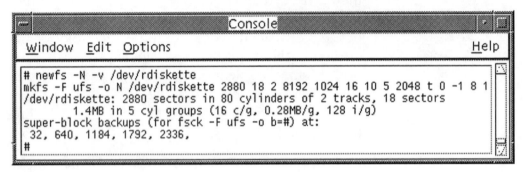

```
# newfs -N -v /dev/rdiskette
mkfs -F ufs -o N /dev/rdiskette 2880 18 2 8192 1024 16 10 5 2048 t 0 -1 8 1
/dev/rdiskette: 2880 sectors in 80 cylinders of 2 tracks, 18 sectors
        1.4MB in 5 cyl groups (16 c/g, 0.28MB/g, 128 i/g)
super-block backups (for fsck -F ufs -o b=#) at:
 32, 640, 1184, 1792, 2336,
#
```

Viewing the mkfs parameters with the -v and -N options.

Some of the parameters in the preceding example should look familiar. Many of these parameters may be altered by using an option flag to newfs instead of manually invoking mkfs. The following table lists each option reading from left to right across the mkfs command line. The first column lists the mkfs parameter, the second column lists the newfs option flag used to modify the parameter, and the third column describes the parameter.

Options cross reference for newfs to mkfs

mkfs parameter	newfs option flag	Description
-F ufs		Indicates the type of file system to be created.
-o N	-N	No change flag; show what would be done, but does not actually create the file system.
/dev/rdiskette		Raw device on which the file system should be created (in this case, a floppy diskette).

mkfs parameter	newfs option flag	Description
2880	-s	Total size of the file system (in blocks).
18		Number of sectors per track.
2	-t	Number of tracks per cylinder.
8192	-b	Size of the allocation block.
1024	-f	Size of a file fragment.
16	-c	Number of cylinders per cylinder group.
10	-m	Minimum free space required on the file system.
5	-r	Rotational speed of disk specified in rpm for newfs, and in revolutions per second for mkfs.
2048	-i	Number of bytes allocated per inode.
t	-o	Optimization method (space or time).
0	-a	Alternate blocks per cylinder.

When newfs is invoked, it prompts the operator for confirmation before proceeding. The creation of a file system is a destructive process. The administrator must ensure that the data on the partition has been copied to tape before allowing newfs to operate.

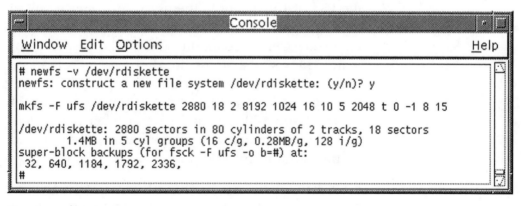

```
# newfs -v /dev/rdiskette
newfs: construct a new file system /dev/rdiskette: (y/n)? y

mkfs -F ufs /dev/rdiskette 2880 18 2 8192 1024 16 10 5 2048 t 0 -1 8 15

/dev/rdiskette: 2880 sectors in 80 cylinders of 2 tracks, 18 sectors
       1.4MB in 5 cyl groups (16 c/g, 0.28MB/g, 128 i/g)
super-block backups (for fsck -F ufs -o b=#) at:
32, 640, 1184, 1792, 2336,
#
```

Creating a file system.

After asking the operator for confirmation, newfs invokes mkfs to create the file system. The mkfs command reports the location of each of the super-block backups as it creates them. It creates a set of index nodes (inodes) and the linked data structure used to access them. Typically, about 10 percent of the total data blocks available are also set aside to avoid the greatly reduced performance a *ufs* file system encounters as the last few data blocks are allocated.

If mkfs reports an error, it usually means that the disk has been improperly partitioned or formatted incorrectly. The mkfs command expects a disk partition to contain a number of blocks which is an integer multiple of the number of blocks in a single disk cylinder. Consult Chapter 9, "Disk Subsystem Hardware," for help in correctly formatting and partitioning a disk.

Tuning Space Usage with the newfs Command

The default newfs parameters will typically result in the loss of 10 percent of the usable disk space on the partition. Part of the space loss is due to the amount of reserved over-subscription space. The newfs command allows alterations in the amount of space reserved for over-subscription storage. The newfs utility also allows tuning of the number of copies of alternate super-blocks, and several other parameters which affect the amount of storage space consumed by the partitioning process.

Refer to Chapter 11, "Disk and File System Maintenance," for more information on using newfs for file system tuning to reduce overhead and improve disk subsystem performance.

Working with Local File Systems

Once a file system has been created, it must be added to the UNIX file tree. At selected spots called mount points, the file tree spans from one file system to the next. File systems are often referred to by the names of their mount points. One could say the system shown below contains the /, /tmp, /usr, and /export/ home file systems. Before mounting a new file system, the administrator should determine which file systems are already in use on the system.

A file tree spanning four file systems.

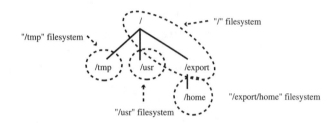

Identifying File Systems to be Mounted

Solaris provides several ways for the administrator to identify the mounted file systems. The **/etc/mount** command invoked with no arguments, the **df** command, and examination of the */etc/mnttab* file are the most common methods.

df Command

The **df** command offers several different ways of displaying information about file systems mounted on a system. Running the command with no argument results in the display of information for all mounted file systems. It is possible to restrict the df output to a single file system by giving a directory in that file system as the last argument to the df command.

```
glenn% df
/               (/dev/dsk/c0t3d0s0):        11690 blocks        9292 files
/usr            (/dev/dsk/c0t3d0s6):        786776 blocks       322309 files
/proc           (/proc):                    0 blocks            204 files
/dev/fd         (fd):                       0 blocks            0 files
/tmp            (swap):                     218816 blocks       3288 files
/opt            (/dev/dsk/c0t3d0s5):        91236 blocks        73801 files
/var/mail       (mail.astro.com:/var/mail): 309376 blocks       96776 files
glenn%
```

The output produced by df shows the mount point, the device mounted (in parentheses), the number of free 512-byte disk blocks, and the number of free files that can be created on the file system. Note that *nfs* mounted file systems list the name of the server which is exporting the file system, and the name of the file system on the server. In the example, it is evident that the *glenn* machine obtains the */var/mail* file system from the *mail.astro.com* machine. When df is invoked with a directory argument, the output format changes as seen below.

```
glenn% df /
File system             kbytes      used     avail    capacity    Mounted on
/dev/dsk/c0t3d0s0       20664       14819    3785     80%         /
glenn%
```

This listing shows the device name first, followed by the total, used, and available space in 1024-byte (kbyte) units, the proportion of the file system already in use, and finally, the mount point. To see this listing for all mounted file systems, use the **-k** option flag with no directory argument.

The most complete df listing is obtained by using the -**g** option. This option formats and displays the *statvfs* structure, which details some of the parameters used to create the file system and provides a complete description of the available space. The -g option flag can be used with or without a directory argument to display all file systems or a single file system.

```
glenn% df -g /
/                        (/dev/dsk/c0t3d0s0):    8192 block size    1024 frag size
41328 total blocks    11690 free blocks        7570 available     11520 total files
9292 free files       8388632 filesys id
ufs fstype            0x00000004 flag           255 file name length
glenn%
```

➥ ***NOTE:*** *It is important to remember the differences among the three different block sizes discussed in this chapter. A disk block is the basic unit of storage on the physical medium. This unit is often 512 bytes long, but not always. An allocation block is the basic unit of space in a file system. It is either 8,192 or 4,096 bytes in length. A fragment, often 1,024 bytes in length, is a portion of an allocation block.*

/etc/mnttab

Another way to determine which file systems are mounted is by examination of the */etc/mnttab* file. This file is created and maintained by the */etc/mount* command. Every time a file system is mounted, an entry is added to the *mnttab* file. When a file system is unmounted, the entry is removed from the *mnttab* file. A typical *mnttab* file might resemble the following example.

```
# cat /etc/mnttab
/dev/dsk/c0t3d0s0 / ufs rw,suid,dev=800018 869622929
/dev/dsk/c0t3d0s6 /usr ufs rw,suid,dev=80001e 869622929
/proc /proc proc rw,suid,dev=2740000 869622929
fd /dev/fd fd rw,suid,dev=27c0000 869622929
/dev/dsk/c0t1d0s6 /var ufs rw,suid,dev=80000e 869622929
swap /tmp tmpfs dev=0 869622932
mail.astro.com:/var/mail /var/mail nfs soft,bg,rw,dev=2700002 869764852
```

mount Command

The **mount** command allows the administrator to view which file systems are mounted, as well as providing a method of mounting file systems. When invoked without arguments, the mount command lists mounted file systems by their mount points, showing the device mounted at each mount point, the mount options used, and the time the file system was mounted.

```
glenn% /etc/mount
/ on /dev/dsk/c0t3d0s0 read/write/setuid on Tue Aug 9 05:58:19 1997
/usr on /dev/dsk/c0t3d0s6 read/write/setuid on Tue Aug 9 05:58:19 1997
/proc on /proc read/write/setuid on Tue Aug 9 05:58:19 1997
/dev/fd on fd read/write/setuid on Tue Aug 9 05:58:19 1997
/tmp on swap read/write on Tue Aug 9 05:58:23 1997
/opt on /dev/dsk/c0t3d0s5 setuid/read/write on Tue Aug 9 05:58:24 1997
/var/mail on mail.astro.com:/var/mail soft/bg/read/write/remote on Tue Aug 9
12:20:52 1997
glenn%
```

Mounting File Systems

File systems can be mounted via several methods: manual mounting by invoking the **/etc/mount** command, the system can automatically mount the file system at boot time after consulting the */etc/vfstab* file, or the volume manager may be used to mount the file system.

General Procedure for Mounting a File System

The general procedure for mounting a file system is outlined below.

1. Format the disk drive.

2. Partition the disk drive.

3. newfs the partition.

4. Create a mount point for the file system.

5. Determine which method will be used to mount the file system (manual mount, *vfstab* mount, automounter, volume manager).

6. Mount the file system.

mount Command Revisited

The **mount** command has several options that control the way in which a file system is mounted. The default action is to mount the file system and base its read and write access on the permissions of the files in the file system. The syntax of the mount follows.

```
# mount -o option,option,...,option disk_device mount_point
```

Options are given as a comma-separated list following the **-o** flag (e.g., **mount -o rw,quota**). Some of

the more common options for the mount command
are summarized in the next table.

Mount command options

Option	Description
quota	Activates the user disk quota system when the file system is mounted.
ro	Mount the file system read-only.
-F fstype	Mount a file system of *fstype*.
-a	Mount all file systems with "mount at boot" (in the *vfstab* file) set to yes.
-o	Use specific *fstype* options when mounting file system.
-p	Provide a list of mounted file systems in *vfstab* format.
-v	Provide verbose output when mounting file system(s).
-O	Overlay the file system on an existing mount point.
-m	Mount file system without making an entry in */etc/mnttab*.
nosuid	Disallow setUID execution for files in this file system.

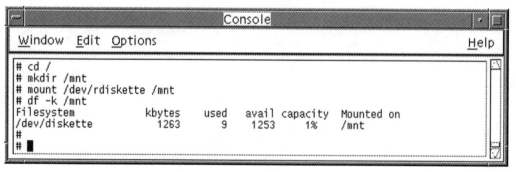

Mounting a file system.

Automatic Mounting

If the administrator needed to type in the **mount** command every time it was necessary to mount a disk, it would quickly become a tiresome chore. Solaris provides several methods for automating disk mounting. The *vfstab* file provides the ability for file systems to be mounted automatically at boot time. The **automounter** allows network file systems to be mounted on demand. The automounter is discussed in detail in Chapter 20, "Automating NFS with Automount." The volume manager provides a means for users to mount/unmount removable media without granting them root privileges.

vfstab

One method of automating the process of mounting file systems is to add the file system to the */etc/vfstab* file. This file is read by the **/etc/mountall** command when it is run as part of the system boot sequence. The vfstab file lists file systems to be mounted when the system is booted. A quick way to add items to vfstab is to use the **-p** option to the mount command.

An example using a floppy disk file system follows. First, the file system is mounted, then the current *vfstab* is saved, and finally, the **mount** and **grep** commands are used to append the mount instructions to the */etc/vfstab* file.

✓ **TIP:** *Saving a copy of the current* vfstab *is good practice as it provides an easy way to recover from any mistakes made when adding new file systems.*

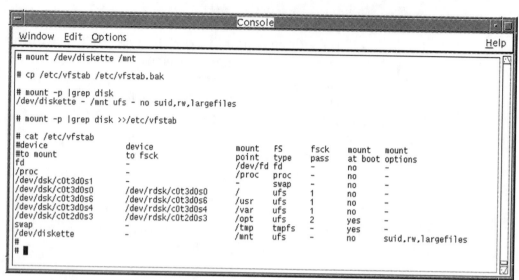

Adding a file system to /etc/vfstab.

It is also possible to add new file systems by editing the *vfstab* file and entering the required information manually. The *vfstab* file format is described in the comment lines at the top of the file. The fields are separated by a tab or spaces. A typical *vfstab* file is shown below. The fields and respective functions of those fields appear in the next table.

#device	device		mount	FS	fsck	mount	mount
#to mount	to fsck		point	type	pass	at boot	options
#							
#/dev/dsk/c1d0s2	/dev/rdsk/c1d0s2		/usr	ufs	1	yes	–
/proc	–		/proc	proc	–	no	–
fd	–		/dev/fd	fd	–	no	–
swap	–		/tmp	tmpfs	–	yes	–
/dev/dsk/c0t3d0s0	/dev/rdsk/c0t3d0s0		/	ufs	1	no	–
/dev/dsk/c0t3d0s6	/dev/rdsk/c0t3d0s6		/usr	ufs	2	no	–
/dev/dsk/c0t3d0s5	/dev/rdsk/c0t3d0s5		/opt	ufs	5	yes	–
/dev/dsk/c0t3d0s1	–		–	swap	–	no	–
/dev/dsk/c1t3d0s1	–		–	swap	–	no	–
mail.astro.com:/var/mail	–		/var/mail	nfs	no	yes	soft,bg,rw

Fields and functions for vfstab

Field name	Description
Device to mount	Name of device to be mounted.
Device to fsck	Raw device to be checked by the fsck utility.
Mount point	Directory where the device should be added to the UNIX file tree.
FS type	File system type.
fsck pass	Number indicates the ordering method by which the file system will be checked.
Mount at boot	Yes to cause file system to mount at boot. No to prevent file system mount at boot.
Mount options	Options to be passed to the mount command.

fsck Pass Field

UNIX file systems (*ufs*) are the only file systems on which the file system checker (*fsck*) operates. If the fsck pass field contains a minus sign (-) or a zero (0), no file system integrity checking is performed. A file system with an *fsck* pass of one (1) indicates that the file system is to be checked sequentially (in the order it is listed in the *vfstab* file). Note that the / (root) file system is always checked first.

File systems with an *fsck* pass number greater than one (1) are checked in parallel (simultaneously). For systems with a large number of disks spanning multiple disk controllers or disk busses, parallel file system checking is generally faster than sequential checking. For efficiency, use *fsck* on file systems of similar size on different disks simultaneously.

Once the appropriate information has been entered in the *vfstab* file, the administrator can cause the system to mount all file systems by invoking the following command:

```
# /etc/mount -a
```

Unmounting a File System

The complement to mounting a file system is to unmount it by using the **/etc/umount** command. Exercise care when unmounting file systems. Some file systems are required for proper system operation. Other file systems may be unmounted while allowing the system to continue to operate, but the system will not perform as expected. The syntax for the umount command follows:

```
# umount mount_point (i.e. umount /mnt)
```

or

```
# umount device (i.e. umount /dev/dsk/c0t2d0s1)
```

If you are uncertain about the effect that unmounting a file system may have on the system, it is best to bring the system to the single-user state before invoking the **umount** command. In the case of a file server, unmounting a file system may also have an effect on the *nfs* client machines on which the file system is mounted.

Unmounting a file system.

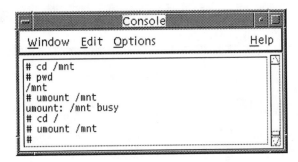

```
┌─────────────────────Console─────────────────┐
│ Window  Edit  Options                  Help  │
├──────────────────────────────────────────────┤
│ # cd /mnt                                    │
│ # pwd                                        │
│ /mnt                                         │
│ # umount /mnt                                │
│ umount: /mnt busy                            │
│ # cd /                                       │
│ # umount /mnt                                │
│ #                                            │
└──────────────────────────────────────────────┘
```

Volume Manager

Adding entries to the *vfstab* file works well for hard disks but is not suitable for removable media such as floppies and CD-ROMs. These devices tend to be mounted and unmounted much more frequently than hard disks. To handle such devices, use the *volume manager*. The volume manager automatically detects the presence of removable media, determines the file system type, and then mounts the media appropriately.

Automatic detection of the media is limited to the CD-ROM device. The Solaris File Manager accomplishes this with a combination of a daemon (**vold**) and a configuration file (*/etc/vold.conf*) which specifies the actions to be taken for various removable devices and file system types. For floppy disks, the volume manager is unable to detect the presence of new disks and must be informed that a new disk is present. This is accomplished by invoking the **/bin/volcheck** command as shown in the next illustration.

```
┌─────────────────────────────────────────────────────────────────┐
│ ─                            Console                        ▫ □   │
├─────────────────────────────────────────────────────────────────┤
│  Window   Edit   Options                                    Help  │
├─────────────────────────────────────────────────────────────────┤
│ # df -k                                                           │
│ Filesystem           kbytes    used   avail capacity  Mounted on  │
│                                                                   │
│ /dev/dsk/c0t3d0s0     30702   23310    7362    76%    /           │
│ /dev/dsk/c0t3d0s6    200515  181176   19139    91%    /usr        │
│ /proc                     0       0       0     0%    /proc       │
│ fd                        0       0       0     0%    /dev/fd     │
│ /dev/dsk/c0t3d0s4     15342    5705    9622    38%    /var        │
│ /dev/dsk/c0t2d0s3    148167   10969  137050     8%    /opt        │
│ /dev/dsk/c0t2d0s6    240463  201981   38242    85%    /usr/openwin│
│ swap                  56920     196   56724     1%    /tmp        │
│ # volcheck                                                        │
│                                                                   │
│ # df -k                                                           │
│                                                                   │
│ Filesystem           kbytes    used   avail capacity  Mounted on  │
│ /dev/dsk/c0t3d0s0     30702   23313    7359    77%    /           │
│ /dev/dsk/c0t3d0s6    200515  181176   19139    91%    /usr        │
│ /proc                     0       0       0     0%    /proc       │
│ fd                        0       0       0     0%    /dev/fd     │
│ /dev/dsk/c0t3d0s4     15342    5705    9622    38%    /var        │
│ /dev/dsk/c0t2d0s3    148167   10969  137050     8%    /opt        │
│ /dev/dsk/c0t2d0s6    240463  201981   38242    85%    /usr/openwin│
│ swap                  56864     200   56664     1%    /tmp        │
│ /vol/dev/diskette0/unnamed_floppy                                 │
│                        1423       0    1423     0%    /floppy/unnamed_floppy│
│ #                                                                 │
│ # ▮                                                               │
└─────────────────────────────────────────────────────────────────┘
```

Mounting a floppy disk using the volume manager.

The volume manager always mounts floppy disks under the */floppy* directory which it controls. CD-ROMs are likewise mounted under the */cdrom* directory. Note that in the example the **volcheck** command was issued from an ordinary user account and not the super-user. The volume manager removes the need for users to have root privileges in order to mount removable media.

The Solaris File Manager communicates with the volume manager and will automatically access a new File Manager window when a removable device is mounted. The volume manager may also activate the File Manager. Under the File menu on the main File

Manager window a menu item allows the user to check for new floppies. This allows the File Manager to perform the same action as typing the **volcheck** command. The newly created File Manager window also has an Eject button which will unmount the file system and eject the disk from the drive.

File Manager detection of volume manager mounted disks.

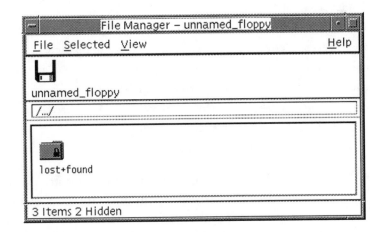

Configuring the Volume Manager

Changing the actions taken and devices under control of the volume manager is a simple matter of modifying the */etc/vold.conf* file. The configuration information in the file is formatted by section. A sample */etc/vold.conf* file appears below.

```
#
# @(#)VOLD.CONF = 6   93/05/17 SMI
#
# Volume Daemon Configuration file
#
# Database to use (must be first)
db db_mem.so
# Labels supported
```

```
label dos label_dos.so floppy
label cdrom label_cdrom.so cdrom
label sun label_sun.so floppy
# Devices to use
use cdrom drive /dev/dsk/c0t6 dev_cdrom.so cdrom0
use floppy drive /dev/diskette dev_floppy.so floppy0
# Actions
insert /vol*/dev/diskette[0-9]/* user=root /usr/sbin/rmmount
insert /vol*/dev/dsk/* user=root /usr/sbin/rmmount
eject /vol*/dev/diskette[0-9]/* user=root /usr/sbin/rmmount
eject /vol*/dev/dsk/* user=root /usr/sbin/rmmount
notify /vol*/rdsk/* group=tty /usr/lib/vold/volmissing -c
# List of file system types unsafe to eject
unsafe ufs hsfs pcfs
```

The lines beginning with a pound sign (#) are considered comments and are used here to help delineate the various sections of the configuration information. The first two sections, beginning with the comments *Database* and *Labels*, describe the database routines and disk label types the volume manager recognizes. These two sections should not be modified.

Devices to Use Section

The third section, marked by the comment *Devices to use*, lists the names and types of the removable media devices the volume manager should monitor. Each line in this section starts with the keyword *use*.

```
use cdrom drive /dev/dsk/c0t6 dev_cdrom.so cdrom0
```

The use keyword is followed by the type of device, either CD-ROM or floppy, and the keyword *drive*.

```
use cdrom drive /dev/dsk/c0t6 dev_cdrom.so cdrom0
```

Following the device type is the Solaris name for the device. Note that the CD-ROM device name specifies only the first five characters of the full special device name. Because the volume manager will monitor and mount all available slices it finds on a CD-ROM disk, the only information items needed are the specific controller and target portions of the device name.

```
use cdrom drive /dev/dsk/c0t6 dev_cdrom.so cdrom0
```

Following the special device name is the name of the shared object used to manage the device. This must match up with the device type specified (e.g., if the device type is *cdrom*, the shared object must be *dev_cdrom.so*). Finally, the symbolic name used in the */device* directory is listed. The first device of a given type has a 0 (zero) appended to its name, while the second is appended with a 1, and so on. For instance, a second CD-ROM drive located at target 5 on the built-in SCSI controller would be placed under volume manager control by adding a line like the following to the devices section of the *vold.conf* file:

```
use cdrom drive /dev/dsk/c0t5 dev_cdrom.so cdrom1
```

Actions Section

The next section, which begins with the comment *Actions*, specifies the actions to be taken when certain events occur. The basic events are the insertion of media into a drive (*insert*), removal of media from a drive (*eject*) and notification of problems (*notify*). An example entry in the actions section appears below.

```
eject /vol*/dev/diskette[0-9]/* user=root /usr/sbin/rmmount
```

Each line lists an event followed by a regular expression. When an event occurs, each line which begins with the name of the event (insert, eject or notify) is checked. If the volume on which the event occurs matches the regular expression, then the remainder of the action line comes into play. The remainder of this line includes the name of the user or group identification to be used to run the listed command with the listed arguments. In the example line above, when the eject event occurs and the volume matches the regular expression */vol*/ dev/diskette[0-9]/**, the command **/usr/sbin/rmmount** would be run with the root user permissions.

rmmount.conf Configuration File

The **/usr/sbin/rmmount** command has its own configuration file called */etc/rmmount.conf.* Although not often modified, this file allows the specification of additional actions to occur when a disk is mounted. A common use of this feature is to allow CD-ROMs which are mounted by the volume manager to be automatically shared, or made accessible to other workstations on the network via NFS. To accomplish this, a share line is added to the bottom of the */etc/rmmount.conf* file as follows:

```
share cdrom*
```

This line would share any CD-ROM mounted by the volume manager without any restrictions. To control access the administrator can add options to the share line in a form similar to the share command. The share command is covered in more detail in Chapter 19, "Accessing Network Resources with NFS."

Mounting Non-ufs File Systems with the Volume Manager

The volume manager is also able to handle file system types other than *ufs*. For instance, inserting an MS-

DOS formatted floppy and using the File Manager or the **volcheck** command results in the disk being mounted under the */floppy* directory on a mount point which bears the floppy volume name. Thus, inserting an MS-DOS floppy whose volume name is "transfer" and selecting Check For Floppy from the File menu in the File Manager will cause the window shown in the following illustration to appear. Note the DOS label on the floppy disk icon which indicates the file system type.

File Manager mounting an MS-DOS floppy.

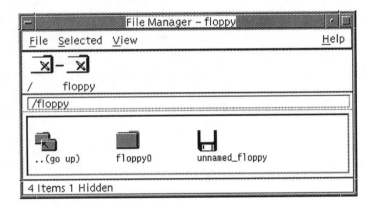

Starting and Stopping the Volume Manager

The CD-ROM and floppy disk drives on a Solaris system are usually under the control of the volume manager. Some CD-ROM and floppy operations require that these devices not be under the influence of the volume manager. Solaris makes it possible to disable the volume manager and work directly with the CD-ROM and/or floppy drive. To disable the volume manager, invoke the **/etc/init.d/volmgt** script with the **stop** option. To restart the volume manager, invoke the **/etc/init.d/volmgt** script with the **start** option.

*Stopping the
volume manager.*

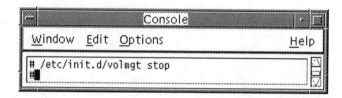

Summary

Local file systems and their relationships to the UNIX
file tree constitute a basic building block of a Solaris
system. Understanding how file systems are created,
mounted, checked, and tuned enables the administrator to effectively manage corporate disk space.
More importantly, understanding how to mount, create, and unmount file systems on an active system can
result in minimization of down-time for corporate
users.

Disk and File System Maintenance

Previous chapters explored the naming conventions of devices under Solaris 2, disk subsystem hardware, and disk subsystem software. These chapters covered the information the system administrator needs in order to install and manage system disk drives. Unfortunately, disk drives break down. Disk drives are also throughput-limited devices. The bits can only be written to, and retrieved from, the magnetic platters at a finite speed.

This chapter presents information to help the system administrator recognize disk problems, and methods which may be used to avoid or recover from disk failures. In addition, the discussion focuses on file system performance tuning and alternate disk subsystem technology which may be employed for high throughput file servers.

File System Repair

Although UNIX file systems are fairly rugged, problems may occur which cause damage to the index structure. Solaris keeps some of the file system structure in memory and periodically updates that structure to disk as a means to improve file system performance. However, an inadvertent shutdown, such as a power failure, might result in the index structure on the disk not being completely updated. This type of file system damage requires repair before the file system can be mounted.

Automatic File Checking

The integrity of file systems automatically mounted at boot time is checked by the **fsck** command. The volume manager will also use the fsck command to check any ufs file systems it mounts. In both of these cases, fsck is run using the *p* (preen) option. The preen option automatically fixes minor file system problems, but will exit if major file system damage is encountered.

Using the fsck preen options to check a file system.

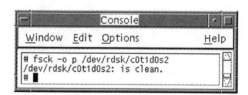

In the event of major damage to the index structure, fsck must be run manually without the preen option. Running fsck in this mode provides more detail about the checks performed by fsck. The following illustration is typical of fsck output for an intact file system.

```
 ┌────────────────────────────Console──────────────────────────┐
 │ Window  Edit  Options                                   Help │
 ├──────────────────────────────────────────────────────────────┤
 │ # fsck /dev/rdsk/c0t1d0s2                                     │
 │ ** /dev/rdsk/c0t1d0s2                                         │
 │ ** Last Mounted on /book                                     │
 │ ** Phase 1 - Check Blocks and Sizes                          │
 │ ** Phase 2 - Check Pathnames                                 │
 │ ** Phase 3 - Check Connectivity                              │
 │ ** Phase 4 - Check Reference Counts                          │
 │ ** Phase 5 - Check Cyl groups                                │
 │ 2 files, 9 used, 95412 free (20 frags, 11924 blocks, 0.0% fragmentation) │
 │ #                                                            │
 └──────────────────────────────────────────────────────────────┘
```

Sample fsck output for a clean file system.

Recoverable File System Damage

Typically, file system damage occurs when a system is shut down incorrectly (e.g., power failure or system crash). As a result, the file system index information in memory may not have been written to disk. A disk with this type of file system damage can usually be repaired with **fsck**.

The following example shows a file system which was damaged when information was written over the primary super-block. To recover from this problem, fsck needs to know where to find an alternate super-block. To specify the alternate super-block, invoke fsck with the **-o b=#** command line options. This allows fsck to reconstruct the primary super-block by using the information stored in an alternative super-block.

```
 ┌────────────────────────────Console──────────────────────────┐
 │ Window  Edit  Options                                   Help │
 ├──────────────────────────────────────────────────────────────┤
 │ # fsck /dev/rdsk/c0t1d0s2                                    │
 │ BAD SUPER BLOCK:  MAGIC NUMBER WRONG                         │
 │ USE AN ALTERNATE SUPER-BLOCK TO SUPPLY NEEDED INFORMATION;   │
 │ eg. fsck -o -b=# [special ...]                              │
 │ where # is the alternate super block.  SEE fsck_ufs(1M).    │
 │ #                                                           │
 │ # fsck -o b=3248 /dev/rdsk/c0t1d0s2                         │
 │ Alternate super block location:  3248                       │
 │ ** /dev/rdsk/c0t1d0s2                                        │
 │ ** Last Mounted on /book                                    │
 │ ** Phase 1 - Check Blocks and Sizes                         │
 │ ** Phase 2 - Check Pathnames                                │
 │ ** Phase 3 - Check Connectivity                             │
 │ ** Phase 4 - Check Reference Counts                         │
 │ ** Phase 5 - Check Cyl groups                               │
 │ 1595 files, 13393 used, 80105 free (377 frags, 9966 blocks, 0.4 │
 │ % fragmentation)                                            │
 │ #                                                           │
 └──────────────────────────────────────────────────────────────┘
```

Disk with a bad super-block.

✓ **TIP:** *The alternate super-block locations are listed by the* newfs *command when the file system is created. The locations can be listed again with* newfs -N *on the raw disk device, as shown in the following illustration.*

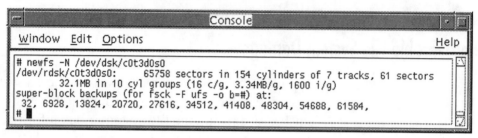

```
┌──────────────────────────────Console──────────────────────────────┐
│ Window  Edit  Options                                        Help  │
├────────────────────────────────────────────────────────────────────┤
│ # newfs -N /dev/dsk/c0t3d0s0                                        │
│ /dev/rdsk/c0t3d0s0:    65758 sectors in 154 cylinders of 7 tracks, 61 sectors │
│       32.1MB in 10 cyl groups (16 c/g, 3.34MB/g, 1600 i/g)          │
│ super-block backups (for fsck -F ufs -o b=#) at:                    │
│  32, 6928, 13824, 20720, 27616, 34512, 41408, 48304, 54688, 61584,  │
│ # ■                                                                 │
└────────────────────────────────────────────────────────────────────┘
```

Invoking newfs with the -N flag to identify alternate super-block locations.

⊷ **NOTE:** *The "FILE SYSTEM WAS MODIFIED" warning given by* fsck *should serve to remind the operator to run* fsck *again. Sometimes a single invocation of* fsck *does not correct all problems present in the index structure of a disk. Using* fsck *a second time catches any problems that were not corrected during the first invocation.*

Example of a Recoverable Disk

The disk in the next example contains extensive file system damage. It appears to be recoverable, but successful recovery will take some work. The first problem is a damaged directory. The command will reconstruct the directory structure by replacing mandatory parts of the directory file, but it may not be able to recover the pointers to the individual files. The result is files (and possibly directories) which are no longer contained in any directory. The complete sequence of fsck actions needed to reconstruct the file are illustrated in the following sections.

Using fsck to repair a heavily damaged disk, part 1.

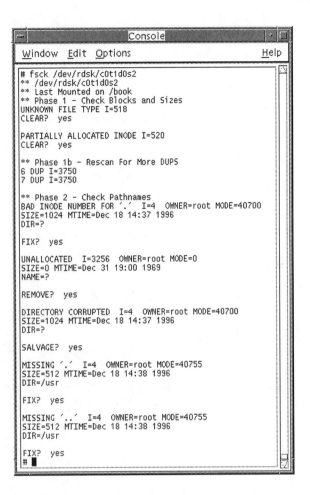

```
# fsck /dev/rdsk/c0t1d0s2
** /dev/rdsk/c0t1d0s2
** Last Mounted on /book
** Phase 1 - Check Blocks and Sizes
UNKNOWN FILE TYPE I=518
CLEAR? yes

PARTIALLY ALLOCATED INODE I=520
CLEAR? yes

** Phase 1b - Rescan For More DUPS
6 DUP I=3750
7 DUP I=3750

** Phase 2 - Check Pathnames
BAD INODE NUMBER FOR '.'  I=4  OWNER=root MODE=40700
SIZE=1024 MTIME=Dec 18 14:37 1996
DIR=?

FIX? yes

UNALLOCATED  I=3256  OWNER=root MODE=0
SIZE=0 MTIME=Dec 31 19:00 1969
NAME=?

REMOVE? yes

DIRECTORY CORRUPTED  I=4  OWNER=root MODE=40700
SIZE=1024 MTIME=Dec 18 14:37 1996
DIR=?

SALVAGE? yes

MISSING '.'  I=4  OWNER=root MODE=40755
SIZE=512 MTIME=Dec 18 14:38 1996
DIR=/usr

FIX? yes

MISSING '..'  I=4  OWNER=root MODE=40755
SIZE=512 MTIME=Dec 18 14:38 1996
DIR=/usr

FIX? yes
#
```

lost+found Directory

Once the directories are repaired, the repair effort moves on to reconnecting files and directories to appropriate directories. The *lost+found* directory is used by **fsck** as a place to deposit files and directories which will be relinked to the directory tree. In the following illustration, the *lost+found* directory itself has been damaged and is recreated.

The files and directories in *lost+found* are listed by the number of the index node (inode) they inhabit. File name information is stored in the directory file structure. When a file system sustains damage to the directory file structure, the names of the files in the directory are frequently damaged (or lost). However, the file's contents, ownership, and permissions are usually left intact because they are stored in the inode for the file. If the inode for the file is intact, fsck can save the file's contents even if the name of the file is lost.

✓ **TIP:** *It is often faster (and easier) not to reconnect zero length files. Zero length files may occur if a program creates a temporary file, opens it, and then removes it while the file is still open. The result is a file which does not belong to any directory, but is still available to be written and read by the program. Under normal operations the file is cleaned up when the program exits.*

File system repair efforts, part 2.

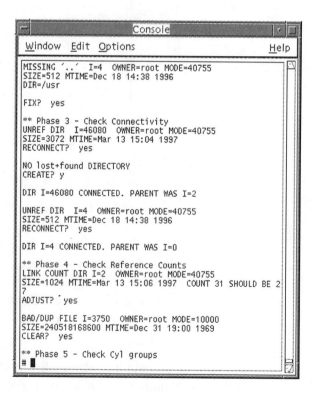

```
MISSING '..'   I=4   OWNER=root MODE=40755
SIZE=512 MTIME=Dec 18 14:38 1996
DIR=/usr

FIX?  yes

** Phase 3 - Check Connectivity
UNREF DIR   I=46080   OWNER=root MODE=40755
SIZE=3072 MTIME=Mar 13 15:04 1997
RECONNECT?  yes

NO lost+found DIRECTORY
CREATE?  y

DIR I=46080 CONNECTED. PARENT WAS I=2

UNREF DIR   I=4   OWNER=root MODE=40755
SIZE=512 MTIME=Dec 18 14:38 1996
RECONNECT?  yes

DIR I=4 CONNECTED. PARENT WAS I=0

** Phase 4 - Check Reference Counts
LINK COUNT DIR I=2   OWNER=root MODE=40755
SIZE=1024 MTIME=Mar 13 15:06 1997  COUNT 31 SHOULD BE 2
7
ADJUST?  yes

BAD/DUP FILE I=3750  OWNER=root MODE=10000
SIZE=240518168600 MTIME=Dec 31 19:00 1969
CLEAR?  yes

** Phase 5 - Check Cyl groups
#
```

With the files and directories reconnected to the file tree in the lost+found directory, fsck repairs the reference count information in the directories and updates free block information in the super-block. The modified directories and super-block are thereby corrected with respect to the actions fsck took earlier.

File system repair efforts, part 3.

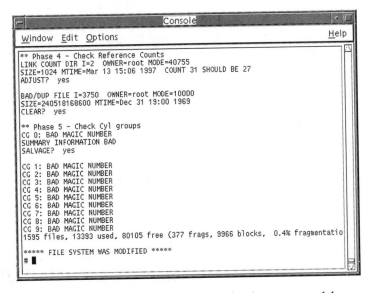

```
                                    Console
Window  Edit  Options                                        Help

** Phase 4 - Check Reference Counts
LINK COUNT DIR I=2  OWNER=root MODE=40755
SIZE=1024 MTIME=Mar 13 15:06 1997  COUNT 31 SHOULD BE 27
ADJUST?  yes

BAD/DUP FILE I=3750  OWNER=root MODE=10000
SIZE=240518168600 MTIME=Dec 31 19:00 1969
CLEAR?  yes

** Phase 5 - Check Cyl groups
CG 0: BAD MAGIC NUMBER
SUMMARY INFORMATION BAD
SALVAGE?  yes

CG 1: BAD MAGIC NUMBER
CG 2: BAD MAGIC NUMBER
CG 3: BAD MAGIC NUMBER
CG 4: BAD MAGIC NUMBER
CG 5: BAD MAGIC NUMBER
CG 6: BAD MAGIC NUMBER
CG 7: BAD MAGIC NUMBER
CG 8: BAD MAGIC NUMBER
CG 9: BAD MAGIC NUMBER
1595 files, 13393 used, 80105 free (377 frags, 9966 blocks,  0.4% fragmentatio

***** FILE SYSTEM WAS MODIFIED *****
# 
```

Finally, fsck is run again to verify that no problems remain from the first run of fsck. With the file system structure intact, the file system is mounted. Now comes the tough job of looking through the files in the *lost+found* directory. This is not easy because the file names are missing. Fortunately, in this case the only items in the *lost+found* directory are directories, and their contents provide clues to respective names. The directories can be returned to their proper places and the file system can be put back into service.

Restoring Files from Backup versus Repairing a File System

Fortunately, repairing file systems is not a common task for the system manager. When confronted with repairing a file system, keep in mind the balance between restoration of the file system from backups versus the repair effort. If a recent backup is available it may be quicker to restore the file system than attempt repair. Repairing a file system can be very time-consuming, and it is not always easy to determine what to do with the remnants **fsck** leaves behind.

For example, the damaged example file system generated over 5,000 lines of fsck output, and required almost an hour to manually repair the damage. A full restore of the 14 Mb of data on this file system would have been a much more productive use of the administrator's time.

File system repair efforts, part 4.

```
                              Console
Window  Edit  Options                                        Help
# mount /dev/dsk/c0t1d0s2 /book
# ls -lR /book
total 2
    drwx-----T   4 root      root         1024 Mar 13 12:09 .

/book/lost+found:
total 18
    drwx------   2 root      root         8192 Mar 13 12:09 #004
    drwxr-xr-x   2 bin       lp           8192 Mar 13 12:09 #3256

/book/lost+found/#003:
total 0

/book/lost+found/#3256
total 112
    drwxr-xr-x   9 root      other         512 Mar 13 12:09 .
    drwxr-xr-x  29 root      other        1024 Mar 13 15:35 ..
    drwxr-xr-x   2 root      other         512 Mar 13 12:09 alerts
    drwxr-xr-x   2 root      other         512 Mar 13 12:09 classes
    drwxr-xr-x   2 root      other         512 Mar 13 12:09 fd
    drwxr-xr-x   2 root      other         512 Mar 13 12:09 forms
    drwxr-xr-x   2 root      other         512 Mar 13 12:09 interfaces
    drwxr-xr-x   2 root      other         512 Mar 13 12:09 printers
    drwxr-xr-x   2 root      other         512 Mar 13 12:09 pwheels
#
```

Unrecoverable File System Damage

Some of the problems in a damaged file system can be repaired, while others cannot. Disk drives which have suffered media damage may contain unrecoverable data sectors. The **fsck** command will notify the user about blocks that are no longer readable, and then exit. Disks which have sustained media damage are often beyond repair. It may be possible to copy the undamaged files to another disk (or tape), and manually select critical files which have not been backed up. Damaged floppy diskettes can often be reused once they are reformatted, but it is best to replace hard disks if you suspect media damage. The next section explores a few typical disk failure modes.

Disk with unrecoverable errors.

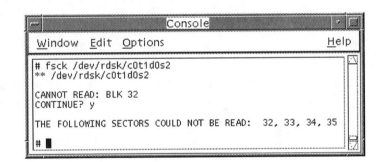

```
# fsck /dev/rdsk/c0t1d0s2
** /dev/rdsk/c0t1d0s2

CANNOT READ: BLK 32
CONTINUE? y

THE FOLLOWING SECTORS COULD NOT BE READ:   32, 33, 34, 35

#
```

Recognizing Failing Disks

Over time all disk drives will experience damage of one sort or another. Damage is usually limited to repairable file system problems due to unexpected shutdowns. Hardware failure is also a possible (albeit less frequent) cause of disk subsystem failure. What are the symptoms of a failing disk drive? Can anything be done to prevent such failures? Once the telltale signs of an impending failure are present, what can be done to avoid catastrophe?

Disk Failure Modes

Most disk drive failures are not characterized by immediate catastrophic failure. More typically, bad spots will appear gradually until the system can no longer read/write major portions of the drive. The system will usually print messages on the console informing the user that something unexpected has occurred.

Soft Errors

In many cases, bad spots on the disk surface can be read without losing data because the disk drive contains logic to detect and correct multi-bit information errors. As the system encounters these spots, the **syslog** utility will be informed (which will create console error messages) that a retryable error has been encountered.

These messages mean that the system had to try several times to read a portion of the disk. System administrators should not be concerned if they notice occasional retry errors. A few soft errors are to be expected from time to time. If the errors become more frequent, you should study the problem and take corrective action before it becomes critical.

Typical retryable disk error message.

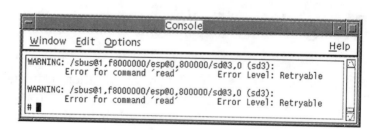

```
┌─────────────────────────────Console────────────────────────────────┐
│  Window  Edit  Options                                        Help  │
│ ┌─────────────────────────────────────────────────────────────────┐ │
│ │WARNING: /sbus@1,f8000000/esp@0,800000/sd@3,0 (sd3):             │ │
│ │        Error for command 'read'         Error Level: Retryable  │ │
│ │                                                                  │ │
│ │WARNING: /sbus@1,f8000000/esp@0,800000/sd@3,0 (sd3):             │ │
│ │        Error for command 'read'         Error Level: Retryable  │ │
│ │# █                                                               │ │
│ └─────────────────────────────────────────────────────────────────┘ │
└──────────────────────────────────────────────────────────────────────┘
```

Hard Errors

Sometimes the system cannot read a bad spot. The drive logic that detects these errors also fails to correct them. In this case the system will report that it has encountered a fatal disk error. Whenever a fatal disk error is reported, the administrator should be very concerned about the integrity of the files on the disk drive. Steps must be taken to rectify the problem immediately, or data may be destroyed.

Typical hard disk error message.

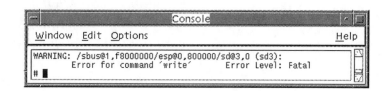

Electronics Failures

A less typical disk subsystem failure involves problems with disk drive electronics. When a drive electronics failure occurs, the system may not be able to open the drive for read/write access. This type of failure is not easily repaired by the end user. In the unfortunate event that the drive had not been backed up to tape, it is sometimes possible to replace the drive circuit boards so the contents of the failing drive may be copied to tape. This operation usually voids the manufacturer's warranty, and should not be attempted by the system administrator. Most disk manufacturers and third party disk repair centers will perform this service for a fee.

✓ **TIP:** *With SCSI disk systems, failure resulting from loose connections or faulty terminators is common. Checking all cable connections, power supplies, and terminators before replacing a drive is recommended.*

☛ *WARNING: The administrator should always ensure that there is a usable, recent backup of the data on a disk before attempting to reformat a failing drive. If a backup has not been performed, contact your disk vendor or data recovery service before attempting to work on the disk. Data recovery services can often salvage a large percentage of the data from a failing disk drive. If the administrator has written format information or surface test patterns over the data, the task of recovering the data are much more difficult and expensive.*

Repairing Damaged Disks

There are several courses of action that may be followed when the administrator suspects that a disk is failing. Basic procedures for repairing bad disks are listed below.

❏ Reboot the system to determine if the failure is due to corrupted system information.

❏ If the block number of a bad spot is known, use the **repair** portion of the **format** utility to map out the spot on the drive.

❏ If there are several bad spots on the drive, use the **analyze** portion of the format utility to perform non-destructive surface analysis of the drive to search for and optionally repair bad spots.

❏ Reformat the defective disk. Remember to perform a file system backup of all file systems on the drive before you reformat it. A simple reformat will often allow the drive to be placed back into operation.

❏ Replace the drive with a known good drive. Copy all data to the new drive. Note that this option can be time-consuming and expensive.

Using format to Repair Bad Spots

If a disk drive is generating a large number of error messages on a few disk blocks, consider repairing the bad spots with the **format** utility **repair** command. The bad spot will be marked as unusable so that the system will not attempt to access this sector in the future. The repair facility will also "map" the bad spot to a spare sector on the disk drive, and copy the data from the bad spot to the replacement sector.

If the **analyze** facility locates a large number of bad spots on the disk, it may be prudent to reformat the entire disk. Before formatting the disk, verify that current backups are available for file systems on the failing disk. Once the drive has been reformatted, reload the file systems from the dump tapes.

Using repair to fix a bad spot on a disk.

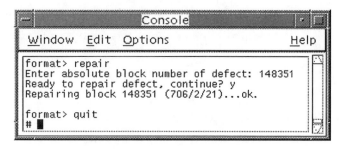

In some cases, **format** cannot recover the data from a bad spot on a disk. In this case, the system will print a warning message stating that it has encountered a fatal error. If an attempt is made to repair such a sector the bad spot will be mapped out, but the file which contained the bad sector may be destroyed by the mapping process. If this occurs, you will need to restore the damaged file from a backup tape.

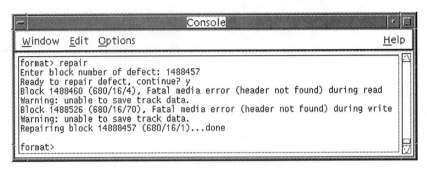

Repairing a fatal error with the repair command.

If the block number of the bad spot is not known, or if there appear to be multiple bad spots on the disk, configure the format utility **analyze** routine to scan the entire drive to find and repair all bad spots.

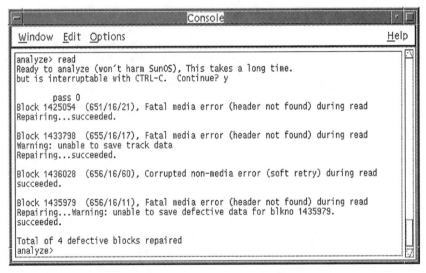

Using analyze to locate and repair bad spots on a disk.

If analyze continues to find errors on successive passes, it may be time to replace the disk drive with a new one. If another drive (of the same model and geometry) is available, a simple shell script may be used to copy files from a failing drive to a new drive as shown below.

```
┌──────────────────────────── Console ────────────────────────────┐
│ Window   Edit  Options                                      Help │
├──────────────────────────────────────────────────────────────────┤
│ # for i                                                          │
│ > in 0 6                                                         │
│ > do                                                            │
│ > dd if=/dev/rdsk/c0t1d0s"$i" of=/dev/rdsk/c0t2d0s"$i" bs=100b  │
│ > fsck /dev/rdsk/c0t2d0s"$i"                                    │
│ > done                                                          │
│ read: I/O error                                                 │
│ 162+0 records in                                                │
│ 162+0 records out                                               │
│ ** /dev/rdsk/c0t2d0s0                                           │
│ ** Last Mounted on /usr/tools                                  │
│ ** Phase 1 - Check Blocks and Sizes                            │
│ ** Phase 2 - Check Pathnames                                   │
│ ** Phase 3 - Check Connectivity                                │
│ ** Phase 4 - Check Reference Counts                            │
│ ** Phase 5 - Check Cyl groups                                  │
│ 1669 files, 4946 used, 2562 free (26 frags, 317 blocks,  0.3% fragmentation) │
│ read: I/O error                                                 │
│ 1603+0 records in                                               │
│ 1603+0 records out                                              │
│ ** /dev/rdsk/c0t2d0s6                                           │
│ ** Last Mounted on /usr/lang                                   │
│ ** Phase 1 - Check Blocks and Sizes                            │
│ ** Phase 2 - Check Pathnames                                   │
│ ** Phase 3 - Check Connectivity                                │
│ ** Phase 4 - Check Reference Counts                            │
│ ** Phase 5 - Check Cyl groups                                  │
│ 1420 files, 12788 used, 61934 free (222 frags, 7714 blocks,  0.2% fragmentation) │
│ #                                                               │
└──────────────────────────────────────────────────────────────────┘
```

Using dd to copy files between disks.

If the new disk drive is a different model, or if the geometry of the failing disk differs from that of the new one, use **ufsdump** to copy the bits from the failing drive to another disk or directly to tape. Once the failing disk has been backed up to tape, remove it from the system and install the replacement drive. With the new drive installed and formatted, use **ufsrestore** to restore the file system contents to the new drive.

```
# ufsdump 0f - /dev/rdsk/c0t2d0s6 | (cd /mnt ; ufsrestore rf -)
  DUMP: Writing 32 Kilobyte records
  DUMP: Date of this level 0 dump: Sat Mar 15 23:44:06 1997
  DUMP: Date of last level 0 dump: the epoch
  DUMP: Dumping /dev/rdsk/c0t2d0s6 (grumpy:/usr/lang) to standard output.
  DUMP: Mapping (Pass I) [regular files]
  DUMP: Mapping (Pass II) [directories]
  DUMP: Estimated 27952 blocks (13.65MB).
  DUMP: Dumping (Pass III) [directories]
  DUMP: Dumping (Pass IV) [regular files]
Warning: ./lost+found: File exists
  DUMP: 27902 blocks (13.62MB) on 1 volume at 97 KB/sec
  DUMP: DUMP IS DONE
# sync
# umount /mnt
# fsck /dev/rdsk/c0t1d0s6
** /dev/rdsk/c0t1d0s6
** Last Mounted on /mnt
** Phase 1 - Check Blocks and Sizes
** Phase 2 - Check Pathnames
** Phase 3 - Check Connectivity
** Phase 4 - Check Reference Counts
** Phase 5 - Check Cyl groups
1420 files, 12788 used, 57326 free (230 frags, 7137 blocks,  0.3% fragmentation)
#
```

Using ufsdump/ufsrestore to copy files between disks.

Improving Disk Subsystem Performance

The input/output (I/O) subsystem of any computer is one of the major pieces of the system performance puzzle. An inadequate I/O subsystem can severely limit the performance of the system. On personal computers, this I/O limitation is not (usually) a significant problem. On the other hand, such limitations in large computer and file server systems can bring the corporation to its knees.

Because of the importance of high-performance I/O in the corporate computing environment, several new technologies have emerged to address this need. Not all performance problems, however, are the result of insufficient I/O subsystems. Sometimes the file system may be tweaked, or tuned, to provide significantly better system performance. The following sections touch upon selected aspects of file system tuning and technologies available to improve I/O subsystem performance.

File System Optimization

The file system created by **newfs** will provide acceptable performance for most applications. There are a few situations where adjusting or tuning the basic file system parameters will result in better performance. One such situation involves applications which have very specific file system needs. For instance, file system tuning usually improves performance of special disks, disk controllers, and applications which write files of a particular size (e.g., many small files, few large files).

The various parameters which affect file system performance can be set using option flags when the file system is created, or they can be adjusted later using the **tunefs** command.

tunefs Command

The **tunefs** command provides a way to adjust various file system parameters after a file system has been created. The tunefs command can also be used to correct mistakes made when the file system was created with **newfs** or to adjust for changing disk usage. Some file system parameters cannot be adjusted without recreating the file system. In those cases it will be necessary to dump the file system to tape and recreate the file system using different parameters.

Similar to tuning a musical instrument or a small engine, the best approach to file system tuning is to adopt a cycle of test, adjust, and re-test. The testing should resemble the application which will be in use on the file system. In fact, the application itself should be used as the test if possible. File system performance measurements are helpful, such as the amount of time it takes for a basic operation to complete (to judge the effect of tuning efforts). With the measurement defined and a test method developed, it is possible to evaluate file system changes to determine whether they improve disk subsystem performance.

➥ **NOTE:** *A growing number of modern SCSI disks are "smart" in that they have a sophisticated internal controller and memory buffer which hide certain performance characteristics. These disks provide high performance, but are not readily tunable. The internal controller and memory buffer mask the effects of selected adjustments such as rotational delay factors.*

What tunefs Can and Cannot Do

File system tuning can only provide limited performance improvements. As servers became faster and jobs more demanding, the roles of the file server and file server disk subsystem have come under increasing scrutiny.

Some file servers provide large file systems for users to access over the network. These file systems are full of current information (files) and are frequently accessed. Other file servers are used for archival storage; files consist of historical information and are not used very often. Both types of file servers could be referred to as "disk farms."

Performance tuning is critical to the first type (active files), and probably not very important for the archive server. Unfortunately, **tunefs** can only do so much to improve performance for either type of server. When tunefs is no longer sufficient, you may wish to investigate other available technologies.

In most instances, disk farms are set up to provide high speed access to important data. Typical applications include database systems, medical images, and other space intensive information services. In a few instances, the disk farm is used for long-term archival storage. Because the information on these systems is not accessed very often, high performance is not always the driving criteria in file system tuning.

While tunefs is capable of changing many characteristics of the disk subsystem, it is not a do-all–save-all I/O subsystem fix-it tool. Certain performance parameters are beyond repair with utilities such as tunefs. A few of these parameters are the rate at which the disk can transfer data, rotational speed of the drive, and the amount of storage space available on the drive.

On the other hand, tunefs can change several important file system parameters. The following sections will outline several typical changes made with the tunefs command. Many of these changes can be made without the need for backups and restoration of files in the file system. In order to implement a few types of changes, the file system will need to be dumped to tape, then reloaded from the backup media once the changes are implemented. In all cases, adjustment of file system parameters requires that the file system must be unmounted.

➥ **NOTE:** *It is always a good idea to verify that the file system has been backed up to tape before making changes. If something goes wrong, the backup media could be used to reload the files once a new file system is created with the desired parameters.*

Tuning for Special Disks and Disk Controllers

By default, **newfs** uses the information in the disk label to calculate several factors related to the ability of the disk, controller, and CPU to read or write information. As the disk rotates, each block in the track moves past the disk heads. If the controller and CPU are fast enough, blocks may be written or read in the order they come under the heads.

Slower controller/CPU combinations must skip one or more blocks between read/writes in order to keep up with the I/O demands of the system. The newfs and **tunefs** commands allow the rotational interleave aspect of the file system to be adjusted by calculating the required time for a single block read to be processed. The proper ordering of read/write operations and skip operations is calculated from this value in combination with the rotational speed of the disk.

Some disk controllers are capable of writing or reading multiple blocks in a single operation due to high speed buffer memory located in the controller. The tunefs command allows specification of the number of buffers the controller hardware provides. This improves file system performance by avoiding unnecessary block skipping. The option flags used to adjust these parameters are listed in the following table.

↪ **NOTE:** *Setting the proper parameter values requires detailed knowledge of the disk hardware. These values are generally set by* newfs.

tunefs parameters associated with high performance disk controllers

newfs option	tunefs option	Parameter
-d gap	**-d gap**	The "gap" is the time (in milliseconds) required for a disk service interrupt to be completed and a new transfer started. This value will be used along with rotational speed of the drive, and the number of sectors per track, to determine the number of blocks to skip between disk transfers.
-C maxconfig	**-a maxconfig**	The maximum number of blocks that can be written contiguously before a block must be skipped. This number depends on disk and controller characteristics.

Tuning for Small and Large Files

Applications which consistently read and write very small (or very large) files can often benefit from file system tuning. Large files are often split between cylinder groups on the disk. A cylinder group is a collection of cylinders used by the disk I/O routines in the SunOS kernel to improve disk performance by grouping a file's data blocks close together on the disk.

When a large file is split over two or more cylinder groups, the file's data blocks are spread across the disk. Consequently, extra seek time is required when reading or writing large files. Adjusting the maximum number of blocks that can be allocated to a single file within a cylinder group may ameliorate this problem.

It is also possible to adjust basic allocation block and fragment sizes on the file system. Larger allocation block and fragment sizes favor large files by reducing the time required for file allocation at the expense of reduced space efficiency. If an application stores small files (exclusively), a smaller allocation block will improve speed, and a smaller fragment size will improve space utilization efficiency by avoiding the allocation of blocks that are much larger than the data to be stored.

Tuning for Usenet-news

Applications which use many small files, such as Usenet-news, typically require more index nodes (inodes) than are normally created by the default **newfs** operation. This value can be changed by adjusting the parameters that control the number of bytes per inode.

To find the number of inodes created for the file system, divide the size of the file system by the number of bytes per inode. The default size number of bytes that can be referenced by an inode is 2,048. This assumes that the average file will be at least 2,048 bytes in length. A smaller value may be needed in some cases. When in doubt, err on the side of more inodes (a smaller number of bytes per inode). If a file system runs out of inodes, additional files cannot be created even if additional storage blocks are available.

```
# newfs /dev/rdsk/c0t1d0s6
newfs: construct a new file system /dev/rdsk/c0t1d0s6: (y/n)? y
/dev/rdsk/c0t1d0s6:      160230 sectors in 763 cylinders of 6 tracks, 35 sectors
        78.2MB in 48 cyl groups (16 c/g, 1.64MB/g, 768 i/g)
super-block backups (for fsck -F ufs -o b=#) at:
 32, 3440, 6848, 10256, 13664, 17072, 20480, 23888, 26912, 30320, 33728, 37136,
 40544, 43952, 47360, 50768, 53792, 57200, 60608, 64016, 67424, 70832, 74240,
 77648, 80672, 84080, 87488, 90896, 94304, 97712, 101120, 104528, 107552,
 110960, 114368, 117776, 121184, 124592, 128000, 131408, 134432, 137840,
 141248, 144656, 148064, 151472, 154880, 158288,

# newfs -i 1024 /dev/rdsk/c0t1d0s6
newfs: construct a new file system /dev/rdsk/c0t1d0s6: (y/n)? y
/dev/rdsk/c0t1d0s6:      160230 sectors in 763 cylinders of 6 tracks, 35 sectors
        78.2MB in 48 cyl groups (16 c/g, 1.64MB/g, 1536 i/g)
super-block backups (for fsck -F ufs -o b=#) at:
 32, 3440, 6848, 10256, 13664, 17072, 20480, 23888, 26912, 30320, 33728, 37136,
 40544, 43952, 47360, 50768, 53792, 57200, 60608, 64016, 67424, 70832, 74240,
 77648, 80672, 84080, 87488, 90896, 94304, 97712, 101120, 104528, 107552,
 110960, 114368, 117776, 121184, 124592, 128000, 131408, 134432, 137840,
 141248, 144656, 148064, 151472, 154880, 158288,
#
```

Using newfs to create a Usenet-news file system with more inodes than default file systems.

newfs and tunefs options for adjusting drive block sizes

newfs option	tunefs option	Parameter
-b size	N/A	Basic allocation block size is 8,192 bytes. Only 4,096 or 8,192 are valid entries for this parameter. Use 4,096 only for file systems which will contain very small files.
-f frag	N/A	Size of a block fragment. Acceptable values are 512, 1,024, 2,048, 4,096, and 8,192. (8,192 is valid only if the basic allocation block value is also 8,192.) The smaller values use disk space efficiently at the expense of speed. Larger values provide better read/write speed at the expense of storage space efficiency. The default is 1,024.
N/A	-e maxbpg	Maximum number of blocks a single file may be allocated within a single cylinder group. The default is approximately 25% of the blocks in the cylinder group.
-i nbpi	N/A	Number of bytes per inode. Lower values result in more inodes, and larger values result in fewer inodes. The default is 2,048 bytes.

Tuning for Storage Space Efficiency versus Speed

The disk storage routines in the SunOS kernel have two strategies available for disk allocation: time or space. Time efficiency refers to the time required to allocate space and write files. Optimizing time is wasteful of space because transfers often result in gaps being created in lieu of long continuous disk writes.

Space efficiency refers to the efficient use of scattered blocks on the disk. Optimizing space wastes time because a file is allocated to blocks scattered around the disk, and the disk heads must move more frequently to read or write a file. The administrator can use **tunefs** to set the preferred method on each file system.

newfs and tunefs options related to space versus time argument

newfs option	tunefs option	Parameter
-o strategy	**-o strategy**	Strategy is either space or time.
-m free	**-m free**	Minimum free space percentage. A reserve of less than 10% will adversely affect performance.

> ↦ **NOTE:** *The space versus time storage strategy routines are automatically set by* newfs *depending on the allocated minimum free space. The defaults are to use space optimization when the minimum free space is less than 10% of the disk blocks, and to use time optimization when the minimum free space is 10% or more of the disk blocks.*

Using the space strategy on a small disk may be desirable when efficient utilization of disk space is more important. On a large file system which has a small value for the minimum free space, it may desirable to use the time strategy to improve performance. Another factor related to the strategy routines is the free space reserve. The SunOS file system routines operate most efficiently when at least 10% of the basic allocation blocks are free. Routines must do much more work to allocate space for a file under the 10% threshold. File systems are normally created with a 10% reserve, but tunefs can be employed to adjust the reserve.

Using tunefs to change the file system optimization method.

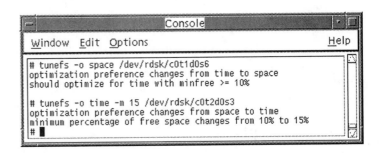

Alternate Disk Subsystem Technologies

For most applications, the standard disk subsystem would provide adequate performance and storage space efficiency. However, in some cases I/O subsystem demands require more throughput or higher availability than is available with standard interfaces. The following sections explore a few of these high-throughput, high-availability technologies.

Redundant Arrays of Inexpensive Disks (RAID)

Servers which provide disk farm service are prime candidates for the use of alternate disk subsystem technology. One increasingly popular technology is Redundant Array of Inexpensive Disks (RAID). RAID disk arrays consist of a SCSI interface, a specialized controller, and an array of disk drives. RAID systems ship with specialized software to control and administer the operation of the disk array. All standard Solaris commands work for the RAID disk array because the driver software takes care of the interface to the RAID subsystem.

RAID disk systems allow for high-throughput, high-availability disk I/O subsystems. RAID technology offers several new options to consider in I/O subsystem design and implementation. In particular, RAID I/O subsystems provide several "levels" of service. Each RAID level describes a different method of data storage on the disk drive(s). Each level has unique characteristics, and some of these levels are more suitable to certain applications than other I/O implementations. The most frequently encountered RAID implementations are discussed in the following sections.

Fiber Channel

In order to provide maximum I/O subsystem performance, many RAID arrays include a fiber channel host interface. This interface provides a 25 megabit per second (or faster) interface between the disk array and the file server. The noise immunity and improved signal quality of the fiber optic based interconnect provide a virtually error free I/O channel for high-traffic file servers. For high-end file servers such fiber based links can provide excellent disk performance.

For a small to medium file server, the standard SCSI interface may be used to connect a RAID array to the system. The 20 megabit/second fast/wide SCSI interface on many of today's workstations provides very good performance in all but the most demanding environments.

RAID Level 0

RAID level 0 implements a striped disk array. This array contains several disk drives. The data are broken down into small segments, and each data segment is written to a separate disk drive. Disk I/O performance is improved by spreading the I/O load across many channels and drives. A striped disk does not offer data redundancy, nor does it provide better fault tolerance than a standard I/O subsystem.

Typical RAID LEVEL 0 implementation.

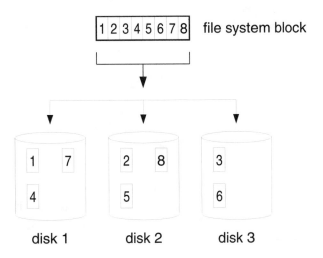

RAID Level 1

RAID level 1 implements a mirrored disk array containing two disk modules. The information written to one disk is also written (mirrored) to the other disk. This ensures that the data are available even if one of the disk drives fails. RAID level 1 offers better data availability, but does not offer high-throughput I/O subsystem performance. It is possible to mirror a striped disk array, thereby providing better I/O performance as well as high data availability. The primary disadvantage of using RAID level 1 is the cost of purchasing duplicate disk drives.

Typical RAID level 1 implementation.

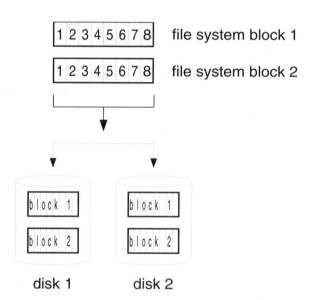

RAID Level 2

RAID level 2 interleaves data sectors across many disks. The controller breaks up the blocks and spreads them over a small group of disk drives. An error detection and correction process is built into RAID level 2 systems to provide some measure of fault tolerance. RAID level 2 is not found on many systems because other RAID levels offer better performance and fault tolerance.

RAID Level 3

RAID level 3 interleaves data sectors across many disks much like RAID level 2. However, RAID level 3 uses a single parity disk per group of drives to implement a fault tolerant disk system. All drive spindles are synchronized such that the read/write heads on all drives in the array are active at the same time. This synchronization also ensures that the data are written to the same sector on every drive in the array. If a disk fails, the system continues to operate by recreating the data from the failed disk to the parity disk.

RAID Level 4

RAID level 4 implements striping with the addition of a parity disk. This provides for some of the fault tolerance missing in RAID level 0. RAID level 4 is not found on many systems, because other RAID levels can provide for better throughput and fault tolerance.

Typical RAID level 3 implementation.

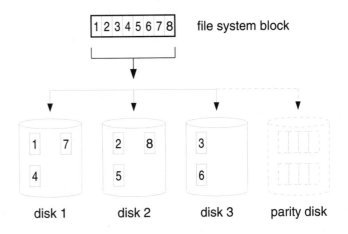

RAID Level 5

RAID level 5 implements a large disk storage array. Similar to a level 0 RAID array, the disk array uses striping to improve system performance. RAID level 5 also implements a parity protection scheme. This scheme allows the system to continue operation even if one of the disk drives in the array fails. The data that are normally stored on the failed drive are recreated from the parity information stored elsewhere in the array. Once the failed disk drive is replaced, the data are restored automatically by the RAID controller. RAID level 5 provides improved system availability and I/O subsystem throughput.

Typical RAID level 5 implementation.

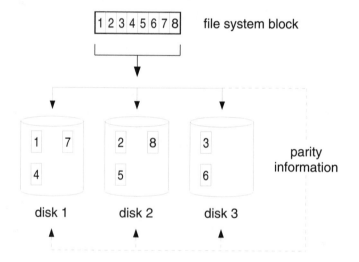

Software Which Emulates RAID Arrays

Hardware solutions are not the only means of implementing disk mirroring and striping. Some software vendors offer applications packages that allow standard SCSI disks connected to standard SCSI controllers to act as a disk array. One such package is Sun Microsystems' Solstice Disk Suite product.

The Disk Suite software allows the administrator to group standard SCSI disk drives into array-like structures. These pseudo arrays may be used to create logical volumes which span the multiple disk drives in the array. For example, it is possible to build a single 10 Gb file system out of five 2.1 Gb disk drives using the Disk Suite software. The performance of these pseudo arrays will not be as good as that of a RAID array, but the cost differential may make the pseudo array more cost effective for some applications.

Summary

The I/O subsystem of any computer system is one of the most important pieces of the system performance puzzle. The system administrator must monitor the system for disk errors and other signs of problems in order to ensure the safety of users' files. At the same time, the administrator must strive to provide file systems which meet the needs of the corporate computing environment. This chapter covered some typical disk failure modes, file system problems, and methods to repair (or circumvent) these problems. Various alternative (high-performance/high-availability) disk subsystem technologies were also explored.

Managing System Software

Previous chapters of this book focused on disk geometry and file systems. The discussion centered on how disk drives are divided into file systems, and how these file systems track files and directories. With these physical setup tasks complete, the time has come to move on to a discussion of how to install and manage the software that will occupy the file systems.

Software Maintenance Concepts

This chapter introduces the concepts of software packages, software clusters, and software configurations. Under Solaris, application software distributions are divided into three general categories discussed in the following sections.

Packages

Software for Solaris 2 is distributed in what is commonly referred to as software packages. A software package is a collection of files that performs a function, such as the on-line manual pages, or the OpenWindows demonstration programs. There are over 80 software packages in Solaris 2. Package names are prefixed with the letters SUNW. For example, the package name for the on-line manual pages is *SUNWman*, and the OpenWindows demos, *SUNWowdem*.

The software packages in Solaris 2 conform to a software standard known as the applications binary interface (ABI). ABI compliant software can be easily managed (i.e., installed and/or removed) with standard systems administration utilities which will be discussed later in this chapter.

Clusters

Related software packages are grouped into software clusters. Cluster names do not use the *SUNW* prefix. Instead they use logical names found in the Sun environment. For example, the OpenWindows Version 3 cluster contains 13 software packages related to OpenWindows version 3. Individual packages may be added to or deleted from a cluster depending on the specific needs at each site. **SunInstall** stores a list of the packages that are installed and removed from the system. Other tools in the package maintenance suite allow the administrator to determine which software packages and clusters are installed on the system.

Configurations

The **SunInstall** utility groups software packages and clusters into categories known as software configurations. There are four software configurations in Solaris 2. Each configuration supports different levels of system sophistication. The four configurations are *core*, *end user, developer,* and *entire.*

•◦ **NOTE:** *Recent releases of Solaris 2 also include the "entire plus OEM support" configuration. This configuration provides support for third party SPARC compatible systems.*

The core software configuration contains the minimum software required to boot and operate a standalone host. The core does not include the OpenWindows software nor the on-line manual pages. It does, however, include sufficient networking software and OpenWindows drivers to run OpenWindows from a server sharing the OpenWindows software. A server would not be built from the core configuration. The core software configuration requires approximately 112 Mb of disk space.

The end user software configuration contains the core configuration software and additional software typically employed by end users. It includes Open-Windows and the end user version of AnswerBook. It does not, however, contain the on-line manual pages. The end user software configuration requires approximately 312 Mb of disk space.

The developer software configuration contains the core and end user configuration software and additional software typically used by systems and software developers. It includes the on-line manual pages and the full implementation of OpenWindows and compiler tools, but does not include compilers nor debugging tools. The developer's software configuration requires approximately 561 Mb of disk space.

↦ **NOTE:** *In Solaris 1 and earlier releases of Sun operating systems, compilers were included (bundled) with the release media. Compilers and debuggers are not bundled with Solaris 2.*

The entire software configuration contains the complete Solaris 2 release, and requires approximately 636 Mb of disk space. Unbundled products are not included in the entire release configuration.

It is important to note that the software configurations can be modified. For example, the on-line manual pages cluster could be added to the end user configuration. Likewise, the on-line manual pages could be removed from the developer configuration. These modifications can be made when the system is being built or afterwards.

Software Maintenance Utilities

Solaris 2 provides two methods for use during software installation: the command line oriented *pkg* utilities, and the graphical interface of the Software Manager Tool (*swmtool*). Because the command line utilities provide the basis for the graphical tools, they will be discussed first.

Command Line Utilities

The command line software maintenance utilities provide the system administrator with the ability to perform software installation, verification, and removal over a terminal based interface. The command line utilities can also be used from a graphics console if the administrator so desires.

pkginfo Command

The **pkginfo** command is used to identify the packages installed on the system.

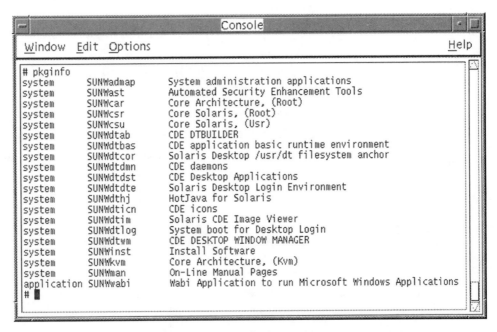

```
# pkginfo
system      SUNWadmap    System administration applications
system      SUNWast      Automated Security Enhancement Tools
system      SUNWcar      Core Architecture, (Root)
system      SUNWcsr      Core Solaris, (Root)
system      SUNWcsu      Core Solaris, (Usr)
system      SUNWdtab     CDE DTBUILDER
system      SUNWdtbas    CDE application basic runtime environment
system      SUNWdtcor    Solaris Desktop /usr/dt filesystem anchor
system      SUNWdtdmn    CDE daemons
system      SUNWdtdst    CDE Desktop Applications
system      SUNWdtdte    Solaris Desktop Login Environment
system      SUNWdthj     HotJava for Solaris
system      SUNWdticn    CDE icons
system      SUNWdtim     Solaris CDE Image Viewer
system      SUNWdtlog    System boot for Desktop Login
system      SUNWdtwm     CDE DESKTOP WINDOW MANAGER
system      SUNWinst     Install Software
system      SUNWkvm      Core Architecture, (Kvm)
system      SUNWman      On-Line Manual Pages
application SUNWwabi     Wabi Application to run Microsoft Windows Applications
#
```

Using pkginfo to identify the packages installed on a system.

The pkginfo command can also be used to display more detailed information on particular packages on your system. Command line options are summarized below.

❏ -q—Do not list any information.

❏ -x—Designate an extracted listing of package information.

❏ -l—Specify long format output.

❏ -p—Display information on partially installed packages.

❏ -i—Display information for fully installed packages.

❏ -r—List the installation base for relocatable packages.

❏ -a arch—Specify the package architecture as arch.

❏ -v version—Specify the version of the package as *version.*

❏ -c category—Display all packages which match category.

❏ -d device—Specify the device which contains the package to be checked.

❏ -R root_path—Specify the path to the package to be checked.

For example, to cause pkginfo to display more detailed information on a specific package, use the **-l** option.

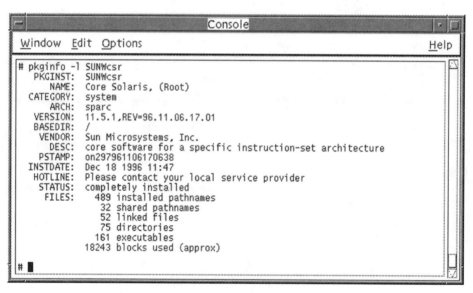

```
# pkginfo -l SUNWcsr
   PKGINST:  SUNWcsr
      NAME:  Core Solaris, (Root)
  CATEGORY:  system
      ARCH:  sparc
   VERSION:  11.5.1,REV=96.11.06.17.01
   BASEDIR:  /
    VENDOR:  Sun Microsystems, Inc.
      DESC:  core software for a specific instruction-set architecture
     PSTAMP:  on297961106170638
  INSTDATE:  Dec 18 1996 11:47
   HOTLINE:  Please contact your local service provider
    STATUS:  completely installed
     FILES:     489 installed pathnames
                 32 shared pathnames
                 52 linked files
                 75 directories
                161 executables
              18243 blocks used (approx)
#
```

Using pkginfo to display detailed information about packages on a system.

➠ **NOTE:** *The descriptions of command line options (flags) in this and subsequent sections are con-*

densed synopses of flag operations. Consult the on-line manual pages for more information on these commands.

pkgadd Command

The **pkgadd** utility allows the operator to copy a package from the distribution media to a location on one of the system's disks. Applications are typically distributed on magnetic or optical media such as cartridge tape, floppy diskette, or CD-ROM. The pkgadd utility copies the package from the media, and installs it in the specified directory. Some of the more useful command line options for the pkgadd command are listed below.

❏ -n—Non-interactive mode.

❏ -a admin—Employ the user-supplied admin file in place of the default admin file.

❏ -d device—Define the device which contains the package to be installed.

❏ -R root_path—Define the full path to where the package should be installed.

❏ -r response—Use the response file for answers to interactive mode installations.

❏ -s spool—Install package in the spool directory.

❏ pkginst—Package name and instance of the package to be installed.

A typical invocation of the pkgadd command follows:

```
pkgadd -d device SUNWpkgA
```

The above statement instructs pkgadd to search the device specified by the name device to locate the *SUNWpkgA* package. Part of the information included with the distributed package is a default installation map. The default installation map specifies installation directory, file modes, and file ownership for all files in the package. Once the system locates the package on the distribution media, the user is prompted for more information. These prompts allow the user to override the default selections.

➜ **NOTE:** *In releases subsequent to Solaris 2.4, certain unbundled and third party applications are no longer compatible with the* pkgadd *utility. These applications require interaction throughout the installation process. To install these packages (released prior to Solaris 2.4), the* NONABI_SCRIPTS *environment variable must be defined as* TRUE. *This can be accomplished with the following commands:* setenv NONABI_SCRIPTS=TRUE *and* export NONABI_SCRIPTS.

Using pkgadd to install a package under Solaris 2.

```
Console

Window   Edit  Options                                      Help

grumpy# pkgadd -d `pwd` SUNWsmaud

Processing package instance <SUNWsmaud> from </cdrom/showme_video_2_0_
1>

SunSolutions ShowMe Audio
(sparc) 2.0.1,REV=1.0.1
        Copyright 1994 Sun Microsystems, Inc. All Rights Reserved.
            Printed in the United States of America.
2550 Garcia Avenue, Mountain View, California, 94043-1100 U.S.A.

Using </opt> as the package base directory.
## Processing package information.
## Processing system information.
## Verifying package dependencies.
WARNING:
    The <SUNWsm-ml> package "SunSolutions Motif" is a
    prerequisite package and should be installed.

Do you want to continue with the installation of <SUNWsmaud> [y,n,?] y
## Verifying disk space requirements.
## Checking for conflicts with packages already installed.
## Checking for setuid/setgid programs.

Installing SunSolutions ShowMe Audio as <SUNWsmaud>

## Installing part 1 of 1.
/opt/SUNWsunsol/ShowMe/bin/bin-SUNOS/showmeaud
/opt/SUNWsunsol/ShowMe/bin/bin-SVR4/showmeaud
/opt/SUNWsunsol/bin/showmeaud <symbolic link>
[ verifying class <none> ]

Installation of <SUNWsmaud> was successful.
grumpy# ▮
```

pkgrm Command

The **pkgrm** command is used to undo the operation of the **pkgadd** command, or to remove a software package from the system. Before removing a package, pkgrm checks other packages on the system to identify interdependencies. If interdependencies are found, the pkgrm command consults the *admin* file to determine the action to be taken. Selected useful command line options for the pkgrm command appear below.

❑ -n—Non-interactive mode. The command exits if interaction is required.

❑ -R—Defines the full path name to the package to be removed.

❑ -admin—Employs the user-supplied admin file in place of the default admin file.

❑ -spool—Removes the package from the spool directory.

❑ pkginst—Specifies the name and instance of the package to be removed.

Using pkgrm to remove a package under Solaris 2.

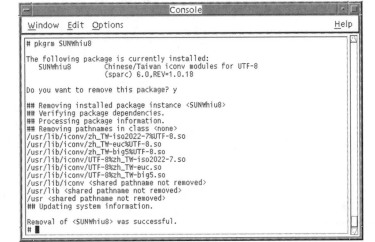

```
# pkgrm SUNWhiu8

The following package is currently installed:
   SUNWhiu8        Chinese/Taiwan iconv modules for UTF-8
                   (sparc) 6.0,REV=1.0.18

Do you want to remove this package? y

## Removing installed package instance <SUNWhiu8>
## Verifying package dependencies.
## Processing package information.
## Removing pathnames in class <none>
/usr/lib/iconv/zh_TW-iso2022-7%UTF-8.so
/usr/lib/iconv/zh_TW-euc%UTF-8.so
/usr/lib/iconv/zh_TW-big5%UTF-8.so
/usr/lib/iconv/UTF-8%zh_TW-iso2022-7.so
/usr/lib/iconv/UTF-8%zh_TW-euc.so
/usr/lib/iconv/UTF-8%zh_TW-big5.so
/usr/lib/iconv <shared pathname not removed>
/usr/lib <shared pathname not removed>
/usr <shared pathname not removed>
## Updating system information.

Removal of <SUNWhiu8> was successful.
#
```

↝ **NOTE:** *In releases subsequent to Solaris 2.4, certain unbundled and third party applications are no longer compatible with the* pkgrm *utility. These applications require interaction throughout the process of removal. To remove these packages (released prior to Solaris 2.4), the* NONABI_SCRIPTS *environment variable must be defined as* TRUE. *This can be accomplished with the following commands:* setenv NONABI_SCRIPTS=TRUE *and* export NONABI_SCRIPTS.

pkgchk Command

The **pkgchk** utility checks a package to determine if it is properly installed, or if it has been properly removed. A few command line options for this utility appear below.

❐ -l—Lists information on the files comprising the package. This option cannot be used with a, c, f, g, and v options.

❐ -a—Audits the attributes, but does not check file contents.

❐ -c—Audits file contents, but does not check attributes.

❐ -f—Corrects file attributes when possible.

❐ -q—Quiet mode.

❐ -v—Verbose mode.

❐ -x—Searches exclusive directories.

❐ -n—Prevents checking of volatile (editable) files.

❐ -p path—Checks only the listed path(s) for accuracy.

❐ -i file—Reads a list of path names from a file.

❐ -d device—Specifies the device containing the spool directory.

❐ -R root_dir—Specifies the path to the package(s).

❐ -m pkgmap—Checks the entire package against the package map.

❐ -e envfile—Uses the *envfile* to resolve parameters listed in the *pkgmap* file.

❐ pkginst—Specifies the package and instance to be checked.

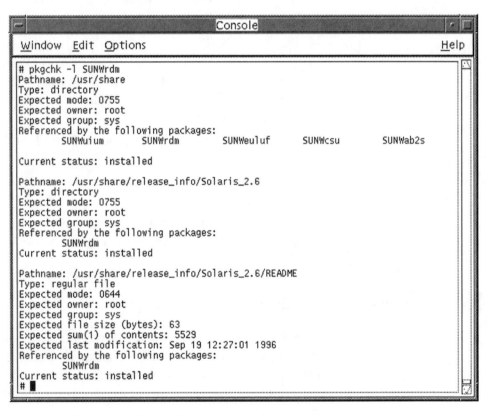

```
# pkgchk -l SUNWrdm
Pathname: /usr/share
Type: directory
Expected mode: 0755
Expected owner: root
Expected group: sys
Referenced by the following packages:
        SUNWuium      SUNWrdm      SUNWeuluf      SUNWcsu      SUNWab2s

Current status: installed

Pathname: /usr/share/release_info/Solaris_2.6
Type: directory
Expected mode: 0755
Expected owner: root
Expected group: sys
Referenced by the following packages:
        SUNWrdm
Current status: installed

Pathname: /usr/share/release_info/Solaris_2.6/README
Type: regular file
Expected mode: 0644
Expected owner: root
Expected group: sys
Expected file size (bytes): 63
Expected sum(1) of contents: 5529
Expected last modification: Sep 19 12:27:01 1996
Referenced by the following packages:
        SUNWrdm
Current status: installed
# ▮
```

Using pkgchk to check the installation of a package under Solaris 2.

Other pkg Commands

Solaris also provides a series of commands which can be used by applications developers to create package distributions. Several of these commands are briefly described below.

❑ pkgask—Stores answers to a request script.

❑ pkgmap—Packages contents description file.

❑ pkgmk—Produces an installable package.

❑ pkgparam—Displays package parameter values.

❑ pkgproto—Generates prototype file entries for input to *pkgmk*.

❑ pkgtrans—Translates the following package formats: (1) file system to data stream; (2) data stream to file system; or (3) file system type 1 to file system type 2.

❑ installf—Adds a file to the installed software database.

❑ removef—Removes a file from the installed software database.

For more information on the use of these commands, consult the Solaris 2 AnswerBook, and the on-line manual pages.

Software Manager Tool (swmtool)

The second method of software maintenance is to use the Software Manager Tool (**swmtool**). The swmtool can be used to install, remove, and perform an integrity check on the software on local or remote systems.

swmtool Interface

To install or remove packages with swmtool, the operator must be operating in a windowing environment, and have root permission. To view the installed packages on a system, use the following statement; root permission is not required.

```
/usr/sbin/swmtool &
```

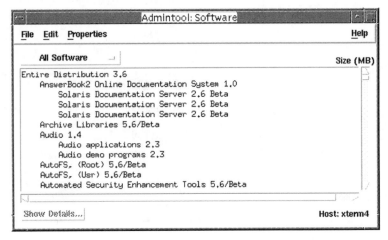

Starting swmtool.

The above command statement causes swmtool to open a new window on your display. Buttons available within this window allow the operator to set the parameters for package installation or removal.

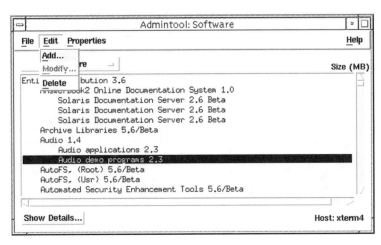

Setting parameters for package installation.

By clicking on the Props button, the operator can set the source media and the directory on the source media which contains the package to be installed.

↦ **NOTE:** *Beginning with Solaris 2.6, the* swmtool *buttons have been redefined. The Props button is now called Properties.*

Setting the source media and package directory.

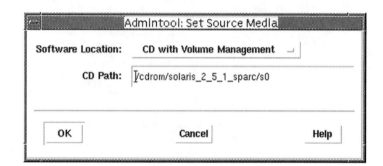

The swmtool will read the source volume and display a description of each package available on the media. Select the package of interest by clicking on the icon to the left of the package description. More information on the package or individual clusters within a package is available by double-clicking on the icon, or by choosing the Expand button on the software menu.

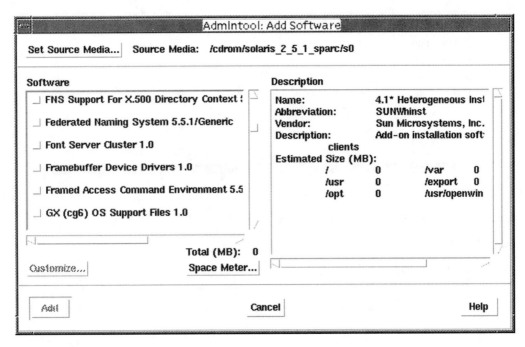

Packages available for installation displayed by the swmtool command.

Once you have chosen the software to be installed, click on the Begin Installation button to start the installation process. The swmtool command may request input to allow it to properly install the selected packages.

➡ **NOTE:** *Beginning with Solaris 2.6, button names have been redefined. The Begin Installation button is now called the Add button.*

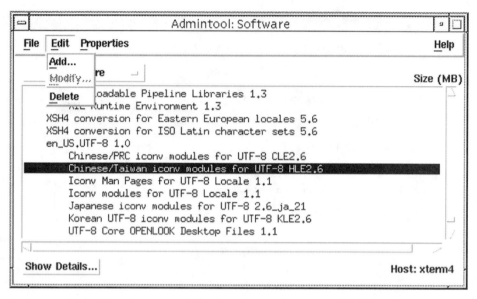

In Remove(Delete) mode, swmtool displays the packages available for removal.

To remove a package with swmtool, the operator must first select the Remove mode of operation. The swmtool command will display a list of packages available on the system. Select the package to be removed by clicking on the icon for the desired package. Once the operator has selected all packages to be removed, click on the Begin Removal button. The swmtool will ask for confirmation of the removal operation.

➼ **NOTE:** *Beginning with Solaris 2.6, the names of the buttons have been redefined. The Remove button is now located in the Edit menu, and is labeled as the Delete button.*

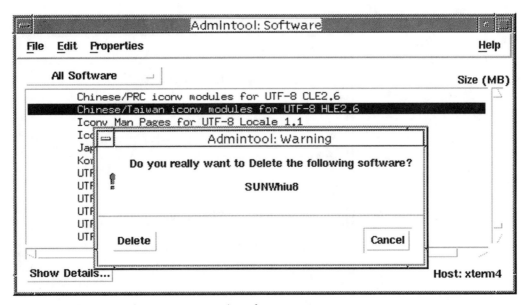

The swmtool command requests removal confirmation.

Upon completion of swmtool tasks, the operator should exit by selecting the box in the upper left corner of the swmtool window, and moving the pointer to the Exit item on the menu.

Software Maintenance Log Files

The **pkgadd**, **pkgrm**, and **swmtool** utilities create several files in the */var/sadm* directory to inform the operator of the success or failure of respective operations. These files may be browsed with the **more** command to monitor the steps followed during a package installation or removal.

If the installation of a package fails, the reason for the failure will be written to the appropriate package log file in */var/sadm*. In the event that problems are encountered during the installation or removal of a particular package, it may be helpful to peruse the installation log file for the package to determine the source of the problems.

*Typical contents of a /
var/sadm/system
package installation log
file.*

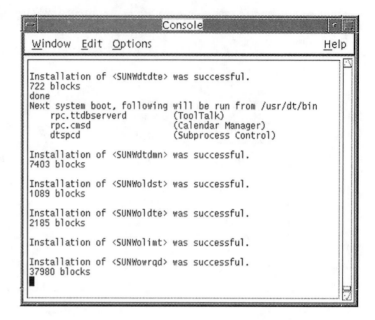

```
┌──────────────────────────Console───────────────────────┐
│  Window  Edit  Options                              Help │
├─────────────────────────────────────────────────────────┤
│ Installation of <SUNWdtdte> was successful.             │
│ 722 blocks                                              │
│ done                                                    │
│ Next system boot, following will be run from /usr/dt/bin│
│      rpc.ttdbserverd         (ToolTalk)                 │
│      rpc.cmsd                (Calendar Manager)         │
│      dtspcd                  (Subprocess Control)       │
│                                                         │
│ Installation of <SUNWdtdmn> was successful.             │
│ 7403 blocks                                             │
│                                                         │
│ Installation of <SUNWoldst> was successful.             │
│ 1089 blocks                                             │
│                                                         │
│ Installation of <SUNWoldte> was successful.             │
│ 2185 blocks                                             │
│                                                         │
│ Installation of <SUNWolimt> was successful.             │
│                                                         │
│ Installation of <SUNWowrqd> was successful.             │
│ 37980 blocks                                            │
│ ▮                                                       │
└─────────────────────────────────────────────────────────┘
```

➤ **NOTE:** *While it may appear that log files are super-fluous and could be removed, it is generally rec-ommended that these files should be left on the system for future reference.*

Summary

This chapter expanded on the Solaris 2 definition of a software package. Command line and graphical interface software installation, removal, and mainte-nance commands were examined. The chapter also covered the strengths and weaknesses of each method of software management, and how to deter-mine which method would be most efficient for a particular package.

Now that the system has an operating system installed, the mass storage system is ready for use and the end user packages have been installed, it is time to prepare the system for users. The next several chapters discuss activities aimed at enabling the system, and include tips on how to perform daily system administration functions to keep the system running smoothly.

Adding Terminals and Modems

With an understanding of the mass storage subsystem and file system and software management, you are ready to tackle other daily administrative tasks. One of the more common tasks on any computer system is the installation, setup, and policing of network services and serial lines.

The serial ports on most systems are used for dial-up modem access, tty terminal access, and line printer output. A discussion of line printers on serial ports is found in Chapter 14, "Managing Printers." This chapter is focused on serial ports and how they should be configured if used for dial-up modems and tty terminal access. Examples show how to set up ports for particular services, and how to monitor system access through the serial ports.

In many applications, the system administrator will be asked to monitor and control access to the systems via network links. As you may recall, system security was discussed in Chapter 7, "Managing System Security." This chapter will re-examine several security topics, including how the system software should be configured to set up and manage the network services for optimum system security.

Working with Serial Devices

Before examining the methods used to add and control serial ports on Solaris systems, a few basic terms and concepts are defined below.

❏ A serial port is an input/output channel between a serial device (terminals, printers, modems, and plotters, to name a few) and the operating system.

❏ The data are transmitted through a serial port in a bit-by-bit (serial) manner.

❏ The speed at which the data are transferred is referred to as the baud rate.

The baud rate is a measure of bits per second. For instance, a baud rate of 300 means that the data are being sent at 300 bits per second. There are numerous standard baud rates provided by computer systems. The most common are 300, 600, 1200, 2400, 4800, 9600, 19,200, and 38,400. Most modern terminals and printers operate at transmission speeds of 9600 baud or faster. Solaris provides the means to set the serial port speed as required by the device, or as desired by the user.

⟿ **NOTE:** *Although the terms "baud" and "bits-per-second" do not necessarily equate to the same* *thing, it is common practice to refer to them as the same. To be more accurate, baud refers to the number of times per second that a signal changes its value. Bits per second refers to the data rate between the computer and the device. The terms are treated as interchangeable in this chapter.*

Most of the terminals on the market offer a set of unique features in addition to those found on all similar type units. In order for the operating system to allow the use of different terminal types, a table of terminal capabilities must be available to the operating system. This table is used to determine how to set the serial port appropriately for the connected terminal. Solaris contains such a database of terminal capabilities for many common terminals. In addition, Solaris allows the addition of new terminal definitions via this database.

Because serial ports allow access to the operating system, the administrator must be aware of the use of all such ports on the system. Unattended terminals are often used by unauthorized users attempting to access computer systems. The administrator should be aware of all serial port connections to the system, and all ports which are not currently in use should be disabled. The administrator should also remind authorized system users about the dangers of leaving a log-in session unattended.

Using the Service Access Facility

Under Solaris there are two methods available to the system administrator for managing serial port facilities. First is the **admintool** utility, which graphically interfaces to the Service Access Facility (SAF), and second, manual invocation of utilities provided by the SAF. Both methods of serial port administration use the same underlying set of utilities.

With the SAF you can manage serial ports and network devices, which includes the tasks below.

❏ Manage and troubleshoot TTY devices.

❏ Manage and troubleshoot network print service requests.

❏ Manage and troubleshoot the Service Access Controller.

❏ Add and manage listen port monitor services.

❏ Add and manage ttymon port monitor services.

The SAF is not a single program, but rather a series of background processes and commands which control those processes. Selected components of the SAF are described below.

❏ The top level SAF program is the service access controller (SAC). The SAC controls port monitors administered through the **sacadm** command.

❏ The sacadm command is used to administer the SAC, which controls the ttymon and listen port monitors.

❏ The **pmadm** command administers the ttymon and listen services associated with ports.

❏ The **ttymon** facility is used to monitor serial line service requests.

❏ The **listen** facility monitors requests for network services.

The following sections examine each component of the SAF in detail to see how these facilities interact, and what each facility does for system administrators.

Service Access Controller (SAC)

The service access controller (SAC) program manages the operation of all port monitors. These port monitors are actually programs which continuously watch for any request to access devices connected to the serial ports. The SAC process is started automatically when the system is booted to the multi-user init state. When invoked, the SAC initializes the system environment by interpreting the **/etc/saf/_safconfig** start-up script. The configuration files allow local customization of serial port parameters.

After the SAC has created a custom environment as directed by the start-up scripts, it reads the */etc/saf/_sactab* file and starts port monitors as required.

Contents of a typical /etc/saf/_sactab file.

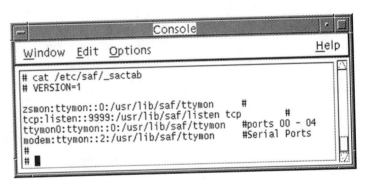

The SAC starts up a new process for each port monitor as directed by the */etc/saf/_sactab* file. The SAC maintains control of new processes, and can therefore be considered the "parent" process. Thus, the new processes are called "child" processes.

✓ **TIP:** *The process hierarchy can be viewed using the* ps -ef *command. The output column labeled PID is the process ID. The output column labeled PPID shows the parent process ID.*

Each child process started by the SAC invokes the appropriate **/etc/saf/port_monitor/_config** script to customize its own environment and start the processes as determined by the configuration file. In most cases, the process started for the port is the login process.

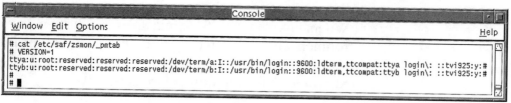

```
# cat /etc/saf/zsmon/_pmtab
# VERSION=1
ttya:u:root:reserved:reserved:reserved:/dev/term/a:I::/usr/bin/login::9600:ldterm,ttcompat:ttya login\: ::tvi925:y:#
ttyb:u:root:reserved:reserved:reserved:/dev/term/b:I::/usr/bin/login::9600:ldterm,ttcompat:ttyb login\: ::tvi925:y:#
#
# █
```

Contents of a typical /etc/saf/zsmon/_pmtab file.

Port Monitors

Solaris implements the ttymon and listen monitors. Every time a user attempts to log in to a system through a directly connected terminal or modem, the ttymon monitor is invoked. Once invoked, ttymon monitors serial port lines for incoming data. When data are present, the ttymon monitor determines the proper line disciplines and baud rate by consulting the */etc/ttydefs* file.

Typical contents of the / etc/ttydefs file.

```
┌─────────────────────────────────────────────────┐
│                    Console              · □      │
├─────────────────────────────────────────────────┤
│  Window  Edit  Options                    Help   │
├─────────────────────────────────────────────────┤
│ # cat /etc/ttydefs                               │
│ # VERSION=1                                      │
│ 460800:460800 hupcl:460800 hupcl::307200         │
│ 307200:307200 hupcl:307200 hupcl::230400         │
│ 230400:230400 hupcl:230400 hupcl::153600         │
│ 153600:153600 hupcl:153600 hupcl::115200         │
│ 115200:115200 hupcl:115200 hupcl::76800          │
│ 76800:76800 hupcl:76800 hupcl::57600             │
│ 57600:57600 hupcl:57600 hupcl::38400             │
│ 38400:38400 hupcl:38400 hupcl::19200             │
│ 19200:19200 hupcl:19200 hupcl::9600              │
│ 9600:9600 hupcl:9600 hupcl::4800                 │
│ 4800:4800 hupcl:4800 hupcl::2400                 │
│ 2400:2400 hupcl:2400 hupcl::1200                 │
│ 1200:1200 hupcl:1200 hupcl::300                  │
│ 460800E:460800 hupcl evenp:460800 evenp::307200  │
│ 307200E:307200 hupcl evenp:307200 evenp::230400  │
│ 230400E:230400 hupcl evenp:230400 evenp::153600  │
│ 153600E:153600 hupcl evenp:153600 evenp::115200  │
│ 115200E:115200 hupcl evenp:115200 evenp::76800   │
│ 76800E:76800 hupcl evenp:76800 evenp::57600      │
│ 57600E:57600 hupcl evenp:57600 evenp::38400      │
│ 38400E:38400 hupcl evenp:38400 evenp::19200      │
│ 19200E:19200 hupcl evenp:19200 evenp::9600       │
│ 9600E:9600 hupcl evenp:9600 evenp::4800          │
│ 4800E:4800 hupcl evenp:4800 evenp::2400          │
│ 2400E:2400 hupcl evenp:2400 evenp::1200          │
│ 1200E:1200 hupcl evenp:1200 evenp::300           │
│ auto:hupcl:sane hupcl:A:9600                     │
│ console:9600 hupcl opost onlcr:9600::console     │
│ console1:1200 hupcl opost onlcr:1200::console2   │
│ console5:19200 hupcl opost onlcr:19200::console  │
│ contty:9600 hupcl opost onlcr:9600 sane::contty1 │
│ contty1:1200 hupcl opost onlcr:1200 sane::contty2│
│ contty5:19200 hupcl opost onlcr:19200 sane::contty│
│ 4800H:4800:4800 sane hupcl::9600H                │
│ 9600H:9600:9600 sane hupcl::19200H               │
│ 19200H:19200:19200 sane hupcl::38400H            │
│ 38400H:38400:38400 sane hupcl::2400H             │
│ 2400H:2400:2400 sane hupcl::1200H                │
│ 1200H:1200:1200 sane hupcl::300H                 │
│ 300H:300:300 sane hupcl::4800H                   │
│ conttyH:9600 opost onlcr:9600 hupcl sane::contty1H│
│ contty1H:1200 opost onlcr:1200 hupcl sane::contty2H│
│ contty5H:19200 opost onlcr:19200 hupcl sane::conttyH│
│ #                                                │
│ # ■                                              │
└─────────────────────────────────────────────────┘
```

The ttymon facility uses the information in the */etc/tty-defs* file to determine the default line settings and proceeds to set the line accordingly. Once this is done, ttymon passes control to the log-in process.

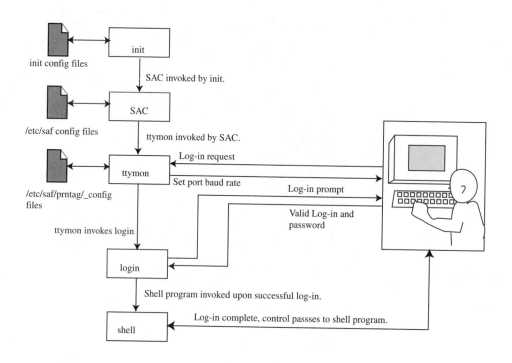

Steps involved in invoking the log-in process.

TTY Port Monitor (ttymon)

The ttymon facility is a streams-based port monitor. The streams interface is a standardized input/output scheme for use with serial port devices. The streams facility allows the system to communicate with serial port devices through standard system calls and facilities used to communicate with other system devices. Streams are defined by the System V Release 4 interface description (SVID).

The ttymon facility uses the streams interface to monitor serial ports and set serial line modes, line disciplines, and baud rates. The ttymon facility also invokes the log-in process.

Ttymon runs under the control of the SAC program and is configured via the **sacadm** command. Each invocation of ttymon can monitor multiple ports. The ports for each ttymon process are specified in the port monitor's administrative file. Configuration files specify the ports managed by the administrative file. The configuration files, in turn, are created by the **pmadm** and **ttymadm** commands.

The ttymadm command, a ttymon-specific administrative command, formats ttymon information and writes it to standard output. In brief, ttymadm provides the ability to present such information to the sacadm and pmadm commands. The ttymadm command does not administer ttymon, but rather is a means for the sacadm and pmadm commands to provide such administration.

Network Listener Monitor (listen)

The listen port monitor monitors the network for service requests. As requests arrive and are accepted, listen invokes the proper server to process the request. The listen monitor is configured via sacadm. Each invocation of the listen monitor may provide multiple services. The services provided by listen are specified in the port monitor administrative file, which is in turn configured via the **pmadm** and **nlsadmin** commands.

The listen monitor may be used with any connection-oriented provider that conforms to transport layer interface (TLI) specifications. Under Solaris, listen monitors provide some services not managed by **inetd**, such as the network printing service.

The **nlsadmin** command, a listen monitor-specific administrative command, formats listen information and writes it to standard output. This command provides the ability to present such information to the sacadm and pmadm commands. The nlsadmin command does not administer listen, but rather supplies a means for the sacadm and pmadm commands to provide such administration.

The **sacadm** command provides administration of the port monitor for the SAC. With the use of command line options, the system administrator can use sacadm to accomplish the tasks listed below.

❐ Add or remove port monitors.

❐ Start or stop port monitors.

❐ Enable or disable port monitors.

❐ Install or replace a configuration script.

❐ Install or replace a port monitor configuration script.

❐ Print port monitor information.

Non-privileged users may request the status of port monitors, and print port and system configuration scripts. All other commands require root access.

The **pmadm** command provides administration of services for the SAC. By using command line options, the system administrator can use pmadm to execute the tasks below.

❐ Add or remove a network service.

❐ Enable or disable a network service.

❐ Install or replace a service configuration script.

❐ Print service information.

The pmadm command may be used by non-privileged users to list the status of the service, or to print out configuration scripts. All other access to the pmadm command requires root privileges.

terminfo Database

The *terminfo* database built by **tic** describes the capabilities of terminals and printers. The devices are described in *terminfo* files by specifying a set of capabilities for the device. Entries in the *terminfo* database files consist of comma-separated fields. These fields contain capabilities of the device represented with Boolean, numeric, and string values. The database is located in */usr/share/lib/terminfo/*.

Information in the database is used by the **curses** library routines to define terminal cursor movement capabilities. By using the terminfo database, these utilities can work with a variety of terminals without the need to modify the programs for each type of terminal on the system. The information in these database files is compiled from the ASCII definition files in */usr/share/lib/termcap*.

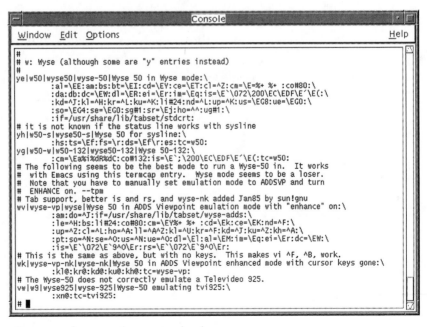

Contents of a typical termcap database entry.

~ **NOTE:** *The complexities of* terminfo *entries are beyond the scope of this chapter. For more information on the data contained in the* terminfo *database, consult the on-line* terminfo *manual page.*

Setting Up Terminals

With an understanding of the functions performed by each portion of the SAF, it is time to examine how the SAF may be used to manage terminal setup tasks. As previously mentioned, there are two interfaces to the SAF. The **admintool** program employs a graphical user interface for user interaction. The individual SAF facility commands accomplish the same tasks, but the operator must supply these commands with the proper arguments in order to accomplish the same functions.

A few simple port setup examples will show how each of these facilities operates, while also highlighting the strengths and weaknesses of each method of interaction. One of the more common tasks is to add a terminal to a system. This requires that the port be configured via the **sacadm** utility or admintool.

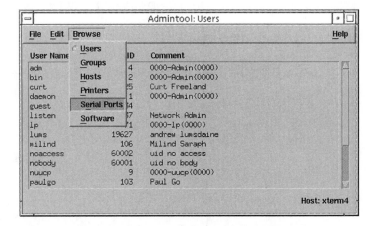

Serial Ports manager in Admintool.

The admintool offers a Serial Ports manager selection which is a screen-oriented process that leads the user through serial port setup. The sacadm command is a program that uses command line options to perform the same tasks. Using the admintool, select the name of the port that is to be modified.

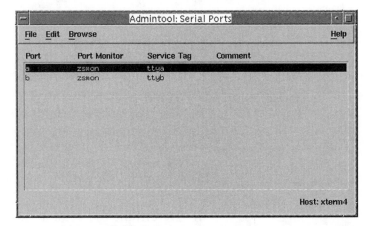

Selecting the port to be modified.

Once the appropriate port is selected, the Modify Serial Port menu of admintool's Serial Ports manager allows modification of parameters such as the baud rate, terminal type, and port enable or disable. Standard templates to simplify the port setup are provided in a pop-up menu activated by clicking on the Template button.

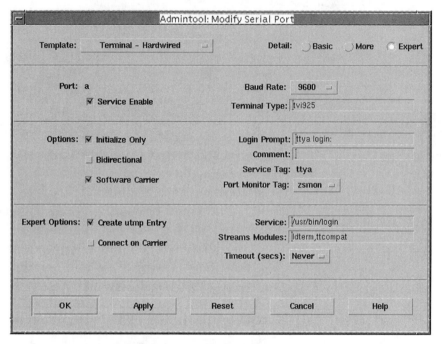

Setting port parameters with the Modify Serial Port menu.

Once the appropriate parameters have been set for a serial port, apply them by clicking the Apply button on this menu. The proper files will be updated, and the appropriate port monitors started.

When using the sacadm command to add a terminal to a system, you would use the following basic steps.

1. Determine the version number of the port monitor by using the **ttyadm -v** command. The numeric response is needed in the next step.

```
# ttyadm -v
```

2. Once the version of the port monitor program is known, it is possible to enable the port via the **sacadm** command. Substitute the version number from above for *version*, and substitute the port monitor tag for *pmtag*.

```
# sacadm -a -p pmtag -t ttymon -c /usr/lib/saf/ttymon -v version
```

3. When the port is enabled, start up a log-in process. Add a log-in service for the port which was just enabled by using the sacadm command. Substitute the port monitor tag for pmtag, the version number for version, the device name for *dev_path*, and the tty label for *ttylabel*:

```
# sacadm -a -p pmtag -s svctag -i root -fu -v version \
-m "`ttyadmin -S y -d dev_path -l ttylabel -s /usr/bin/login -l ldterm, ttcompat`"
```

As illustrated by the above example, adding a terminal is much easier to accomplish using the Serial Port Manager.

Setting Up Modems

Another common request encountered by a system administrator is to set up and administer dial-up modems. Again, this requires the use of either the **sacadm** command or the **admintool** command. Using admintool, follow the same steps for adding a terminal to the system. This time, select the appropriate modem type from the Template menu in the Modify Serial Port dialog box. Once the appropriate parameters for the modem have been set, apply the changes by clicking on the Apply icon.

⊷ **NOTE:** *To use UNIX to UNIX copy protocol (UUCP) with the newly installed modem, set the line disci-*

pline to use eight data bits, with no parity. This is accomplished by selecting the Baud Rate menu item and clicking on the Other menu item. Set the appropriate baud rate (as defined in /etc/ttydefs) that sets the parity and the correct number of data bits. One such entry would be 9600E for a 9600 baud modem.

The steps for adding a dial-in or bidirectional modem to a system using the sacadm command appear below.

1. Display all port monitors.

```
# sacadm -l -t ttymon
```

If the port already has a port monitor, skip the next step.

2. If there is no port monitor on the chosen port, use the sacadm command to start one.

```
# sacadm -a -p pmtab -t ttymon -c /usr/lib/saf/ttymon -v `ttyadm -V`   -y "comment"
```

3. Determine if the chosen port already has a configured service by using the **pmadm** command.

```
# pmadm -l -s svctag
```

If there is no service configured on the port, skip the next step.

4. Use the pmadm command to delete the service attached to the chosen port.

```
# pmadm -r -p pmtag -s svctag
```

5. Create the desired type of modem service. Use the port service tag listed in the */dev/term* directory for the *svctag* variable.

Modem Services

At this point the modems should be configured according to their intended use. If the modem will be used exclusively to allow remote users to log in to the system, it is referred to as a "dial-in" modem. Conversely, if the modem will be used to allow users on this system to contact other computers, the modem is referred to as a "dial-out" modem. Modems which allow incoming and outgoing connections are referred to as "bidirectional" modems.

The port setup for the three types of modems are different. The following commands initialize the serial port for the appropriate type of modem.

Setting Up a Dial-in Modem

A dial-in modem requires a log-in session manager so that users may log in to the system on this port. The following command provides a typical port setup for a dial-in modem:

```
# pmadm -a -p pmtag -s svctag -i root -v `ttyadm -V` -fu \
-m "`ttyadm -S n -d dev_path -s /usr/bin/login -l ttylabel -m ldterm,ttcompat`" \
         -y "comment"
```

Setting Up a Dial-out Modem

A dial-out modem has no need for a log-in manager, but it does require a method by which the user can "open" the device for communication. To add a dial-out modem to a system, take the following steps.

1. Add the modem to */etc/uucp/Devices*. The format of this file is ACU *cua/svctag - speed type*. The *svctag* will be the port identifier as listed by an *ls - lsa* of the */dev/term* directory.

```
# vi /etc/uucp/Devices
```

2. Disable log-ins on the chosen port:

```
# pmadm -d -p pmtag -s svctag
```

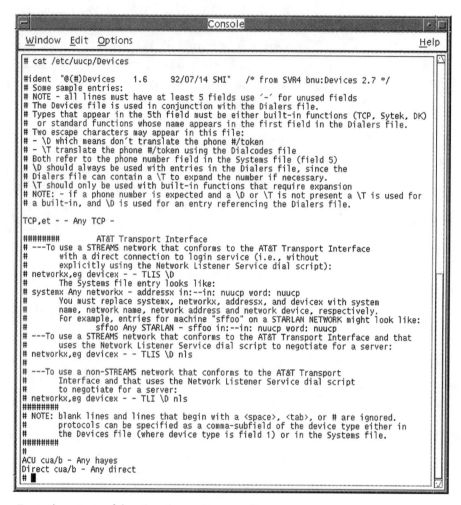

```
                              Console

Window  Edit  Options                                        Help

# cat /etc/uucp/Devices

#ident  "@(#)Devices    1.6    92/07/14 SMI"   /* from SVR4 bnu:Devices 2.7 */
# Some sample entries:
# NOTE - all lines must have at least 5 fields use '-' for unused fields
# The Devices file is used in conjunction with the Dialers file.
# Types that appear in the 5th field must be either built-in functions (TCP, Sytek, DK)
#  or standard functions whose name appears in the first field in the Dialers file.
# Two escape characters may appear in this file:
# - \D which means don't translate the phone #/token
# - \T translate the phone #/token using the Dialcodes file
# Both refer to the phone number field in the Systems file (field 5)
# \D should always be used with entries in the Dialers file, since the
# Dialers file can contain a \T to expand the number if necessary.
# \T should only be used with built-in functions that require expansion
# NOTE: - if a phone number is expected and a \D or \T is not present a \T is used for
# a built-in, and \D is used for an entry referencing the Dialers file.

TCP,et - - Any TCP -

########         AT&T Transport Interface
# ---To use a STREAMS network that conforms to the AT&T Transport Interface
#        with a direct connection to login service (i.e., without
#        explicitly using the Network Listener Service dial script):
# networkx,eg devicex - - TLIS \D
#        The Systems file entry looks like:
# systemx Any networkx - addressx in:--in: nuucp word: nuucp
#        You must replace systemx, networkx, addressx, and devicex with system
#        name, network name, network address and network device, respectively.
#        For example, entries for machine "sffoo" on a STARLAN NETWORK might look like:
#                sffoo Any STARLAN - sffoo in:--in: nuucp word: nuucp
# ---To use a STREAMS network that conforms to the AT&T Transport Interface and that
#        uses the Network Listener Service dial script to negotiate for a server:
# networkx,eg devicex - - TLIS \D nls
#
# ---To use a non-STREAMS network that conforms to the AT&T Transport
#        Interface and that uses the Network Listener Service dial script
#        to negotiate for a server:
# networkx,eg devicex - - TLI \D nls
########
# NOTE: blank lines and lines that begin with a <space>, <tab>, or # are ignored.
#        protocols can be specified as a comma-subfield of the device type either in
#        the Devices file (where device type is field 1) or in the Systems file.
########
#
ACU cua/b - Any hayes
Direct cua/b - Any direct
# ■
```

Typical contents of the /etc/uucp/Devices file.

Setting Up a Bidirectional Modem

A bidirectional modem requires both the log-in session manager and a method by which users can open it for outbound communications. Steps to add a bidi-rectional modem to a system appear below.

1. Determine whether services have been config-ured for the port.

```
# sacadm -l -t ttymon
```

2. If no port monitor is present, start one with the fol-lowing command.

```
# sacadm -a -p pmtag -t ttymon -c /usr/lib/saf/ttymon -v `ttyadm -V` \  -y "comment"
```

3. Determine whether services have been config-ured for the port.

```
# pmadm -l -t ttymon
```

4. If services are associated with the port, delete them.

```
# pmadm -r -p pmtag -s svctag
```

5. Create a bidirectional port service.

```
# pmadm -a -p pmtag -s svctag -i root -v `ttyadm -V` -fu \                   -m
"`ttyadm -b -S n -d /dev/term/svctag -s /usr/bin/login -l ttylabel -m \ ld-
term,ttcompat`" -y "comment"
```

6. Add the modem to */etc/uucp/Devices*. For exam-ple, to use a ttya as a 9600 baud Telebit Trailblazer modem line, you could enter the statements below.

```
  grumpy% ls -lsa /dev/term
total 16
  2 drwxrwxr-x 2 root  root  512 Jul 2 1996 .
 10 drwxrwsr-x 17 root sys  4608 May 2 11:10 ..
  2 lrwxrwxrwx  1 root  root  38 Jul 2 1996 a -> ../../devices/sbus@1f,0/
```

```
zs@f,1100000:a
   2 lrwxrwxrwx  1 root  root   38 Jul 2 1996 b -> ../../devices/sbus@1f.0/
zs@f,1100000:b
```

Point to Point Protocol (PPP)

The point to point protocol (PPP) allows serial port devices to become network interfaces. Solaris implements the PPP protocol via the three optional software packages listed below.

❏ SUNWpppk—PPP kernel modules.

❏ SUNWapppu—Link manager and log-in service modules.

❏ SUNWapppr—Configuration files for PPP.

The PPP packages consist of the components listed below.

❏ */etc/asppp.cf* configuration file—Contains setup information used by the link manager as required to manage the PPP connection.

❏ */usr/sbin/aspppd* link manager—Package which handles the dynamic link setup and shutdown for PPP connections.

❏ */usr/sbin/aspppls* log-in service—Package that manages the PPP log-in services.

❏ */var/adm/log/asppp.log* log file—Log file which contains information about the PPP link usage.

❏ */tmp/.asppp.info* FIFO file—Named pipe file used for communications between the components of the PPP package.

❏ */etc/init.d/asppp* file—Script run by init to start and stop the PPP software.

❏ */etc/uucp/Dialers* file—Contains information about the dialer used on this system. Used only with dial-out connections.

❏ */etc/uucp/Devices* file—File contains information about the serial port devices used for PPP connections. Used only with dial-out connections.

❏ */etc/uucp/Systems* file—Describes the systems that can communicate with the system via PPP connections. Used only for dial-out connections.

The PPP version shipped with Solaris allows for several modes of operation. Refer to the *Solaris Administrator's Answerbook* to identify the configuration most appropriate for individual applications. The Solaris version of PPP also allows for dynamic or static IP address assignment, and inbound or outbound connections. Security options include the challenge handshake authentication protocol (CHAP), and the password authentication protocol (PAP). The use of PAP and CHAP is discussed in more detail in Chapter 18, "Network Security."

Customizing PPP

In order to use PPP on a Solaris system, the administrator must customize several configuration files. The information required prior to beginning customization is listed below.

❏ Addressing method—Will the link employ static or dynamically assigned addresses? This information is required for the */etc/asppp.cf* file.

❏ Network interface—Which serial port will be used? Available options are *ipdptpN* and *ipdN*, where *N* is the link number. The *ipdN* interface is used for multipoint links. This information is required for the */etc/asppp.cf* file.

❐ Type of connection supported—Single end point to single end point or multiple end point to single end point? This information is required for the */etc/asppp.cf* file.

❐ Dial-in or Dial-out—What type of service will this system provide? This information is required for the */etc/asppp.cf* and */etc/uucp/** files.

❐ Type of name service used—Will the link employ NIS+ (Network Information Service Plus), DNS (Domain Name Service), FNS (Federated Naming System), or other name services? The name service, routing, and authentication methods are required in the */etc/asppp.cf* file.

❐ Type of authentication to be used—Will the link use the challenge handshake authentication protocol (CHAP) or password authentication protocol (PAP)? This information is required for the */etc/asppp.cf* file.

❐ Routing requirements—Will data from this link be routed to other network destinations via dynamic or static routing?

Refer to the following figure for the typical contents of the */etc/asppp.cf* file in a single end point to single end point setup. Substitute the local net mask for the net mask value, and the host names for the "remote" and "local" references in the illustration. Refer to the "Solaris TCP/IP and Data Communications Administration Guide" in the *Answerbook* for more details about the information contained in the */etc/asppp.cf* file.

```
┌─────────────────────────────────────────────────────────────┐
│ ═                        Console                      · □     │
├─────────────────────────────────────────────────────────────┤
│  Window  Edit  Options                                 Help   │
├─────────────────────────────────────────────────────────────┤
│ # cat /etc/asppp.cf                                         ▲ │
│                                                               │
│ #ident  "@(#)asppp.cf   1.10    93/07/07 SMI"                 │
│ #                                                             │
│ # Copyright (c) 1993 by Sun Microsystems, Inc.                │
│ #                                                             │
│ # Sample asynchronous PPP /etc/asppp.cf file                  │
│ #                                                             │
│ #                                                             │
│ ifconfig ipdptp0 plumb LOCAL REMOTE netmask 255.0.0.0 up      │
│ path                                                          │
│         inactivity_timeout 120     # Approx. 2 minutes        │
│         interface ipdptp0                                     │
│         debug_level 8                                         │
│         peer_system_name REMOTE # The name we log in with (also in │
│                                 # /etc/uucp/Systems          ▼ │
│ # █                                                           │
└─────────────────────────────────────────────────────────────┘
```

Typical contents of the /etc/asppp.cf file.

Starting and Stopping PPP

Once all configuration files are customized, the PPP software can be started by manual start-up or automatically at boot time.

To manually start PPP, enter the following command:

```
# /etc/init.d/aspppd start
```

To manually stop PPP, enter

```
# /etc/init.d/aspppd stop
```

To arrange for automatic start-up of the PPP software the administrator needs to add links from the appropriate */etc/rc.d* directory to the */etc/init.d/aspppd* file. Refer to Chapter 5, "Boot and Shutdown Procedures," for more information about configuring services for automatic start-up.

Checking for PPP Connections

The administrator can use the **netstat** command to determine whether a PPP connection is currently active. The **snoop** command can also be used to determine the type of traffic currently flowing across the PPP link. Refer to Chapter 17, "Network Configuration and Management," for more information about the use of the netstat and snoop commands. In the following figure, the *ipdptp0* link is active, indicating that an active PPP connection is in progress.

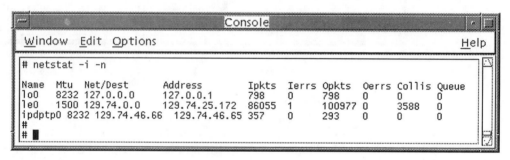

Using netstat to determine if PPP links are active.

Disabling and Removing Serial Services

There may be instances when disabling serial port access is desirable. Again, the **sacadm** or admintool command may be used to achieve this goal. Using admintool, select the Modify Serial Port menu, click on the Disable Port button, and then click on the Apply icon to disable the port.

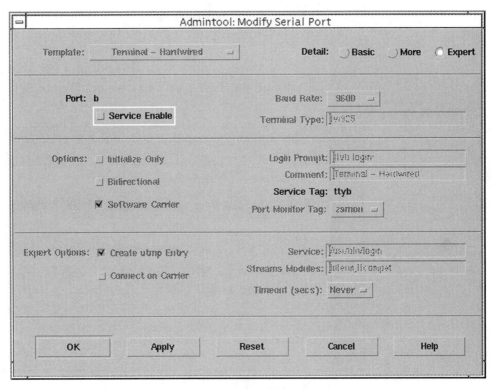

Using the serial port monitor to disable a serial port.

Under most circumstances, however, it is simpler to disable or enable a port by manually invoking the appropriate commands. The basic steps to manually disable a port appear below.

1. Determine the monitor tag for the port which is to be disabled by using the **pmadm** command.

```
# pmadm -l
```

2. Disable the port by using the pmadm command. Substitute the information provided in step 1 for the pmtag field.

```
# pmadm -d -p pmtag -s svctag
```

To remove a serial port service entirely, use the Serial Port Manager or the command line interface.

1. To remove a port with the Serial Port Manager, activate the Edit menu and select Delete Service.

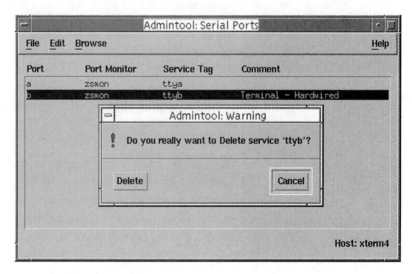

Using the Serial Port Manager to remove a serial port.

2. To delete the port service with the command line interface, disable the port by following step 1, and then proceed to step 3.

3. Delete the port service using the pmadm command.

```
# pmadm -r -p pmtag -s svctag
```

Summary

This chapter examined how serial port and network services requests are administered. Each utility in the service access facility (SAF) suite was examined to determine the function of the utility, and how the utilities interact for administration of serial port devices. Examples demonstrated the use of the SAF utilities and the admintool's serial ports monitor to manage serial ports. This chapter also examined how to use the sacadm and pmadm commands with command line arguments to add, modify, and delete serial port services.

Managing Printers

The Solaris printing service provides a complete network printing environment that allows sharing of printers across machines, management of special printing situations such as forms, and the filtering of output to match up with special printer types such as those which use the popular PostScript page description language. There are two sides to managing printers. On one side is the technical aspect of connecting a printer to a system and configuring the software to work with the printer. On the other is printing policy, including who should be allowed to use a particular printer or certain forms, or whether a particular printer should be shared among workstations on the network. The printing service provides tools to make connections to printers and to manage printing policy.

Printing Terminology

Printing services on Solaris use a set of special commands, daemons, filters, and directories. Files are not directly sent to the printer, but are spooled and then printed, freeing whatever application submitted the print request to move on to other tasks. The term *spooling* refers to the temporary storage of a file for later use. It comes from the early history of computing when files to be printed were saved temporarily on spools of magnetic tape. Here spooling is defined as the process of putting a file in a special directory while it waits to be printed. Spooling puts the SunOS multi-tasking capability to good use by allowing printing to occur at the same time as other activities.

Printing service.

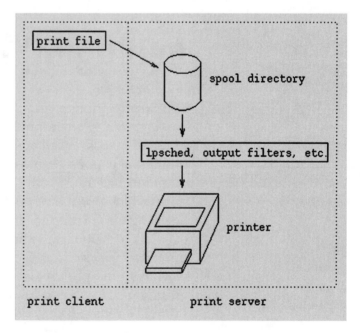

The actual work of printing is executed by a printing *daemon*. Daemon is the nickname given to all system processes running on a UNIX system. Using **ps -e** to take a look at all processes on a workstation, one would see a number of daemons. The main printing daemon is called **lpsched**.

Most output bound for a printer will require filtering before it is printed. The term "filter" is used to describe a program which transforms the contents of one file into another format as the file passes through the program. For instance, when an ASCII file is sent to a printer which accepts the PostScript page description language, the print service first runs the file through a filter to transform the file from ASCII into PostScript. The resulting PostScript file contains complete instructions in a form the printer can use to print the page and is somewhat larger than the original file due to the addition of this information. After filtering, the PostScript file is sent to the printer which reads the description of the pages to be printed and then prints them.

Printing on the Network

Printing can be set up as a network-wide service. Machines with directly connected printers which accept print requests from other machines are called *print servers*. Machines which submit print requests over the network to other machines are called *print clients*. A machine can be both a print server and a print client. Printers directly attached to the workstation are known as *local printers*, while printers attached to other workstations reached via the network are known as *remote printers*.

Many commands are used to set up the printing service, enforce printing policies and handle special devices. Some of these commands, such as **lpadmin**, are used throughout the printing process to manage all aspects of the printing service. Other commands, such as **lpusers**, are used to manage a single aspect of printing. The following table summarizes printing service commands. These commands will be described in detail later in the chapter as the discussion focuses on printer setup and management situations.

Printing service commands

Command	Description
lpadmin	Printer setup and a few management aspects.
admintool	Graphical system management tool that includes printer setup.
lpusers	Manages printer job priorities.
lpmove	Changes the destination of print jobs in the spool area.
lpforms	Manages forms and associated parameters.
lpshut	Stops the entire printing service.
lpshed	Starts the printing service.
lpset	Defines printer configuration in /etc/ printers.conf or FNS.
lp	Submits print jobs.
printtool	Graphical interface for submitting print jobs.
lpstat	Displays the status of the printing service and the status of individual jobs and printers.
cancel	Stops individual print jobs.
enable	Starts a printer printing.
disable	Stops printing on a particular printer.
accept	Starts print jobs accumulating in the print spool for a particular printer.
reject	Stops the print service from accepting jobs to be put in the printer spool area for printing on a particular printer.

Printer Setup

A printing service setup is comprised of four phases. First, the printer must be physically connected to a machine which will become known as the *print server*. The print service software on the print server machine is then configured to match the connection parameters, and type and model of the attached printer. Next, the print service is configured for other machines which will share the printer to act as *print clients*. Finally, configuration of special features such as printwheels, forms, and printer access controls may be required to complete the installation.

Setting Up a Local Printer

The most basic printing service configuration is the local printer. Remote printing and print servers are extensions of the local printing process. To configure a system for local printing, a printer is attached to a workstation. Because there are nearly as many types of printers as system types, the details of connecting a printer will not be covered here. In general, the printer is connected to the workstation or server and configured as necessary to use the connection. Common connection methods include serial connections using the ports available on the rear of many workstations, and parallel connections using special parallel ports or special purpose interface cards such as the S-Bus card used with some SPARC printers. Printers have a variety of configuration methods from small DIP switches to keypads and display panels.

For a common PostScript printer such as an Apple LaserWriter, a serial cable must be purchased or constructed to connect the printer to the workstation. The printer and workstation manuals should provide information on the necessary cable. Changes to the settings on the printer to work with a serial cable may also be required. On a printer such as the LaserWriter the small DIP switches on the rear of the printer should be set as shown in the printer manual under "serial port connections."

Printers connected by a network are becoming increasingly common. For printers treated as if they are print servers, see the "Setting Up Remote Printers" section later in this chapter. Some network attached printers are treated like local printers; the "connection" is obtained over the network by installing communications software provided by the printer manufacturer.

To set up the printing service, the system administrator requires the following information on how the printer is connected.

❏ The port to which the printer is connected.

❏ If the printer is connected via a serial line, the basic serial parameters such as baud rate, parity, data, and stop bits.

❏ The printer type. Solaris supports numerous printers by model name (e.g., HP LaserJet, Apple LaserWriter, NEC Spinwriter).

❏ The input accepted by the printer. Some printers will accept only a certain input language such as PostScript, while others will accept a variety of input languages.

Consider how much disk space is available for the printer spool area. This area is associated with the */var/spool/lp* directory. Depending on how Solaris was initially installed, the spool area may be in a separate disk slice or it may be part of the root directory (/). Sun recommends that at least 8 MB of disk space be reserved to service a single printer for one to three users. The amount needed on a specific system depends on the number and size of the print jobs likely to be run through a particular printer. Factors to consider include the number of users who will submit print jobs, the type of printer, and the speed of the printer. Remember that files with graphical images are bigger than text files which have the same number of pages. If you are setting up a server which will be servicing several printers, consider putting the */var* directory on a separate disk slice with sufficient space for all the printers to be serviced.

✓ **TIP:** *Allocate more disk space for PostScript printers than for other printer types. PostScript print jobs, especially those containing graphics, can be quite large, and are typically much larger than ASCII print jobs. A good rule of thumb is to allow for 1.5 times as much space for a PostScript printer that will be printing mostly text, two to three times as much space for PostScript printers handling grayscale graphics, and five times as much space for printers handling color PostScript images.*

Setting Up an Apple LaserWriter IINT

With the above details in mind, let's walk through an example to demonstrate the process of printer setup. In the example, the printer is an Apple LaserWriter IINT connected via a serial line on port b of the workstation. The LaserWriter accepts only PostScript data, and expects the serial connection to be 9600 baud, 7 data bits, 1 stop bit, and no parity.

To begin the printer setup process, the system administrator should be in OpenWindows with a command window and logged in as the root user. Next, start the **admintool** by typing the following:

```
admintool &
```

The opening screen of the admintool appears in the next illustration.

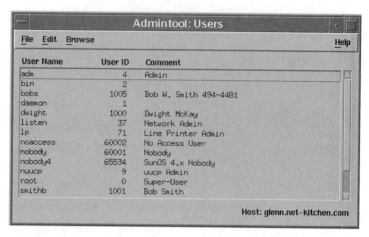

Opening admintool screen.

As seen in Chapter 6, "Creating, Deleting and Managing User Accounts," the admintool can handle several different chores. Select Printers from the Browse menu to begin. The admintool can be used to manage printers on machines across the network. This feature is explored later in the discussion of network sharing of printers. For now, select Local Printer from the Add item in the Edit menu's pull-right selection.

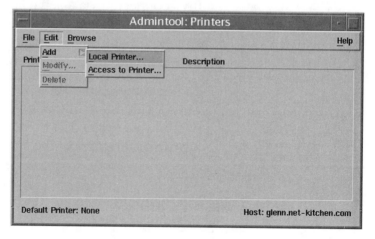

Printer Manager screen.

The resulting window is a form used to specify parameters associated with the printer to be added. A blank form is shown in the next illustration.

Local printer form.

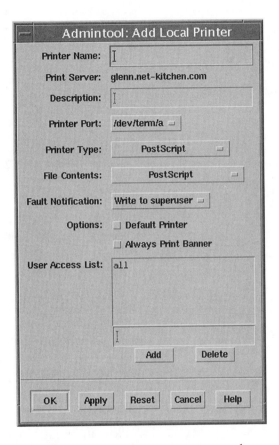

The first item in the form, Printer Name, can be up to 14 characters long. Making the name descriptive of printer usage, location, or ownership to aid users in selecting the correct printer for the job is recommended. Locations with a large number of printers distributed across a network adopt a uniform naming scheme which encodes the location and type of printer in the name. For example, *Lillyps* might be used to name the Post-Script printer in the Lilly building.

The next field in the form is automatically filled in, and lists the name of the local machine as the print server. For the printing service of a local printer, the local machine is both client and server.

The third field is for a comment that can be displayed by various printing commands. You can input additional information about the printer such as the room it is located in and model number or name.

The fourth field asks for the port to which the printer is connected. This is one of the items the system administrator must remember when a printer is connected to a workstation. Select from the menu which appears under the little box to the left of the field. Selecting Other produces a small entry window to enter the SunOS device name for the port the printer is connected to if the printer is not connected to one of the listed ports.

The fifth field for Printer Type refers to the particular manufacturer and model, such as HP Laser or Diablo. Most laser printers which accept the PostScript language can be set to either PostScript or PostScript Reverse. PostScript Reverse causes the printing service to reverse the order of the pages printed. The reversed output order is handy for printers which stack output face up as it eliminates the need to collate the printed pages.

If the printer being installed is not listed in the pop-up menu near the fifth field, check the *terminfo* directories located in */usr/share/lib/terminfo*. A series of directories with single letter and number names is located in the *terminfo* directory. The terminfo entries are grouped by the first letter or number of the entry in the directories. Each entry describes a terminal or printer, and the name of an entry can be used to fill in the printer type field.

If no entry is available for the printer being installed, use an entry for a similar printer. For example, if a dot matrix printer is being installed, try one of the *epson* entries. If this tactic does not meet with acceptable results, try copying the entry for a similar printer and editing it to describe the features of the printer being installed. The manual page for terminfo lists the format and keywords used in terminfo entries.

The sixth field, which looks like a duplicate of the fifth field, refers to the type of information that can be sent to the printer rather than the type of printer. Some printers can automatically detect PostScript and ASCII files, while others can accept only one type of input or the other. Selecting the type of information the printer accepts causes the print manager to configure the printing service to filter the print job as necessary to create the appropriate input type for the printer. For example, if an ASCII text file is sent to a printer whose File Contents field is set to PostScript, the printing service will automatically filter the file to be printed with a program which converts the ASCII file to PostScript and then sends the converted file to the printer.

The seventh field, Fault Notification, indicates where problem messages from the printing service are to be sent. Write to superuser sends the problem messages directly to the terminal of a user logged in as *root*. Alternatively, problem messages can be mailed to the root user or simply not sent by using the None option. The importance of printer problems and who needs to be informed depend on the printer, its location, and local policies.

The eighth field offers two check boxes. If the Default box is checked, the printer described in this form is where output printed on a particular workstation will go when a print command is run without listing an explicit printer name (e.g., **lp file**). The Banner box asks if a separator page should be printed at the start of each job. Because the banner will include the name of the user who printed the job, it is handy at locations with many print jobs being requested by different users.

The final field is an access list for the printer. By default any user can print to any local printer. You may wish to limit printer access to the user who owns the printer, or to members of the local work group. A more detailed discussion of printer access is covered in the "Access Control" section below, and the setup of a remote or network accessible printer appears in "Setting Up Remote Printers."

When the form is complete, click on the OK button. The admintool will install the new printer and make it ready for use.

Printers can also be set up via command lines. The **lpadmin** command covers a super-set of the functions available with the Print Manager. To perform the same printer setup shown in the completed local printer form, use the following lpadmin commands:

```
lpadmin -p 1st-floor -D "Apple LaserWriter" -v /dev/term/b -T ps -I postscript -A
write -o banner

accept 1st-floor
```

The first command creates the printer *1st-floor*. The **-p** option specifies the printer name. The **-D** option includes a comment, and the -**v** option specifies the device to which the printer is attached. The **-T** and **-I** options specify the printer type and file content, respectively. The **-A** option sets the alert type for announcing printer problems. The **-o** option disables the printing of a banner page unless requested by the user. The second command, **accept**, tells the printing system to begin accepting print jobs for the new printer.

Changing printer parameters is a matter of changing the options in the lpadmin command or selecting the printer from the admintool window by double-clicking on it. A form like the one filled in to create a printer will appear allowing you to edit the fields to be changed. To delete a printer, select a printer from the list by clicking on the printer's entry line and selecting Delete from the Edit menu.

The Modify Printer form contains a different set of fields than the Local Printer form. In particular, several items in the Local Printer form are not listed or not changeable. For example, if the printer type needs to be changed, the printer entry must be deleted and re-created with the new type.

The new fields that appear in the Modify Printer form refer to the control of print file printing and queuing, and are duplicated by the **enable/disable** and **accept/reject** commands, respectively.

Modifying a printer's configuration using the modify printer form.

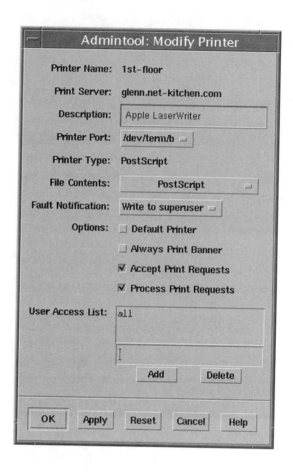

Setting Up Remote Printers

A local printer setup configures the local machine as print server and print client. The default access list created allows other workstations to send jobs to the printer spool to be printed. The only task remaining for remote printing is to provide information about the printer to each print client. Three ways of accomplishing this task are listed below.

❒ If the systems sharing the printer are running Solaris 2.6 and using NIS+ as described in Chapter 21, "Network Name Services," sharing the information is easy. The **admintool** will automatically add the printer information to the NIS+ database. If **lpadmin** was used to set up the printer, the NIS+ databases must be manually updated.

❏ If the systems sharing the printer are running Solaris 2.6, but not using NIS+, the */etc/printers.conf* file must be distributed to each print client. A painless way of accomplishing the distribution is to use the **rdist** command as described in Chapter 16, "Automating Routine Administrative Tasks."

❏ If the print server is not a Solaris 2.6 system, access to the printer via the common Berkeley *lpd* printer protocol can be set up by selecting Access to Printer from the pull-right Add item in the Edit menu of **admintool**. The next illustration provides an example of the ensuing form.

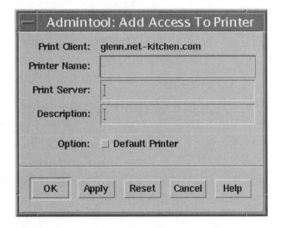

Setting up remote printer access.

The printer name refers to the printer on the print server, and the printer server is the name of the machine the printer is physically connected to. A description can be included as well, and a remote printer can be set as the default printer.

When using **lpadmin** to set up remote printer access, the **-s** option is used to specify the remote system. On *grissom*, the following lpadmin command line would be used to specify access to the *basement* printer on *glenn*.

```
lpadmin -p basement -s glenn
```

Access Control

The printing service provides access control for both local and remote printers. Access control is implemented through *allow* and *deny* lists.

You can select a printer through the Printer Manager window, and edit the access list at the bottom of the Modify Printer form. Take a look at the form in the illustration appearing in the previous section. The list managed by the form is an *allow* list, meaning that users on this list and only users on this list can print on the printer. This technique is handy for situations where the printer is in an individual's office or is used by a small work group.

However, sometimes you need the ability to prevent a certain user or two from using a printer. This is done with a *deny* list. These lists can be created only by using the **lpadmin** command. The following command line denies access to the *basement* printer to user *mary*.

```
lpadmin -p basement -u deny:mary
```

Printing

Solaris provides many tools to submit print jobs to the printing service. There are three basic commands: **lp** for submitting jobs, **cancel** for halting print jobs, and **lpstat** for checking on the progress of a submitted job and the general status of the printing system. There is also an OpenWindows interface called **printtool** which combines aspects of both lp and lpstat, as does the CDE printer icon found on the right side of the CDE "dashboard."

lp and cancel Commands

The **lp** command allows a user to submit print jobs with a large number of options, and to change a print job's options while it is in the print spool waiting to be printed. An example of the simplest form of the lp command follows:

```
lp my-stuff
```

The above command will print the file called *my-stuff* to the default printer. If no file is listed, **lp** reads from the standard input, making it suitable for sending the output of other commands via a pipe. To better control how *my-stuff* or other print jobs are printed, **lp** has a variety of option flags. Some of the most frequently used options are listed in the following table.

Frequently used lp command options

Options	Description
-c	Copy before printing. Normally the printing service makes a link to the file to be printed and reads it at the time printing takes place. However, if you plan to remove or overwrite the file before it is printed, using this option forces the printing service to make a copy of the file. The printing service removes this temporary copy at the successful conclusion of the print job.
-d destination	To print on a specific printer or class of printers, provide the name of the printer or class as the *destination*. A destination of *any* will submit the job to any printer which can satisfy other options that may have been set.
-f form	This option specifies the form or paper to be used when printing the job. With a destination of *any* (see previous option), the form option will route the print job to any printer which has the specified form mounted on it. See the discussion of forms below.

Options	Description
-H handling	Specifies job handling. By using *hold* as the *handling* request, a job can be put in the print spool but not printed. Using *resume* releases a held job to be printed. Specifying *immediate* allows users in the *lpadmin* group to get a print job processed ahead of jobs waiting in the printer spool.
-m	This option causes the print service to send the user electronic mail when a print job concludes.
-n copies	Specifies the number of copies of the file to be printed.
-o option	This flag allows a number of printer-specific options. These include options which modify the line and character spacing and even send special control sequences to certain printers. A common usage is to specify the *nobanner* option to prevent a banner page from being printed. Such option saves paper for printers which handle few print jobs, and where individual jobs are easily sorted out.
-q priority	Sets the print job *priority* in the print queue. The default print priority and the priority limits can be set and controlled on a per user basis as shown later in this chapter.
-t title	Puts the *title* of the print job on the banner page.
-T content	Specifies the *content* type of the print job, such as *PostScript*. Print jobs are routed to printers that can either directly handle the content type, or printers for which a filter is available to convert the content type into a form the printer can handle.
-w	This option causes the print service to write a message on the terminal, commandtool or shelltool window when a submitted print job finishes printing.

With such a wide variety of options, a job is occasionally submitted which specifies incorrect options. When this happens, the printing service cannot satisfy the request and the job will be rejected. If the job is accepted, but the options should be corrected, you need to know the *request id* of the print job. This is the magic code that **lp** provides when a print job is submitted. The next illustration shows an example of a print job which could not be satisfied, followed by one in which the options were wrong and then corrected by using lp. Note how the request id is used.

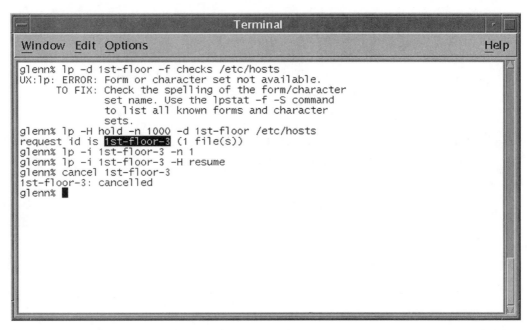

```
glenn% lp -d 1st-floor -f checks /etc/hosts
UX:lp: ERROR: Form or character set not available.
      TO FIX: Check the spelling of the form/character
              set name. Use the lpstat -f -S command
              to list all known forms and character
              sets.
glenn% lp -H hold -n 1000 -d 1st-floor /etc/hosts
request id is 1st-floor-3 (1 file(s))
glenn% lp -i 1st-floor-3 -n 1
glenn% lp -i 1st-floor-3 -H resume
glenn% cancel 1st-floor-3
1st-floor-3: cancelled
glenn% ▊
```

Correcting an incorrect print job using the request id.

cancel Command

The **cancel** command accepts either a *request id* as shown in the previous example, or the name of a printer. Given a request id, cancel removes the print job from the printer spool. Given a printer name, cancel stops the currently printing job. The cancel command can be used only to stop a user's own print jobs. However, the root user can cancel any job submitted by any user. But what if the request id for a print job has been forgotten?

lpstat Command

The **lpstat** command provides information on the status of print jobs, printers, and the printing service daemons. Similar to **lp**, lpstat has numerous options. By default, keying in lpstat will list all print jobs submitted to the printing service. Every job is listed along with the request id of the job so that a user or the system administrator can cancel or modify the request as needed.

```
glenn% lpstat
1st-floor-3              glenn.net-kitchen.com!dwight      121    Jun 30 14:43
1st-floor-4              glenn.net-kitchen.com!dwight      121    Jun 30 14:43
1st-floor-5              glenn.net-kitchen.com!dwight      121    Jun 30 14:43
glenn% █
```

Checking on print jobs with lpstat.

More information is available upon using one of the commonly used options listed in the following table.

Frequently used lpstat command options

Option	Description
-a list	Check to see if destinations or classes in the *list* are accepting print jobs. If no list is present, all destinations and classes are checked.
-c list	List the members of a class. If no *list* is given, all classes and respective members are listed.
-d	Report the default print destination.
-f list -l	Checks to determine if the form in the *list* is available. Adding the *-l* flag will list the description of the form.
-o *list*	Report on print requests in the *list*.

CDE Printer Icon

The CDE printer icon provides drag and drop access to the default printer. It is located in the CDE dashboard to the right of the center set of screen switch buttons. Dragging a file from the CDE File Manager onto the printer icon submits a print job. Clicking on the icon brings up a small window which graphically displays printer status and lists jobs in the print queue. Options are described in the built-in help system found in the menu at the upper right corner of the window.

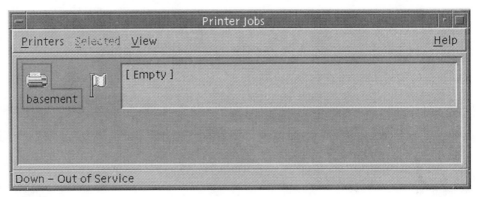

CDE print screen.

Printer and Print Job Management

Sometimes printing problems develop that require printer queues to be managed in more sophisticated ways. In larger organizations with many printers and users there is a constant need to shift printing resources around, and to route print jobs in response to printing demands, hardware failures, and the movement of personnel and equipment. Several commands allow the system administrator to deal with these situations.

Disable and Enable, Reject and Accept

Printers can be stopped by using the **disable** command. By default, disable will stop the currently printing job and save it to be reprinted when the printer is returned to service. The system administrator can cause the printing system to wait for the current job to finish printing and then stop the printer by using the **-W** option, or cancel the current job and not reprint it later by using the **-c** option. She can also leave a message to be displayed when the **lpstat** command is used to explain why the printer is disabled by using the **-r** option flag.

Restarting the printer is a simple matter of using the **enable** command. An example appears in the following illustration.

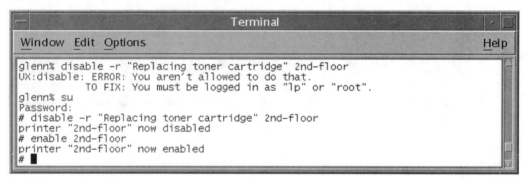

Disabling and enabling a printer.

As the example shows, the **disable** and **enable** commands are reserved for use by the *root* or *lp* users.

While a printer is disabled, print jobs can still be submitted for printing on that printer. Such jobs are held in the printer spool area until printing resumes. The spooling of jobs to be printed is controlled by the

accept and **reject** commands. These two commands start and stop the acceptance of jobs to be spooled. The only option flag, **-r**, provides a means to present a reason when a printer spool rejects jobs. An example appears in the next illustration.

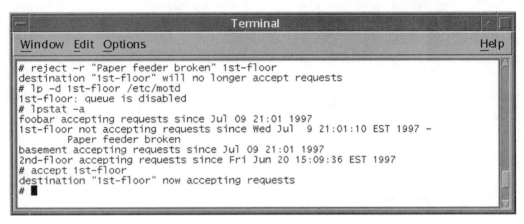

```
# reject -r "Paper feeder broken" 1st-floor
destination "1st-floor" will no longer accept requests
# lp -d 1st-floor /etc/motd
1st-floor: queue is disabled
# lpstat -a
foobar accepting requests since Jul 09 21:01 1997
1st-floor not accepting requests since Wed Jul  9 21:01:10 EST 1997 -
        Paper feeder broken
basement accepting requests since Jul 09 21:01 1997
2nd-floor accepting requests since Fri Jun 20 15:09:36 EST 1997
# accept 1st-floor
destination "1st-floor" now accepting requests
#
```

Rejecting and accepting print jobs.

Once a printer spool has been set to reject, print jobs cannot be sent using **lp**, and the reason indicated by the -r option of the reject command is printed when the **lpstat** command is used to check on the printer.

Moving Print Jobs

The printer and printer spool control commands are used in concert with other printing system commands to handle various situations. Suppose a printer is out of commission or is otherwise unusable for a period of time, and print jobs are pending. Fortunately, the printing service allows print jobs to be moved from one printer spool to another.

The approach to handling this situation follows: (1) reject print jobs for the unavailable printer to prevent more jobs from accumulating; and (2) move the print jobs to an alternative printer. An example appears in the next illustration.

```
┌─────────────────────────────────────────────────────────────────────┐
│ ─                        Terminal                          ▾  □      │
├─────────────────────────────────────────────────────────────────────┤
│  Window  Edit  Options                                       Help    │
├─────────────────────────────────────────────────────────────────────┤
│ # lpstat -o 1st-floor                                                 │
│ 1st-floor-3            glenn.net-kitchen.com!dwight    121   Jul 10 07:34 │
│ 1st-floor-4            glenn.net-kitchen.com!dwight    121   Jul 10 07:34 │
│ 1st-floor-5            glenn.net-kitchen.com!dwight    121   Jul 10 07:34 │
│ # reject 1st-floor                                                    │
│ destination "1st-floor" will no longer accept requests               │
│ # lpmove 1st-floor-3 1st-floor-4 1st-floor-5 2nd-floor               │
│ UX:lpmove: ERROR: Request "1st-floor-3" is done.                     │
│         TO FIX: It is too late to do anything with it.               │
│ total of 2 requests moved to 2nd-floor                               │
│ # lpstat -o 2nd-floor                                                 │
│ 2nd-floor-4            glenn.net-kitchen.com!dwight    121   Jul 10 07:35 │
│ 2nd-floor-5            glenn.net-kitchen.com!dwight    121   Jul 10 07:35 │
│ #                                                                     │
│ #                                                                     │
│ # ■                                                                   │
└─────────────────────────────────────────────────────────────────────┘
```

Moving print jobs to another printer.

Canceled jobs cannot be moved, but any other jobs in a print spool can be directed to another printer.

Removing a Printer

Removing a printer from the print service follows the same general strategy as moving print jobs. First, the printer and spool are stopped and then the printer entry is removed. There are two ways to remove the printer. One is to use the OpenWindows Printer Manager, select the printer from the listing, and pick Delete Printer from the Edit menu. You can also use the **lpadmin** command, the command line equivalent to the Print Manager. The command **lpadmin -x printer-name** removes the printer called *printer-name* from the system.

> **⟶ NOTE:** *Neither the Print Manager nor* lpadmin *will remove a printer with jobs waiting in its printer spool. Before removing a printer, move or cancel all print jobs in the spool.*

Starting and Stopping the Printing Service

The entire printing service can be stopped and started as necessary. With the print service shut off, no jobs are printed or accepted for spooling for any local or remote printer accessible by the workstation. Stopping and starting are accomplished by using the **lpshut** and **lpsched** commands. An example appears in the next illustration.

Stopping and starting the print service.

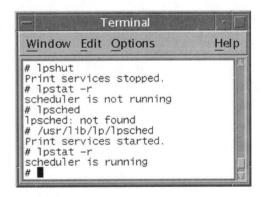

The first command shows **-r**, a special option to the **lpstat** command, that indicates whether the print service scheduler is running. The lpshut command stops the scheduler. Starting the scheduler is a matter of running the scheduler by typing the full path to it as shown in the illustration.

Forms, Printwheels, and Fonts

Some older printers support features such as downloadable fonts, interchangeable printwheels, and font cartridges. In addition, many printers support printing on preprinted forms or paper sources of differing sizes.

Printers with downloadable fonts allow you to change typeface by changing a printwheel or font cartridge, and printing on preprinted forms or different paper sizes can be handled by the printing service. These special situations are handled by the **lpadmin** command and the special forms handling command, **lpforms**.

The general strategy for forms and different fonts or printwheels depends on where the font or form is located. Fonts or forms resident in the printer, such as fonts loaded in the printer's read-only memory and forms located in additional paper feeders, are handled by informing the print service of their presence. Once informed and with a proper *terminfo* entry for the printer, the print service can automatically switch between printer resident fonts and forms based on the selections made for each print request.

Changing Fonts

Modern PostScript printers handle font changes in the page description sent to them; the printing service requires no special processing to handle this. However, older printers which use hardware such as font cartridges or printwheels require special handling. The **lpadmin** command's **-S** option allows the system administrator to specify which printwheels are available. If you are working with older printers, consult the manual page for lpadmin.

Changing Paper

Changing forms on many printers requires changing and aligning paper or switching paper trays. For these situations, the system administrator must load the definitions for the alternative forms into the print service and specify how to announce a change of form. The print service will automatically hold print jobs which require a special form until the system administrator informs the print service that the form has been mounted on the printer. In addition, a user access list can be created to control the users with permission to request and print on a given form such as pre-printed checks.

Configuring the print service to use a form requires a couple of steps. First, the **lpforms** command is used to define the form. This can be executed interactively, or by feeding the answers to lpforms from a file. The next illustration shows two form definitions to describe letter- and legal-size paper for a laser printer.

The first form pertaining to the lpforms command almost completely describes the form by explicitly providing each option. The only option not used here is **comment**, which allows a description of the form to be added to its definition and an alignment pattern which can be printed to aid in aligning pre-printed forms.

In the second form definition, only options that differ from the defaults are used. Results can be viewed by using the **-l** option. Use the **-x** option to remove an incorrect form the print service. Alerts and user lists can also be attached to forms.

Defining forms using the lpforms command.

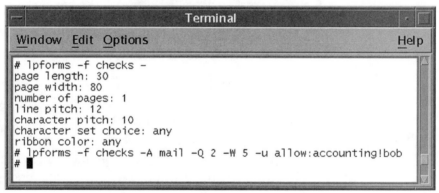

Assigning user lists and alerts to forms.

In the second example, a form is defined for pre-printed checks. The second **lpforms** command assigns an alert message to the form with the **-A, -W,** and **-Q** options. The -A option can be used to send the alert message via mail, writing it on the user's screen (write), running a specified command, or doing nothing (none). The -W option, which specifies the alert message timing, is set to send it every two minutes in the example. The -Q option specifies how many print jobs must be in the printer spool waiting to be printed on this form before the alert is sent.

The **-u** option specifies a user access list. In this case, an allow list is used. As discussed in the previous section on setting up a printer, an allow list specifies the user or users who can print on this form. The notations, *machine!user* and *machine!all* are used to specify a user on a specific machine or all users of a machine, respectively.

Printing Problems and Solutions

The two basic printing problems are not producing output or producing incorrect output. In both cases, a useful approach toward identifying the problem is to walk backwards along the path the print job takes. Start with the printer itself and work toward the command used to submit the print request.

At the printer, start with the obvious. Is the printer on? Is the machine loaded with paper, ribbon, toner, and so forth? Laser printers in particular are intelligent devices and will halt due to a variety of problems including paper jams and other internal faults. Check the printer to be sure it is ready to accept a print job.

Next, check the connection from the printer to the print server. For intelligent printers, such as most laser printers, begin by checking the printer's configuration. Some printers have communications settings that are easily changed by mistake. Verify that the printer is using the correct port and that the parameters for the port are accurate, such as baud rate and parity. Determine whether the cable between the printer and computer is plugged in.

Network parameters should be checked for printers that are directly connected to the network. Checking may be required for the IP address number, subnet mask, and broadcast address. See Chapter 17, "Understanding Basic Local Area Networking," for a discussion of these parameters.

At this point, check the print server. A common source of problems is the incorrect definition of the printer when it was set up. Check each field of the local printer form viewed by using the Printer Manager screen accessed from the **admintool**. Determine whether the communications parameters match the settings on the printer. If incorrect output is the problem, check to determine whether the printer type and file contents fields are correct for the printer in question. In some cases, the system administrator may need to experiment with these settings to find a combination that works.

If the printing problem occurs on a print client, try sending a print job to the printer directly from the print server. This procedure may help to determine the location of the problem (client or server). If the problem is on the client, check to determine whether the printer configuration file, */etc/printers.conf*, is up to date.

Once convinced that the printer is properly connected, the printer server is operating, and the print client is correctly configured, move on to check the network. Use a utility such as **ping** to verify whether communications are working between the print client and server. See Chapter 17, "Understanding Basic Local Area Networking," for more information on setting up, monitoring, and troubleshooting network problems. Finally, check the network log file, */var/lp/*

logs/lpNet. This file lists problems found by the printing service when it sends or receives a print request over the network.

Problems Printing from Windows NT

A common problem when printing from a system running Windows NT to a Solaris printer is the inability to print anything but text files. Documents printed from Microsoft Word and other applications print as a jumble of characters. This is corrected by changing the printer configuration on the Windows NT system as follows. (The menu selections appearing below pertain to Windows NT 4.0.)

1. Select Settings from the START menu.

2. Select Printers.

3. If the Solaris printer does not exist, add the printer according to the instructions that ship with Windows NT.

4. Double-click on the printer icon.

5. Select DOCUMENT PROPERTIES from the FILE menu.

6. Select PostScript Options, and set the following options to NO: Send CTRL-D Before Each Job; Send CTRL-D After Each Job; and Generate Job Control Code.

Summary

The Solaris printing service is comprised of a collection of commands that work together to handle a variety of printing situations. When setting up printing, pay careful attention to the printer's connection to the workstation or server and to the allocation of sufficient printer spool space. Next, employ the printing service's access control mechanisms to implement printer and form access policies.

System Backups and Disaster Recovery

Previous chapters have explored Solaris installation, and managing users, peripherals, security, and network services. One very important area that has yet to be explored is recovering from disaster. Whether the disaster is caused by a user removing a critical file, or loss of data due to a disk drive failure, the system administrator needs to be prepared. One of the simplest ways to protect data is to perform file system backups at regular intervals.

Importance of System Backups

System admininistrators' most important task is to ensure the availability of user data. Every file, every database, every byte of information stored on the system must be available. This requirement dictates the need for backup copies of all data in the event there is a need to retrieve such information at a later date.

However, making a copy of the data and storing it in another office in the same building may not be a solution to the problem. For instance, in the event of a natural disaster the corporation may need to have copies of the data in remote locations to ensure survivability and accessibility of the backup media.

In the event of a hardware failure or some other form of disaster, the system administrator should be able to reload the bulk of user information from backup media. There will almost always be some loss of information. It is impractical to back up every keystroke as it occurs. Therefore, it is possible to lose information entered into the system between backups. The goal of a backup procedure is to minimize data loss and allow the corporation to reload quickly and continue with business.

Which Files Should Be Backed Up?

Backing up all files on the system at regular intervals is good practice. Some files, however, rarely change. The administrator needs to examine contents of system disks to identify the files that should be a part of file system backups, as well as files to be eliminated from regular backups.

Some programs are part of the operating system. Administrators often decide not to back up the operating system binaries, opting to reload them from distribution media in the event of a loss. This practice requires making a backup of any operating system-related files that are created or changed. A few examples might be the NIS+ database, the password file, shadow file, rc files, and customizations which affect those files.

Another set of files that may be eliminated from regular backups are vendor supplied binaries. If the distribution media for commercially purchased software is available, regular backups of these files are not necessary. Again, the administrator should ensure that any customized files or start-up scripts required for those packages are on the backup list.

Files that should be backed up regularly include user files, corporate databases or other important data, and files which have been changed since the time of the last backup. On large systems the amount of data to be backed up may exceed several gigabytes a day. Consequently, the administrator should decide how and when to perform backups in order to minimize impact on the system and loss of important data.

How Often Should Backups Be Performed?

Frequency is another very important factor in successful file system backup strategies. The administrator must determine how much data can be lost before seriously affecting the organization's ability to do business. In many cases losing the data from a single day is deemed an acceptable loss. This is generally the case for activities such as university teaching laboratories, customer service databases, and Internet service providers.

At the other end of the spectrum are situations which do not allow the loss of any data. Corporations which use computers for financial transactions, stock and commodities trading, and insurance activities would be a few examples. In these cases the loss of a single transaction is not acceptable.

A reasonable backup schedule for a typical corporation would call for some form of backup to be performed every day. In addition, the backups must be performed at the same time every day so that the operator knows which tape to use for reload, and to force the administrator into a schedule. Backups are too important to leave to haphazard scheduling. The fact that backups must be maintained on a rigid schedule lends support to the automated backup methods discussed later in this chapter and in Chapter 16, "Automating Administrative Tasks."

⟿ **NOTE:** *The examples in this chapter assume computer system operation seven days a week, and thus, the need for seven backups per week. In some instances, corporate computer systems are idle on weekends, enabling the administrator to use a five day per week backup schedule.*

Backup Strategy and Scheduling

Once the administrator has determined which files to back up, and how often to perform backups, several additional decisions must be made. These decisions include identifying the type of backups to perform, and the devices to use to perform the backups. Such decisions will serve to delineate the parameters of backup scheduling.

Before scheduling the backups, the administrator needs to determine the type of backup strategy to use, the type of backup system to use, and how much media will be required for the chosen methods. The following sections examine several backup schedules, the media requirements for each schedule, and the strengths and weaknesses of each backup method.

Types of Backups

Several types of backups may be performed, and the simplest type is a full dump. A full dump is also referred to as a level 0 (zero) backup. A full dump copies every disk file to the backup media. On systems with large disk storage capacity, it may not be possible to make full dumps very often due to the cost of the tapes, tape storage space requirements, and the amount of time required to copy files to the tapes.

Because full dumps are expensive in terms of time and resources, most programs written to perform backups allow for some form of incremental dump. In an incremental dump, files created or modified since the last dump are copied to the backup media. For instance, if a full dump is performed on a Sunday, the Monday incremental dump would contain only the files created or changed since the Sunday dump.

The type of dump to be performed will typically rely on the chosen backup strategy. The strategy includes factors such as how long to keep copies of the dump media on hand, what type of dump to perform on which day of the week, and how easily information can be restored from the backup media.

Backup Strategies

Because performing dumps at regular intervals is vital, developing a dump schedule is recommended. The dump schedule notes when the operator should perform full and incremental dumps on the systems.

Full dumps allow you to establish a snapshot of the system at a given point in time. All corporate data should be contained on a full dump.

Incremental dumps enable the administrator to recover active files with a minimum of media and time. Incremental dumps capture the data which have changed since the last incremental dump of a lower level. For example, a level 1 dump would back up all files created or changed since the last level 0 dump. A level 2 dump would back up all files since the last level 1 dump, and so on. In general, a level N dump will back up all files which have been altered or created since the last dump at a lower level.

In order to ensure that the backups capture as much of the system data as possible, the interval between backups should be kept to a minimum. With these constraints in mind, it is time to examine a few of the more popular backup strategies.

➡ **NOTE:** *In order to ensure successful dumps, it is recommended that the system be shut down to the single-user* init *state.*

Volume/Calendar Backup

The volume/calendar backup strategy calls for a full system backup once a month. An incremental backup is performed once a week for files which change often. Daily incremental backups catch the files that have changed since the last daily backup.

A typical schedule would be to perform the full (level 0) backup one Sunday a month, and weekly level 3 backups every Sunday of the month. Daily level 5 backups would be performed Monday through Saturday. This would require eight complete sets of media (one monthly tape, one weekly tape, and six daily tapes).

Typical scheduling for the volume/calendar backup strategy

Sun	Mon	Tue	Wed	Thu	Fri	Sat
Date: 1	Date: 2	Date: 3	Date: 4	Date: 5	Date: 6	Date: 7
Level: 0	Level: 5	Level: 5	Level: 5	Level: 5	Level: 5	Level: 5
Tape: A	Tape: B	Tape: C	Tape: D	Tape: E	Tape: F	Tape: G
Date: 8	Date: 9	Date: 10	Date: 11	Date: 12	Date: 13	Date: 14
Level: 3	Level: 5	Level: 5	Level: 5	Level: 5	Level: 5	Level: 5
Tape: H	Tape: B	Tape: C	Tape: D	Tape: E	Tape: F	Tape: G
Date: 15	Date: 16	Date: 17	Date: 18	Date: 19	Date: 20	Date: 21
Level: 3	Level: 5	Level: 5	Level: 5	Level: 5	Level: 5	Level: 5
Tape: H	Tape: B	Tape: C	Tape: D	Tape: E	Tape: F	Tape: G
Date: 22	Date: 23	Date: 24	Date: 25	Date: 26	Date: 27	Date: 28
Level: 3	Level: 5	Level: 5	Level: 5	Level: 5	Level: 5	Level: 5
Tape: H	Tape: B	Tape: C	Tape: D	Tape: E	Tape: F	Tape: G
Date: 29	Date: 30	Date: 31				
Level: 3	Level: 5	Level: 5				
Tape: H	Tape: B	Tape: C				

Recovering from complete data loss with the volume/calendar scheme requires restoring from the most recent full backup, then restoring from the most recent weekly backup, and finally, restoring from each daily backup tape written since the weekly backup.

An advantage to this backup scheme is that it requires a minimum of media. One problem with this backup scheme is that the tapes are immediately reused. For example, every Monday overwrites last Monday's backup information. Consider what would happen if one of the disk drives failed during the second Monday backup. It would not be possible to recover all the data, because the system was in the process of overwriting the backup tape when the drive failed.

Grandfather/Father/Son Backup

The grandfather/father/son backup strategy is similar to the volume/calendar strategy. The major difference between the two schemes is that the grandfather/father/son method incorporates a one-month archive in the backup scheme. This eliminates the problem of overwriting a tape before completing a more recent backup of the file system.

Implementing the grandfather/father/son strategy requires performing a full (level 0) dump once a month to new media. Once a week, an incremental (level 3) backup must be performed which captures all files changed since the last weekly backup. This weekly backup should also be saved on new media. Each day, an incremental level 5 backup must be performed to capture files that have changed since the last daily backup. The daily backups reuse the tapes written one week earlier.

Typical scheduling for the grandfather/father/son backup strategy

Sun	Mon	Tue	Wed	Thu	Fri	Sat
Date: 1	Date: 2	Date: 3	Date: 4	Date: 5	Date: 6	Date: 7
Level: 0	Level: 5	Level: 5	Level: 5	Level: 5	Level: 5	Level: 3
Tape: A	Tape: B	Tape: C	Tape: D	Tape: E	Tape: F	Tape: G
Date: 8	Date: 9	Date: 10	Date: 11	Date: 12	Date: 13	Date: 14
Level: 5	Level: 5	Level: 5	Level: 5	Level: 5	Level: 5	Level: 3
Tape: H	Tape: B	Tape: C	Tape: D	Tape: E	Tape: F	Tape: I

Sun	Mon	Tue	Wed	Thu	Fri	Sat
Date: 15	Date: 16	Date: 17	Date: 18	Date: 19	Date: 20	Date: 21
Level: 5	Level: 5	Level: 5	Level: 5	Level: 4	Level: 5	Level: 3
Tape: H	Tape: B	Tape: C	Tape: D	Tape: E	Tape: F	Tape: J
Date: 22	Date: 23	Date: 24	Date: 25	Date: 26	Date: 27	Date: 28
Level: 5	Level: 5	Level: 5	Level: 5	Level: 5	Level: 5	Level: 3
Tape: H	Tape: B	Tape: C	Tape: D	Tape: E	Tape: F	Tape: K
Date: 29	Date: 30	Date: 31				
Level: 0	Level: 5	Level: 5				
Tape: L	Tape: B	Tape: C				
			Date: 1	Date: 2	Date: 3	Date: 4
			Level: 5	Level: 5	Level: 5	Level: 3
			Tape: D	Tape: E	Tape: F	Tape: M
Date: 5	Date: 6	Date: 7	Date: 8	Date: 9	Date: 10	Date: 11
Level: 5	Level: 5	Level: 5	Level: 5	Level: 5	Level: 5	Level: 3
Tape: H	Tape: B	Tape: C	Tape: D	Tape: E	Tape: F	Tape: G
Date: 12	Date: 13	Date: 14	Date: 15	Date: 16	Date: 17	Date: 18
Level: 5	Level: 5	Level: 5	Level: 5	Level: 5	Level: 5	Level: 3
Tape: H	Tape: B	Tape: C	Tape: D	Tape: E	Tape: F	Tape: I
Date: 19	Date: 20	Date: 21	Date: 22	Date: 23	Date: 24	Date: 25
Level: 5	Level: 5	Level: 5	Level: 5	Level: 5	Level: 5	Level: 3
Tape: H	Tape: B	Tape: C	Tape: D	Tape: E	Tape: F	Tape: J
Date: 26	Date: 27	Date: 28				
Level: 0	Level: 5	Level: 5				
Tape: A	Tape: A	Tape: C				
			Date: 1	Date: 2	Date: 3	Date: 4
			Level: 5	Level: 5	Level: 5	Level: 3
			Tape: D	Tape: E	Tape: F	Tape: K
Date: 5	Date: 6	Date: 7	Date: 8	Date: 9	Date: 10	Date: 11
Level: 5	Level: 5	Level: 5	Level: 5	Level: 5	Level: 5	Level: 3
Tape: H	Tape: B	Tape: C	Tape: D	Tape: E	Tape: F	Tape: M

Sun	Mon	Tue	Wed	Thu	Fri	Sat
Date: 12	Date: 13	Date: 14	Date: 15	Date: 16	Date: 17	Date: 18
Level: 5	Level: 5	Level: 5	Level: 5	Level: 5	Level: 5	Level: 3
Tape: H	Tape: B	Tape: C	Tape: D	Tape: E	Tape: F	Tape: G
Date: 19	Date: 20	Date: 21	Date: 22	Date: 23	Date: 24	Date: 25
Level: 5	Level: 5	Level: 5	Level: 5	Level: 5	Level: 5	Level: 3
Tape: H	Tape: B	Tape: C	Tape: D	Tape: E	Tape: F	Tape: I
Date: 26	Date: 27	Date: 28	Date:29	Date: 30	Date: 31	
Level: 0	Level: 5	Level: 5	Level: 5	Level: 5	Level: 5	
Tape: L	Tape: B	Tape: C	Tape: D	Tape: E	Tape: F	

In order to maintain a one-month archive, the monthly full backup tape should be placed in storage. Each weekly full backup should also be placed in storage. When the time comes to perform the second monthly full backup, new media should be used. When the third monthly backup is due, the first month's full backup media should be reused. The weekly backups are archived in a similar manner.

This scheme requires two sets of monthly backup media (one in storage, one active), five sets of weekly backup media, and six sets of daily backup media. A total of 13 sets of media are required to implement this strategy with a one-month archive of information.

In order to recover from complete data loss, first restore the most recent level 0 backup tape. Next, restore from the most recent of the level 3 backups, if that backup was written after the level 0 backup. When the level 3 backup has been restored, the operator would restore from each of the level 5 backups that were written after the level 3 backup.

This backup strategy requires much more media than the simple volume/calendar strategy. While media cost is increased with this plan, data survivability also increases.

Tower of Hanoi Backup

The Tower of Hanoi backup strategy is a variation of "exponential backup." Both strategies rely on mathematical functions of powers of two. For example, the use of five backup tapes provides for a 32-day schedule. The use of six tapes would provide for a 64-day schedule.

The Tower of Hanoi backup schedule provides outstanding data survivability and a minimum of media. Unfortunately, on a seven-day backup system, the scheduling of full backups as opposed to partial backups can become a problem for the operator.

One way to avoid operator confusion is to perform a special level 0 backup on the first day of each month. This tape would not be one of the five tapes used in the backup cycle. Total media requirements in this scheme would be seven sets of media.

To recover from complete data loss, first restore from the most recent level 0 backup, and then restore from the level 1 backup if that backup was written after the level 0 backup. Next, restore consecutively from the most recent level 3 and 4 backups if both were written after the level 0 backup. Finally, restore each of the level 5 backups that were written after the level 0 backup.

Typical scheduling for the tower of Hanoi backup strategy

Sun	Mon	Tue	Wed	Thu	Fri	Sat
Date: 1	Date: 2	Date: 3	Date: 4	Date: 5	Date: 6	Date: 7
Level: 0	Level: 5	Level: 4	Level: 5	Level: 3	Level: 5	Level: 4
Tape: E	Tape: A	Tape: B	Tape: A	Tape: C	Tape: A	Tape: B
Date: 8	Date: 9	Date: 10	Date: 11	Date: 12	Date: 13	Date: 14
Level: 5	Level: 1	Level: 5	Level: 4	Level: 5	Level: 3	Level: 5
Tape: A	Tape: D	Tape: A	Tape: B	Tape: A	Tape: C	Tape: A
Date: 15	Date: 16	Date: 17	Date: 18	Date: 19	Date: 20	Date: 21
Level: 4	Level: 5	Level: 0	Level: 5	Level: 4	Level: 5	Level: 3
Tape: B	Tape: A	Tape: E	Tape: A	Tape: B	Tape: A	Tape: C
Date: 22	Date: 23	Date: 24	Date: 25	Date: 26	Date: 27	Date: 28
Level: 5	Level: 4	Level: 5	Level: 1	Level: 5	Level: 4	Level: 5
Tape: A	Tape: B	Tape: A	Tape: D	Tape: A	Tape: B	Tape: A
Date: 29	Date: 30	Date: 31				
Level: 3	Level: 5	Level: 4				
Tape: C	Tape: A	Tape: B				
			Date: 1	Date: 2	Date: 3	Date: 4
			Level: 5	Level: 0	Level: 5	Level: 4
			Tape: A	Tape: E	Tape: A	Tape: B
Date: 5	Date: 6	Date: 7	Date: 8	Date: 9	Date: 10	Date: 11
Level: 5	Level: 3	Level: 5	Level: 4	Level: 5	Level: 1	Level: 5
Tape: A	Tape: C	Tape: A	Tape: B	Tape: A	Tape: D	Tape: A
Date: 12	Date: 13	Date: 14	Date: 15	Date: 16	Date: 17	Date: 18
Level: 4	Level: 5	Level: 3	Level: 5	Level: 4	Level: 5	Level: 0
Tape: B	Tape: A	Tape: C	Tape: A	Tape: B	Tape: A	Tape: E
Date: 19	Date: 20	Date: 21	Date: 22	Date: 23	Date: 24	Date: 25
Level: 5	Level: 4	Level: 5	Level: 3	Level: 5	Level: 4	Level: 5
Tape: A	Tape: B	Tape: A	Tape: C	Tape: A	Tape: B	Tape: A
Date: 26	Date: 27	Date: 28				
Level: 1	Level: 5	Level: 4				
Tape: D	Tape: A	Tape: B				

A Reasonable Alternative

The following four-week schedule offers a reasonable backup schedule for most sites. Performing a full dump on the first Sunday of the month provides a monthly snapshot of the system data. Using two sets of dump media allows the operator to store information for two full months.

Note that in the example the Tuesday through Friday incremental backups contain extra copies of files from Monday. This schedule ensures that any file modified during the week can be recovered from the previous day's incremental dump.

To recover from complete data loss, restore the most recent full (level 0) backup tape. Next, restore from the most recent of the weekly (level 3) backups. Once the weekly backups are restored, restore from each of the daily (level 5) backups.

Typical scheduling for a reasonable backup strategy

Sun	Mon	Tue	Wed	Thu	Fri	Sat
Date: 1	Date: 2	Date: 3	Date: 4	Date: 5	Date: 6	Date: 7
	Level: 0	Level: 5	Level: 5	Level: 5	Level: 3	
	Tape: A	Tape: B	Tape: C	Tape: D	Tape: E	
Date: 8	Date: 9	Date: 10	Date: 11	Date: 12	Date: 13	Date: 14
	Level: 5	Level: 5	Level: 5	Level: 5	Level: 3	
	Tape: F	Tape: B	Tape: C	Tape: D	Tape: E	
Date: 15	Date: 16	Date: 17	Date: 18	Date: 19	Date: 20	Date: 21
	Level: 5	Level: 5	Level: 5	Level: 5	Level: 3	
	Tape: F	Tape: B	Tape: D	Tape: D	Tape: E	
Date: 22	Date: 23	Date: 24	Date: 25	Date: 26	Date: 27	Date: 28
	Level: 5	Level: 5	Level: 5	Level: 5	Level: 3	
	Tape: F	Tape: B	Tape: D	Tape: D	Tape: E	

Sun	Mon	Tue	Wed	Thu	Fri	Sat
Date: 29	Date: 30	Date: 31				
	Level: 5	Level: 5				
	Tape: F	Tape: B				

Types of Backup Devices

Once a backup strategy has been chosen, it is time to devote some attention to backup devices. Principal backup device requirements are listed below.

❏   User ability to write data to the device.

❏   Media capable of storing the data for long periods of time.

❏   Supports standard system interconnects.

❏   Supports reasonable input/output throughput.

Most devices used for system backups consist of a magnetic media (tape or disk) and the drive mechanism. The drive mechanism provides the electronics and transport functions to allow the device to store digital signals on the magnetic media. Some backup devices use optical rather than magnetic media to store information. Still other backup devices use a combination of magnetic and optical storage methods. The following sections examine the characteristics of a few of these devices.

Tape Backup Devices

Tape backup devices are probably the most common backup media in use. The media is relatively inexpensive, the performance is reasonable, the data formats are standardized, and tape drives are easy to use. These factors combined make magnetic tape backups an attractive option. One complication of tape backups is deciding which type of tape system to use. Several of the more popular options are explored in the following sections.

1/2-inch 9-Track Tape Drive

Until recently, 1/2-inch tape drives were the mainstay of backup media. The early 1/2-inch tape drives allowed the operator to write information on the tape at 800 bits per inch of tape. Later drives allowed 1600 bits per inch, and the most recent drives allow 6,250 bits per inch. At 6,250 bits per inch, a 1/2-inch tape can hold up to 100 Mb of information.

A typical 1/2-inch tape consists of 2,400 feet of 1/2-inch wide tape on a 10-inch plastic reel. The 1/2-inch tape drives are typically large and expensive, but the tape itself is relatively inexpensive. The 1/2-inch format is still a standard media for many industrial operations.

Older 1/2-inch tape drives required special bus interfaces to connect them to a system, whereas the SCSI interface is common today. Due to the vast quantity of information stored on 1/2-inch tape, the 1/2-inch tape drive will be required for many years to come simply to access the stored information. However, with systems containing literally hundreds of gigabytes of disk storage, 1/2-inch tape becomes impractical for data backup.

Cartridge Tape Drive

Cartridge tape drives store between 10 Mb and several Gb of data on a small tape cartridge. Cartridge tape drives are usually smaller and less expensive than 1/2-inch tape drives. Tape cartridges are typically more expensive than a reel of 1/2-inch tape.

Because the data cartridge stores more information, a cartridge tape must be manufactured to tighter tolerances than 1/2-inch tapes.

Because the media is smaller and the drives less expensive, cartridge tape systems have become a standard distribution media for many software companies. Moreover, cartridge tapes are particularly attractive for disaster recovery situations because many systems still include boot PROM code to boot cartridge distribution tapes.

A cartridge tape distribution of the operating system is easily stored in a remote location. In the event of a failure or disaster at the primary computer location, these distribution tapes could be used to reload/rebuild a system to replace the damaged host.

Most cartridge tape systems use SCSI interconnections to the host system. These devices support data transfer rates up to 5 Mb per second. This transfer rate may be a little misleading, however, as the information is typically buffered in memory on the tape drive. The actual transfer rate from the tape drive memory to the tape media is typically about 500 Kb per second.

8 mm Tape Drive

These tape drives are also small and fast, and use relatively inexpensive tape media. The 8 mm media can hold between 2 and 20 Gb of data. Because of high density storage, 8 mm drives have become a standard backup device on many systems. Several companies also use 8 mm tape as a distribution media for software.

The 8 mm drives use the SCSI bus as the system interconnection. Low density 8 mm drives can store 2.2 Gb of information on tape. These units transfer data to the tape at a rate of 250 Kb per second. High density

8mm drives can store between 5 and 20 Gb of information on a tape.

At the "low" end, the 8 mm drives do not use data compression techniques to store the information on tape. At the "high" end, the drives incorporate data compression hardware used to increase the amount of information that can be stored on the tape. Regardless of the use of data compression, high density 8 mm drives transfer data to tape at a rate of 500 Kb per second. High density drives also allow the user to read and write data at the lower densities supported by the low density drives. When using the high density drives in low density mode, storage capacities and throughput numbers are identical to low density drives.

Digital Audio Tape Drive

Digital audio tape (DAT) drives are small, fast, and use relatively inexpensive tape media. Typical DAT media can hold between 2 and 8 Gb of data. The DAT media is a relative newcomer to the digital data backup market. The tape drive electronics and media are basically the same as the DAT tapes used in home audio systems.

The various densities available on DAT drives are due to data compression. A standard DAT drive can write 2 Gb of data to a tape. By using various data compression algorithms, manufacturers have produced drives which can store between 2 and 8 Gb of data on a tape.

DAT drives use SCSI bus interconnections to the host system. Because DAT technology is relatively new, it offers performance and features not available with other tape system technologies. For instance, the DAT drive offers superior file search capabilities as compared to the 8 mm helical scan drives on the market.

Jukebox System

Jukebox systems combine a "jukebox" mechanism with one or more tape drives to provide a tape system capable of storing several hundred Gb of data. The tape drives in the jukebox systems are typically DAT or 8 mm devices. Many of these systems employ multiple tape drives, and special "robotic" hardware to load and unload the tapes.

Jukebox systems require special software to control the robotics. The software keeps track of the contents of each tape and builds an index to allow the user to quickly load the correct tape on demand. Many commercially available backup software packages allow the use of jukebox systems to permit backup automation.

Digital Linear Tape (DLT)

Digital linear tape backup devices are relative newcomers to the backup market. These tape devices offer huge data storage capabilities, high transfer rates, and small (but somewhat costly) media. Digital linear tape drives can store up to 70 Gb of data on a single tape cartridge. Transfer rates of 5 Mb/second are possible on high-end DLT drives, making them very attractive at sites with large on-line storage systems.

While 8 mm and DAT tapes cost (roughly) $15 per tape, the DLT tapes can run as much as $60 each. But when the tape capacity is factored into the equation, the costs of DLT tapes become much more reasonable. (Consider an 8 mm tape which holds 14 Gb on average versus a DLT cartridge which can hold 60 Gb of data.)

➥ **NOTE:** *Many operators elect not to enable the compression hardware on tape systems, opting instead* *for software compression before the data are sent to the tape drive. In the event of a hardware failure in the tape drive's compression circuitry, it is possible that the data written to tape would be scrambled. By using the software compression techniques, the operator can bypass such potential problems.*

Optical Backup Devices

Magnetic tape has been used as a backup media for many years. Until recently, magnetic tape was the most economical way to back up mass storage devices. Recently, optical storage devices have become another economical means to back up mass storage systems. A few of the more popular optical backup devices are described below.

Compact Disk

Compact disk read-only-memory devices (CD-ROM) are useful for long-term archival of information. Although the name implies that these are read-only devices, recent technology has made it possible to mass market the devices which create the encoded CD-ROM media. These CD-ROM writers (also called CD-recordables) make it possible to consider CD-ROM as a backup device.

One of the major decisions in choosing a backup device is the ability of the media to store information for long periods of time. CD-ROM media offers excellent data survivability. Another advantage to the CD-ROM is the availability of reliable data transportability between systems. This reliability is possible due to the CD-ROM's adherence to industry standardized data formats.

Along with these advantages, the CD-ROM offers a few unique disadvantages. The foremost disadvantage to the CD-ROM as a backup device is the setup cost to create a CD. Not only is the CD-ROM writer an expensive device, setting up and creating a CD is a time-intensive operation. Other disadvantages include the (relatively) high priced media, and the slow speed at which the CD-ROM writers operate compared to tape drives.

WORM Disk

Write once read many (WORM) storage systems have been around longer than CD-ROM backup systems, and ways have been found to overcome some of the throughput problems associated with CD-ROM writers. These devices use standard SCSI system interconnects to interface with the system. Most WORM devices provide data rates similar to SCSI tape drives (500 kilobits per second).

The media used by WORM devices can often hold several Gb of data and is less expensive than the CD-ROM media. Unfortunately, there are no clear industry standards for recording methods on WORM devices. This has led to a plethora of manufacturers who produce drives with proprietary recording standards. Consequently, the data written on one brand of WORM drive (quite often) cannot be read on another brand of WORM drive.

Magneto-Optical Backup Devices

Magneto-Optical Disk

Optical storage systems and associated media are typically expensive. They are also relatively slow devices. Consequently, optical storage systems are rarely used as backup devices. In contrast, magnetic tape (or disk) storage systems are inexpensive and fast. Unfortunately, the media is bulky and susceptible to damage and data loss. By combining the two storage systems into a single system, manufacturers have been able to provide fast, inexpensive, and reliable backup systems.

Many of the magneto-optical systems are hierarchical, meaning that they keep track of how long a file has been in storage since the last modification. Files which are not accessed or modified are often eligible to be stored on the slower optical storage section of the system. Frequently accessed files are maintained on the magnetic storage section of these systems, which allows for faster access to files.

Most magneto-optical storage systems use standard SCSI bus system interconnections. These systems can typically provide the same (or better) data transfer rates as SCSI tape and disk systems.

Disk Systems as Backup Devices

One problem involved in using tape devices for backups is the (relatively) low data throughput rate. If the operator had to back up several gigabytes or terabytes of data daily it would not take long to realize that tape drives are not the best backup method. While optical backup devices offer high storage capacity, the optical devices are often much slower than tape devices. So how can an administrator back up large-scale systems?

One popular method of backing up large-scale systems is to make backup copies of the data on several disk drives. Disk drives are orders of magnitude faster than tape devices, so they offer a solution to one of the backup problems on large-scale systems. But disk drives are much more expensive than tapes. Disk backups also consume large amounts of system resources. For example, you would need 2 Gb disks to back up 100 2 Gb disks. Fortunately, there are software applications and hardware systems available to transparently perform this function.

RAID Disk Arrays

One operating mode of redundant arrays of inexpensive disks (RAID) enables the system to make mirror image copies of all data on backup disk drives. RAID disk arrays also allow data striping for high speed data access. See Chapter 11, "Disk and File System Maintenance," for an in-depth discussion of RAID disk arrays.

Briefly, RAID I/O subsystems provide several "levels" of service. Each RAID level describes a different method of storing the data on the disk drive(s). Each level has unique characteristics, and some of these levels are more suitable to certain applications than other I/O implementations would be.

RAID: A Review

The most common RAID implementations are described below.

RAID LEVEL 0 implements a striped disk array containing several disk drives. The data are broken down into small segments, and each data segment is written to a separate disk drive. This improves disk I/O performance by spreading the I/O load across many channels and drives. A striped disk does not offer data redundancy, nor does it offer better fault tolerance than a standard I/O subsystem.

RAID LEVEL 1 implements a mirrored disk array containing two disk modules. The information written to one disk is also written (mirrored) to the other disk. The procedure ensures that the data are available even if one of the disk drives fails. RAID level 1 offers better data availability, but does not offer high throughput I/O subsystem performance. It is possible to mirror a striped disk array, therefore providing better I/O performance, as well as high data availability. The primary disadvantage of using RAID level 1 is the cost of purchasing duplicate disk drives.

RAID LEVEL 2 interleaves data sectors across many disks. The controller breaks the blocks up and spreads them over a small group of disk drives. An error detection and correction process is built into RAID level 2 systems to provide some measure of fault tolerance. RAID level 2 is not found on many systems, as other RAID levels offer better performance and fault tolerance.

RAID LEVEL 3 interleaves data sectors across many disks much like RAID level 2. But RAID level 3 uses a single parity disk per group of drives to implement a fault tolerant disk system. All drive spindles are synchronized such that the read/write heads on all drives in the array are active at the same time. This synchronization also ensures that the data are written to the same sector on every drive in the array. If a disk fails, the system continues to operate by recreating the data for the failed disk from the parity disk.

RAID LEVEL 4 implements striping with the addition of a parity disk. This provides for some of the fault tolerance missing in RAID level 0. RAID level 4 is not found on many systems, as other RAID levels can provide for better throughput and fault tolerance.

RAID LEVEL 5 implements a large disk storage array. Similar to a level 0 RAID array, the disk array uses striping to improve system performance. RAID level 5 also implements a parity protection scheme, which allows the system to continue operation even if one of the disk drives in the array fails. The data normally stored on the failed drive are recreated from the parity information stored elsewhere in the array. Once the failed disk drive is replaced, the data are restored automatically by the RAID controller. RAID level 5 provides improved system availability and I/O sub-system throughput.

Problems with Disks as Backup Devices

While backing up to disk devices is much faster than backing up to other devices, it should be noted that disk devices present a potentially serious problem. One of the important considerations of backup planning is the availability of the data to the users. In the event of a natural disaster, it may be necessary to keep a copy of the corporate data off-site.

When tape devices are employed as the backup platform, it is a simple matter to keep a copy of the backups off-site. When disk drives are employed as a backup media the process of keeping a copy of the backup media off-site becomes a bit more complicated (not to mention much more expensive). In the case of a RAID disk array, the primary copy of the data is stored on one disk, and the backup copy of the data is stored on another disk. However, both disks are housed in a single box. This makes the task of moving one drive off-site much more complicated.

RAID disk arrays have recently been equipped with fiber channel interfaces. The fiber channel is a high-speed interconnect that allows devices to be located several kilometers from the computer. By linking RAID disk arrays to systems via optical fibers, it is possible to have an exact copy of the data at a great distance from the primary computing site at all times.

In applications and businesses where data accessibility is of the utmost importance, the use of RAID disk arrays and fiber-channel interconnections could solve most, if not all, backup and survivability problems.

Providing high data accessibility, automatic backups, and survivable systems via RAID and fiber channel.

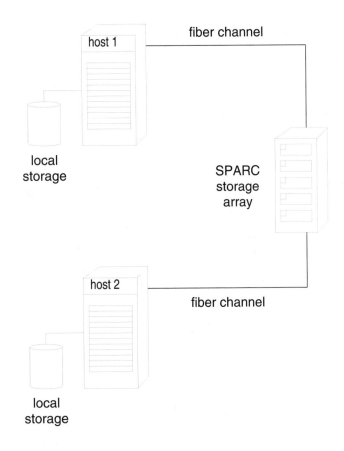

Floppy Disk Backups

Another type of system that offers a challenge for backups is the nomadic computer. These are the laptops and other portable computers that are carried by salespersons, executives, and others in the course of their travels. Because space is at a premium on these systems, they often do not contain tape drives for backup purposes. They do, however, usually contain a floppy disk drive.

The floppy disk drive is also a useful backup device on larger desktop systems. Many users may wish to make backup copies of their own files. Some users will not have access to system tape drives, optical drives, or other more conventional backup devices. But they do have access to floppy disk drives in their desktop workstations.

Floppy disks offer an ideal, low-cost method of backing up files in such instances. While the floppy diskette does not allow storage of very much information, it is an inexpensive and convenient way for users to perform their own backups. And due to the industry standard floppy diskette formats, it also allows users to carry files between systems of different architectures and operating systems.

Backup Commands

As noted in previous sections, several devices are available to use as backup media. There are also several methods available to copy data onto these backup devices. Solaris provides several commands which may be used to perform this task. The most versatile of these commands is ufsdump and its companion, ufsrestore. The **ufsdump** command is a system level backup utility that uses a special format when writing information to the tape.

The **ufsrestore** command understands the special format used by ufsdump, and provides the ability to read ufsdump tapes. The ufsdump/ufsrestore utilities are geared toward providing entire file system backups. These utilities are not particularly useful for building distribution media, nor for performing personal file backups.

In contrast, the **tar** utility is more efficient at building distribution tapes or personal archives. The tar command is not very efficient for use as a general backup utility. The tar command is standard across UNIX platforms, so data written under one UNIX variant are usually accessible under another variant. The tar package uses command line flags to control whether it is writing, or reading from, the tape.

The **cpio** command is a longstanding utility in the UNIX arena. It was available in the earliest releases of UNIX software, and therefore has a large base of users who know and trust it. When coupled with commands which can locate files by attributes such as the modify date, cpio can be used as a backup command.

The **dd** command allows the user to perform high speed copies from one device to another. One caveat must be understood: when dd is used to copy files to tape for backup purposes, the files must be restored to a device which is identical to the original device that was dumped. The dd command (in essence) creates a binary image of the data on the new device. Several filters are built into dd to allow some control of this behavior.

ufsdump Command

This **ufsdump** application was developed to allow the backup of entire systems one at a time. The ufsdump program allows the operator to specify the files to be "dumped" (or backed up to tape), and options to use during the dump. In addition, ufsdump enables scheduling of different levels of dumps on different days. The ufsdump command also allows for dumps which occupy multiple tape reels. Probably the easiest way to learn about the ufsdump command is to examine a few typical instances of how it is used.

In order to perform a full dump of a system, the operator must first know which disk devices are present on the system. To obtain this information, use the **df** command.

Using df to identify the file systems present on a system.

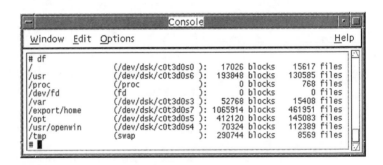

-• NOTE: *The* ufsdump *command requires that the user have read access privileges on the system disks.*

Once the operator has determined which file systems are present on the system, it is easy to determine the command line options to use for backing up the file systems to tape. The syntax of the ufsdump command follows:

```
/usr/sbin/ufsdump [options] [arguments] files_to_dump
```

❧ **NOTE:** *Consult the on-line manual page for more information about the* ufsdump *command, and the* Answerbook *for more complete descriptions of* ufsdump *and available options.*

Some of the commonly used command line options to **ufsdump** are listed below.

❐ 0-9—These numeric values specify the dump level. All files listed in the *files_to_dump* list which have been modified since the last ufsdump at a lower dump level are copied to the *dump_file* destination.

❐ b—Signifies the blocking factor to be used. The default is 20 blocks per write for tape densities of 6,250 BPI (bytes per inch) or less. The blocking factor of 64 is used for tapes with 6,250 BPI or greater density. The default blocking factor for cartridge tapes is 126.

❧ **NOTE:** *The blocking factor is specified in 512-byte blocks.*

❐ c—Signifies that the backup device is a cartridge tape drive. The option sets the density to 1,000 BPI and the blocking factor to 126.

❐ d—Signifies the density of the backup media in BPIs. The default density is 6,250 BPI except when the c option is used. When the c option is used, the density is set to 1,000 BPI per track. Typical values for selected tape devices are listed below.

• 1/2" tape 6,250 BPI

• 1/4" cartridge 1,000 BPI

- 2.3-Gb or 5.0-Gb 8 mm tape 54,000 BPI

❏ D—Signifies that the dump device is a floppy diskette.

❏ f—Signifies the *dump_file*. This option causes *ufsdump* to use *dump_file* as the file to dump to, instead of */dev/rmt/0*.

❏ s—Signifies the size of the backup volume. This option is not normally required because ufsdump can detect end-of-media. When the specified size is reached, ufsdump waits for the operator to change the volume. The size parameter is interpreted as the length in feet for tapes and cartridges, and as the number of 1,024-byte blocks for diskettes. Typical values used with the s option are listed below.

- 1/2" tape 2,300 feet

- 60-Mbyte 1/4" cartridge 425 feet

- 150-Mbyte 1/4" cartridge 700 feet

- 2.3-Gb 8 mm 6,000 feet

- 5.0-Gb 8 mm 13,000 feet

- diskette 1,422 blocks

❏ u—This option causes ufsdump to annotate which file systems were dumped, the dump level, and the date in the */etc/dumpdates* file.

❏ v—This letter signifies that ufsdump should verify the content of the backup media after each tape or diskette is written.

For illustration purposes, assume that an 8 mm tape system is connected to a system as */dev/rmt/0*. To make a full backup of the entire disk *c0t3d0*, the operator could issue the following command:

```
# ufsdump 0fu /dev/rmt/0 /dev/rdsk/c0t3d0s2
```

Using ufsdump to perform a full dump of the system disk.

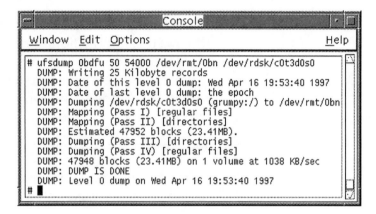

```
                        Console
 Window  Edit  Options                            Help
# ufsdump 0bdfu 50 54000 /dev/rmt/0bn /dev/rdsk/c0t3d0s0
  DUMP: Writing 25 Kilobyte records
  DUMP: Date of this level 0 dump: Wed Apr 16 19:53:40 1997
  DUMP: Date of last level 0 dump: the epoch
  DUMP: Dumping /dev/rdsk/c0t3d0s0 (grumpy:/) to /dev/rmt/0bn
  DUMP: Mapping (Pass I) [regular files]
  DUMP: Mapping (Pass II) [directories]
  DUMP: Estimated 47952 blocks (23.41MB).
  DUMP: Dumping (Pass III) [directories]
  DUMP: Dumping (Pass IV) [regular files]
  DUMP: 47948 blocks (23.41MB) on 1 volume at 1038 KB/sec
  DUMP: DUMP IS DONE
  DUMP: Level 0 dump on Wed Apr 16 19:53:40 1997
#
```

As shown in the illustration, ufsdump sends several messages to the controlling terminal. The messages provide several pieces of information about the status of the dump. Most of the dump information messages are self-explanatory (date of the dump, date of last dump, dump is done). Some of the dump information messages merely echo the command line option settings (files to dump, dump device, block size). Messages that actually provide new information about the status of the current dump include the items below.

❐ *Estimated 47952 blocks (23.41MB).* Estimated size of the dump.

❐ *XX.XX% done, finished in H:MM.* What proportion of the dump is completed, and how long the remainder will take to complete.

❑ *47948 blocks (23.41MB)* on 1 volume at 1038
KB/sec. This message tells how many blocks were
dumped, and how many tapes (volumes) were
required.

ufsrestore Command

Now that file systems have been copied onto a tape,
how is this information retrieved? Solaris provides an
application to restore data from the backup media to
the system mass storage devices. This application is
called **ufsrestore**. The syntax of the ufsrestore command follows:

```
/usr/sbin/ufsrestore options [ arguments ] [ filename ... ]
```

Some of the most useful options to the ufsrestore
commands are listed below.

❑ **i**—Puts ufsrestore in the interactive mode. Commands available in this mode are listed below.

- **add [filename]**—Adds the named file or directory
  to the list of files to extract.

- **cd directory**—Changes to *directory* on the dump
  media.

- **delete [filename]**—Deletes the current directory
  or file from the list of files to extract.

- **extract**—Extracts all files on the extraction list
  from the dump media.

- **help**—Displays a summary of available commands.

- **ls [directory]**—Lists files in *directory* (dump media) or the current directory, which is represented by a period (.).

- **pwd**—Prints the full path name of the current working directory.

- **quit**—Exits immediately.

- **verbose**—Toggles the verbose flag (the program prints a line for every action it takes).

❑ **r**—Restores the entire contents of the media into the current directory.

❑ **x**—Extracts the named files from the media.

❑ **b**—Sets the ufsrestore blocking factor.

❑ **f [dump_file]**—Tells ufsrestore to use *dump_file* instead of */dev/rmt/0* as the file to restore from.

❑ **R**—Resumes restore after volume is changed.

❑ **t**—Prints table of contents for dump file.

❑ **n**—Skips to the nth file when multiple dump files exist on the same tape.

❑ **v**—Displays the name and inode number of each file restored.

Note that the i, r, R, t, and x arguments are mutually exclusive. Only one of these arguments may be used at a time.

Example ufsrestore

This section provides a few examples of **ufsrestore** usage. In the examples, assume that the system has an 8 mm tape drive connected as */dev/rmt/0*.

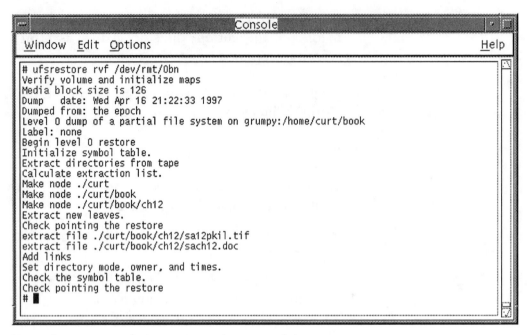

```
# ufsrestore rvf /dev/rmt/0bn
Verify volume and initialize maps
Media block size is 126
Dump    date: Wed Apr 16 21:22:33 1997
Dumped from: the epoch
Level 0 dump of a partial file system on grumpy:/home/curt/book
Label: none
Begin level 0 restore
Initialize symbol table.
Extract directories from tape
Calculate extraction list.
Make node ./curt
Make node ./curt/book
Make node ./curt/book/ch12
Extract new leaves.
Check pointing the restore
extract file ./curt/book/ch12/sa12pkil.tif
extract file ./curt/book/ch12/sach12.doc
Add links
Set directory mode, owner, and times.
Check the symbol table.
Check pointing the restore
# ▮
```

Using ufsrestore to perform a restore to the system disk.

To completely restore an entire disk (*c0t2d0s2*), change directory to the file system to be restored (e.g., */home*) and then issue a command similar to the following, substituting the appropriate disk partition information.

```
# ufsrestore rbf 50 /dev/rmt/0 /dev/rdsk/c0t2d0s2
```

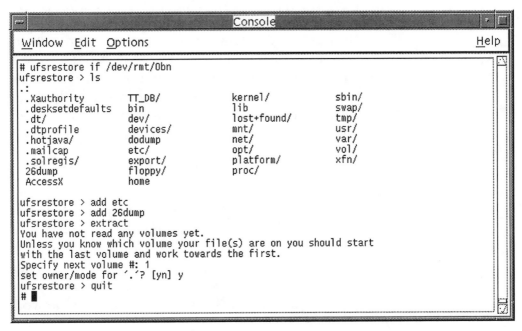

```
# ufsrestore if /dev/rmt/0bn
ufsrestore > ls
.:
 .Xauthority        TT_DB/        kernel/        sbin/
 .desksetdefaults   bin           lib            swap/
 .dt/               dev/          lost+found/    tmp/
 .dtprofile         devices/      mnt/           usr/
 .hotjava/          dodump        net/           var/
 .mailcap           etc/          opt/           vol/
 .solregis/         export/       platform/      xfn/
 26dump             floppy/       proc/
 AccessX            home

ufsrestore > add etc
ufsrestore > add 26dump
ufsrestore > extract
You have not read any volumes yet.
Unless you know which volume your file(s) are on you should start
with the last volume and work towards the first.
Specify next volume #: 1
set owner/mode for '.'? [yn] y
ufsrestore > quit
#
```

Using ufsrestore in interactive mode.

To perform a partial restore to a file system, use the interactive mode of ufsrestore. First, change directory to a temporary storage space (e.g., */tmp/restore*). Next issue a command similar to the string below, substituting the appropriate disk partition information.

```
# ufsrestore ibf 50 /dev/rmt/0
```

✓ **TIP:** *A side effect of using* ufsrestore *is that it creates a "symbol table" file in the file system where the files are restored. This file is named* restore-symtable. *It is wise to remove this file because it is usually quite large. Once the restore is complete, remove the file by issuing the following command:* rm ./restoresymtable.

Remote Backup and Restore

How can a file system dump be performed on a system without a backup device? The **ufsdump** and **ufsrestore** commands allow input and output to be sent to the standard input and output streams. This flexibility allows the ufsdump/ufsrestore output to be sent through the network to a system with a backup device.

The **-f** option for ufsdump and ufsrestore specifies the dump device.

❐ **f dump_file**—This option tells ufsdump to use *dump_file* as the file to dump to. If *dump_file* is specified as a hyphen (-), ufsdump will dump to standard output. If the name of the file is of the form *machine:device*, the dump is carried out from the specified machine over the network using the **rmt** facility.

➡ ***NOTE:*** *Because* ufsdump *is normally run by root, the name of the local machine must appear in the* /.rhosts *file of the remote machine.*

To perform a remote restore using the ufsrestore command, the operator can make use of the -f option again.

❐ **f dump_file**—This option tells ufsrestore to use dump_file as the file to restore from. If *dump_file* is specified as a dash (-), ufsrestore reads from the standard input.

➡ ***NOTE:*** *Because* ufsrestore *is also normally run by root, the name of the local machine must appear in the* /.rhosts *file of the remote machine.*

To copy a home partition to another machine with a remote ufsrestore, consider the following command, substituting the appropriate directory and device names.

```
# ufsdump 0f - /dev/rdsk/c0t2d0s2 | (rsh machine ; cd /home;ufsrestore xf -)
```

tar Command

What happens if the operator does not wish to dump and restore complete file systems? For example, what if a user simply wants to make a tape of the data associated with one project? Is there a simpler method than using ufsdump and ufsrestore? Most UNIX derivatives provide a standardized utility called **tar**. The tar command creates tape archives and provides the ability to add and extract files from these archives.

The syntax for using the tar command was covered in Chapter 4, "Fundamental System Administrator Tools." The following examples outline some typical uses of tar as a backup utility.

To create an archive of */home/project* and */project/data* on the default tar device, the user would issue the following command:

```
% tar c -C /home/project include -C /data/project
```

A user may wish to create an archive of his/her home directory on a tape mounted on drive */dev/rmt/0*. To do so, use the following command:

```
% cd ; tar cvf /dev/rmt/0
```

To read the table of contents of the tar file, issue the following command:

```
% tar tvf /dev/rmt/0
```

To extract files from the tar file, issue the command:

```
% tar xvf /dev/rmt/0
```

It is also possible to get tar to transfer files across the network. To create a tar file on a remote host, use the command:

```
% tar cvfb 20 filenames | rsh host dd of = /dev/rmt/0 obs = 20b
```

To extract files from a remote tar file into the current local directory, issue the command:

```
% rsh -n host dd if = /dev/rmt/0 bs = 20b | tar xvBfb 20 filenames
```

cpio Command

Another command that may be used to back up and restore files is **cpio**. The cpio command copies file archives "in and out." The syntax for the cpio command follows:

```
cpio -key [options] filename
```

The key option to cpio determines the actions to be performed. The flags listed below are mutually exclusive.

❏ **cpio -i** (copy in) extracts files from the standard input, which is assumed to be the product of a previous cpio -o.

❏ **cpio -o** (copy out) reads the standard input to obtain a list of path names. The cpio command then copies those files to the standard output with path name and status information. Output is padded to a 512-byte boundary by default.

❐ **cpio -p** (pass) reads the standard input to obtain a list of path names of files that are conditionally created and copied into the destination directory tree.

❧ **NOTE:** *Consult the on-line manual pages for more complete information on the* cpio *command.*

To make a cpio copy of the files in the current directory in a file called *../newfile*, issue the following command:

```
% ls | cpio -oc >> ../newfile
```

❧ **NOTE:** *The* find, echo, *or* cat *commands can also be used as substitutes for the* ls *command to produce a list of files to be included in the* ../newfile.

To extract the files from *../newfile*, issue the following command:

```
% cat newfile | cpio -icd
```

To copy the contents of the current directory to a new directory called *newdir*, issue the command:

```
% find . -depth -print | cpio -pdlmv newdir
```

❧ **NOTE:** *When using* cpio *in conjunction with* find, *use the* L *option with* cpio *and the* -follow *option with* find.

dd Command

Yet another command available to use as a backup and restore utility is the **dd** utility. The dd command copies the input file to output, applying any desired conversions in the process. When complete, dd reports the number of whole and partial input and output blocks. The syntax of the dd command follows:

```
dd [ option = value ] ....
```

Valid options to the dd command are listed below.

- ❏ **if = filename**—Use *filename* as the input file. Use stdin by default.

- ❏ **of = filename**—Use *filename* as the output file. Use stdout by default.

- ❏ **ibs = n**—Use *n* as the input block size. Use 512 by default.

- ❏ **obs = n**—Use *n* as the output block size. Use 512 by default.

- ❏ **bs = n**—Use *n* as the input and output block size This supersedes the ibn and obn arguments. If no conversion is specified, preserve input block size.

- ❏ **files = n**—Copy and concatenate *n* input files before terminating.

- ❏ **skip = n**—Skip *n* input blocks before performing copy.

- ❏ **iseek = n**—Seek *n* blocks from the beginning of the input file before copying.

- ❏ **oseek = n**—Seek *n* blocks from the beginning of the output file before copying.

- ❏ **count = n**—Copy *n* input blocks.

- ❏ **swab**—Swap every pair of bytes.

- ❏ **sync**—Pad every input block to ibs.

The dd command has a few noteworthy restrictions, listed below.

- ❏ Do not use dd to copy files between file systems with different block sizes.

❐   Using a blocked device to copy a file will result in extra nulls being added to the file, in order to pad the final block to the block boundary.

❐   When dd reads from a pipe, using the ibs = X and obs = Y operands, the output will always be blocked in chunks of size Y. When bs = Z is used, the output block size will be whatever could be read from the pipe at the time.

To copy the files on device */dev/dsk/c1t3d0s2* to tape drive */dev/rmt/2* with a 20 block record size, issue the command:

```
# dd if=/dev/rdsk/c1t3d0s2 of=/dev/rmt/2 bs=20b
```

To extract this information from the tape to a disk named */dev/dsk/c2t1d0s2*, issue the command:

```
# dd if=/dev/rmt/2 of=/dev/rdsk/c2t1d0s2 bs=20b
```

Dealing with Specific Backup Issues

Certain aspects of successful backup and restore strategies require special attention. For instance, how could the operator restore the root file system if the root disk had crashed and there was no way to boot the system? Many administrators are also concerned with how to automate backups to minimize time investment while ensuring successful backups. Next, what happens if a backup requires 2 Gb of backup media, but the backup device can write only 1 Gb to the media?

These are just a few of the file system backup problems that the administrator should be prepared to encounter. Fortunately, they are also topics that have been encountered and solved by the creators of Solaris. With an understanding of the basics of performing backups, it is possible to examine these specific topics in more detail.

Restoring the Root File System

It would appear that one of the most difficult problems faced when using ufsrestore is restoring the root file system. If the root file system is missing, it is not possible to boot the damaged system. Further, it is unlikely that there would be a file system tree to restore to.

One way to accomplish a root file system reload is by booting the system to the single-user state from the CD-ROM distribution media. Once the system is running, the operator could issue the proper commands to reload the root device from backup media.

As discussed in previous chapters, the Solaris operating system can be booted from the distribution media. The following table provides the system type to boot device mapping for a local CD-ROM drive.

System type to Solaris boot device correspondence

System type	Boot command
Sun4	b sd(0,30,1)
SparcEngine	b sd(0,6,5)
Sun4c	boot cdrom
Sun4d	boot cdrom
Sun4m	boot cdrom
Sun4u	boot cdrom

For example, the boot command for a SPARCServer 2000 system would be:

```
ok boot cdrom
```

The above command will load the SunInstall program from the CD-ROM. At this point, follow the instructions in Chapter 3, "Preparing for a Solaris Installation," to load a new copy of the operating system to the new root drive. Alternately, select the menu item which allows SunInstall to bring up a single-user shell on the system. Once the shell prompt appears, reload the root drive from the backup media instead of loading from the distribution media.

Another way to reload the root file system would be to boot the system to the single-user state as a client of another system on the network. At that point, the proper commands could be issued to reload the root device from backup media.

Automated Backups

All commands mentioned in this chapter may be adapted to provide for automated backups. Backup automation requires the creation of a shell script file containing the desired backup commands. When the script files have been created, submit a job to **cron** to execute the command files at the desired time.

For example, place the following commands in a file called */dodump*.

```
#!/bin/sh
# dump /, /usr, /var, /home, /opt, /usr/openwin
#
mt -f /dev/rmt/0b rew
/usr/lib/fs/ufs/ufsdump 0bdfu 50 54000 /dev/rmt/0bn /dev/rdsk/c0t3d0s0
/usr/lib/fs/ufs/ufsdump 0bdfu 50 54000 /dev/rmt/0bn /dev/rdsk/c0t3d0s6
/usr/lib/fs/ufs/ufsdump 0bdfu 50 54000 /dev/rmt/0bn /dev/rdsk/c0t3d0s3
/usr/lib/fs/ufs/ufsdump 0bdfu 50 54000 /dev/rmt/0bn /dev/rdsk/c0t3d0s7
/usr/lib/fs/ufs/ufsdump 0bdfu 50 54000 /dev/rmt/0bn /dev/rdsk/c0t3d0s5
```

```
/usr/lib/fs/ufs/ufsdump 0bdfu 50 54000 /dev/rmt/0bn /dev/rdsk/c0t3d0s4
mt -f /dev/rmt/0b rew
mt -f /dev/rmt/0b offl
```

Place a tape in the */dev/rmt/0* tape drive, log in as root on the system, and type **# sh -x /dodump**. The system will respond by performing a full backup of all file systems listed in the *dodump* file. While this is a simplistic method of automating file system backups, it serves to illustrate the point that it is possible to automate backups. Chapter 16, "Automating Administrative Tasks," will discuss more elegant solutions to the backup automation problem using the **cron** command.

Multi-volume Dumps

Two of the backup commands mentioned in this chapter also allow for multi-volume backups. The **ufsdump** command and the **cpio** command allow a backup to be stored over multiple media sets. The other commands (**tar** and **dd**) will allow the user to split the backups onto several sets of media, but these commands require that the user perform much of the work manually.

The cpio command watches for the "end of medium" event. When cpio detects this event, it stops and prints the following message on the terminal screen.

```
If you want to go on, type device/file name when ready.
```

To continue the dump, the operator must replace the medium and type the character special device name (e.g., */dev/rdiskette*) and press <Enter>. At this point, the operator may choose to have cpio continue the backup to another device by typing the device's name at the prompt.

☞ **WARNING:** *Simply pressing <Enter> at the prompt will cause* cpio *to exit.*

Multi-volume dumps using cpio.

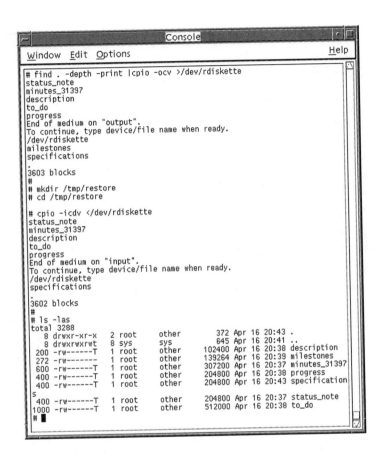

```
                                    Console
 Window  Edit  Options                                      Help
 # find . -depth -print |cpio -ocv >/dev/rdiskette
 status_note
 minutes_31397
 description
 to_do
 progress
 End of medium on "output".
 To continue, type device/file name when ready.
 /dev/rdiskette
 milestones
 specifications
 .
 3603 blocks
 #
 # mkdir /tmp/restore
 # cd /tmp/restore

 # cpio -icdv </dev/rdiskette
 status_note
 minutes_31397
 description
 to_do
 progress
 End of medium on "input".
 To continue, type device/file name when ready.
 /dev/rdiskette
 specifications
 .
 3602 blocks
 #
 # ls -las
 total 3288
    8 drwxr-xr-x  2 root   other      372 Apr 16 20:43 .
    8 drwxrwxrwt  8 sys    sys        645 Apr 16 20:41 ..
  200 -rw------T   1 root   other   102400 Apr 16 20:38 description
  272 -rw-------   1 root   other   139264 Apr 16 20:39 milestones
  600 -rw------T   1 root   other   307200 Apr 16 20:37 minutes_31397
  400 -rw------T   1 root   other   204800 Apr 16 20:38 progress
  400 -rw------T   1 root   other   204800 Apr 16 20:43 specification
 s
  400 -rw------T   1 root   other   204800 Apr 16 20:37 status_note
 1000 -rw------T   1 root   other   512000 Apr 16 20:38 to_do
 # ▮
```

Using **ufsdump** as the backup command is somewhat simpler. Like cpio, ufsdump will detect the end of medium event and stop operation. The ufsdump command will then wait for the operator to change the media before it continues. Unlike cpio, ufsdump does not need the name of the backup device to continue. The operator simply needs to confirm that the

media has been changed and that everything is ready
for ufsdump to continue operation.

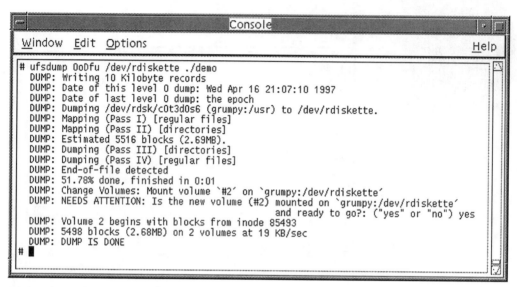

```
# ufsdump OoDfu /dev/rdiskette ./demo
  DUMP: Writing 10 Kilobyte records
  DUMP: Date of this level 0 dump: Wed Apr 16 21:07:10 1997
  DUMP: Date of last level 0 dump: the epoch
  DUMP: Dumping /dev/rdsk/c0t3d0s6 (grumpy:/usr) to /dev/rdiskette.
  DUMP: Mapping (Pass I) [regular files]
  DUMP: Mapping (Pass II) [directories]
  DUMP: Estimated 5516 blocks (2.69MB).
  DUMP: Dumping (Pass III) [directories]
  DUMP: Dumping (Pass IV) [regular files]
  DUMP: End-of-file detected
  DUMP: 51.78% done, finished in 0:01
  DUMP: Change Volumes: Mount volume `#2´ on `grumpy:/dev/rdiskette´
  DUMP: NEEDS ATTENTION: Is the new volume (#2) mounted on `grumpy:/dev/rdiskette´
                                         and ready to go?: ("yes" or "no") yes
  DUMP: Volume 2 begins with blocks from inode 85493
  DUMP: 5498 blocks (2.68MB) on 2 volumes at 19 KB/sec
  DUMP: DUMP IS DONE
#
```

Multi-volume dumps using ufsdump.

Summary

This chapter explored the commands that can be
used to make backup copies of system data, why it is
important to make such backup copies of the data,
and selected methods of avoiding data loss due to
natural or other disasters. While the authors hope that
readers never have to use any of these backup copies
to restore the operation of their systems, such restora-
tions are likely inevitable. Good backups require a lot
of time and attention, but having a reliable copy of the
data are much more acceptable than the time and
expense of rebuilding a system without such backup
copies.

Automating Administrative Tasks

As demonstrated in previous chapters, administering Solaris systems can be a time-consuming process. Simply keeping track of system files could be a full-time job, let alone managing security and user accounts. This chapter explores ways to automate certain mundane system administration chores. The goal of this chapter is to demonstrate ways that the administrator can "free up" time for more urgent tasks, such as helping users.

Updating System Files

One of the most frequent housekeeping jobs an administrator faces is keeping system files current. For example, an administrator at a small site might manage 25 to 30 machines. Each of the machines has a private copy of the **/usr/lib/sendmail** software installed on the local disk. How could the administrator efficiently update the sendmail software on 30 machines?

rdist Command

The **/bin/rdist** command is a remote file distribution utility. The rdist utility allows the administrator to set up a central repository of files which must be updated on multiple machines within the corporate network. To update all machines on the network, the administrator makes changes to the files in the central repository, and then uses the rdist utility to distribute these files to the other hosts on the network.

The rdist utility requires setup before it is used the first time. When the setup is complete, the task of updating files on machines becomes a trivial task.

rdist Setup

The setup for rdist consists of editing two files. First is the */.rhosts* file, which grants permission for users to access the system through the network, and second is the *Distfile*, an ASCII file which lists the commands and files that rdist will use to update the other machines on the network.

/.rhosts File

The */.rhosts* file on each of the machines on the network should contain an entry allowing the distribution server access. For example, if the distribution server were *mercury.astro.com*, and the rdist command required root access, the **rdist** client machines should have the following line in their respective */.rhosts* files.

```
mercury.astro.com root
```

> ⇥ **NOTE:** *The* rcp, rsh, *and* rlogin *commands make use of the* /.rhosts *file to determine access. Access is granted if the machine and user being contacted have a* /.rhosts *file, and if the file contains a line listing the machine from which the connection derives. Exercise great care upon allowing access to the root account via entries in the* /.rhosts *file. See Chapter 18, "Network Security," for more information on root account security.*

Distfile

The rdist command searches for a file named *Distfile* in the current directory. If the file exists, **rdist** executes the commands in the file to update the other hosts on the network. The *Distfile* allows the administrator to list the hosts and respective files to be updated. The syntax of the *Distfile* entries is very similar to the UNIX regular expression syntax. The user may define groups of hosts which require special handling, or macro commands which define a series of commands to be executed on a host.

> ✓ **TIP:** *It is also possible to use* rdist *in a limited mode without all of the setup. To distribute a file from the current machine to another, invoke rdist with the following format:* rdist -c filename host. *For example, the following command would distribute the* /etc/inet/hosts *and* /etc/mail/aliases *files from the current host to host* mercury: rdist -c / etc/inet/hosts /etc/mail/aliases mercury.

Instead of attempting to list the details of how to use the rdist command, the following section will describe typical usage.

rdist Example

Assume that the SpaceCadets Corporation system administrator is attempting to stay current with file maintenance on the corporate hosts. One of the tools which the administrator decides to use is the **/bin/ rdist** command. The following example details the steps that the administrator must follow in order to get rdist running on the corporate Solaris systems.

Challenge

Corporate management has requested that the administrator automate the distribution of several files on the corporate machines. However, not all machines will be updated with the same information. For example, the machines in the engineering department will not allow employees from the marketing department to log in. Conversely, the machines in the marketing department will not allow employees of the engineering department to log in.

Management has also decided that a single e-mail server will suffice for all machines within the corporation. Because the e-mail server is to handle sending and receiving e-mail for all corporate computer systems, the **sendmail** configuration on the server should allow the server to accept mail bound for any system within the corporate domain. Client machines should be configured to send their mail to the server for delivery, and to load the users' e-mail from the mail spool area on the mail server system.

SpaceCadets
Corporation network
configuration.

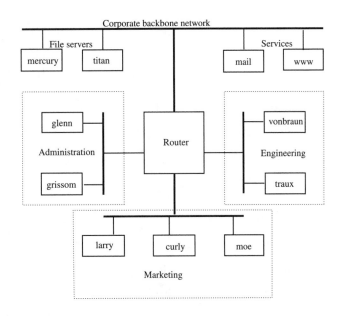

Identifying Files to rdist

One of the first implementation issues to be dealt with is identifying which files to **rdist**. The administrator reviews the list of tasks set forth by corporate management, and develops the following file list.

❏ /.rhosts—Specifies which hosts have privileged network access to the system. All machines must have this file.

❏ /etc/inet/hosts—Contains the mapping of IP addresses to host names for the hosts within the corporation. All hosts must have this file.

❏ /etc/passwd – System password file. The administrator decides that two password files are necessary for engineering and marketing, and that others may be required in the future.

❏ /etc/mail/sendmail.cf—Configuration file for the *sendmail* program. Separate *sendmail.cf* files include one for the e-mail server and another for all other hosts on the network.

❑ /etc/mail/sendmail.cw—Informs the e-mail server of the hosts for which to accept e-mail. The file resides only on the e-mail server system, and lists all corporate host names.

❑ /usr/lib/sendmail—The **sendmail** program. All hosts on the network will have the same sendmail binary.

❑ /etc/mail/aliases—Lists the e-mail aliases used by the corporation. This file resides on all corporate hosts.

❑ /etc/groups—System group file. The administrator decides to make the group files on all machines identical in order to avoid problems in the future.

❑ /etc/shells—Lists the shell programs allowed on the systems. In order to keep users from loading unsupported shells, the administrator decides that this file should be centrally maintained. All hosts will have this file.

Creating a Central Repository

With a list of files in hand, the administrator starts planning the implementation of the **rdist** server. The rdist server contains a central repository that holds the master copy of the files to be distributed to other corporate hosts. The administrator decides to name the directory which contains this central repository of files /*common*. Subdirectories in the /*common* directory with familiar names include the following:

❑ /common/etc—Contains copies of files that will end up in the /*etc* directory of the rdist client machines.

❑ /common/usr—Contains copies of files that will end up in the /*usr* directory of the rdist client machines.

The above scheme allows the administrator to tailor the central repository in an easy to decipher hierarchical storage area. The administrator copies the appropriate files into the central repository and then edits each file to tailor it to the appropriate task.

Identifying Machine Groupings

The next challenge the administrator has to meet is grouping machines into easy to maintain sets. To achieve this objective, the administrator creates several groups of machines.

- ❒ SERVERS—List of file servers to be managed with **rdist**. This group includes the systems that provide the file systems containing applications and user file space for the corporation.

- ❒ SERVICES—List of service machines to be maintained by rdist. This group includes the e-mail server and a Web server.

- ❒ CLIENTS—List of all non-file server machines to be managed under rdist. These machines are desktop workstations. They obtain applications and user file storage space from the corporate file servers, and services (e-mail and Web) from the corporate services machines.

- ❒ ENGR—Subset of the CLIENTS group. These machines are located in the engineering department.

- ❒ MARKET—Another subset of the CLIENTS group. These machines are located in the marketing department.

Implementation

When planning is complete, the administrator creates a *Distfile* within the */common* directory. Appearing below is an example of what the administrator's *Dist-file* might contain.

```
# SpaceCadets Corporation rdist Distfile
# First, define the hosts to be managed.
ALL=(${SERVERS} ${CLIENTS} ${SERVICES} )
SERVERS=(mercury.astro.com titan.astro.com)
CLIENTS=( ${ADMIN} ${ENGR} ${MARKET} )
ADMIN=( glenn.astro.com grissom.astro.com)
ENGR=( vonbraun.astro.com truax.astro.com )
MARKET=( larry.astro.com curly.astro.com moe.astro.com)
SERVICES=( mail.astro.com www.astro.com )
# Next, define the files to be installed on ALL hosts.
hosts:
/common/etc/hosts -> ${ALL}
                install /etc/inet/hosts;
group:
/common/etc/group -> ${ALL}
                install /etc/group;
shells:
/common/etc/shells -> ${ALL}
                install /etc/shells;
sendmail:
/common/usr/lib/sendmail -> ${ALL}
                install /usr/lib/sendmail;
          special "/usr/local/bin/killj sendmail";
                special "/usr/lib/sendmail -bd -q1h";
aliases:
/common/etc/aliases -> ${ALL}
                install /etc/aliases;
                special "/usr/ucb/newaliases";
# Define the files to be installed on the ENGineeRing hosts.
prtconf:
/tmp/trash -> ${ENGR}
                install /tmp/trash;
```

```
                special "rm /tmp/trash";
                special "/etc/prtconf";
passwd.engr:
/common/etc/passwd.engr -> ( ${ENGR} )
                install /etc/passwd.engr;
                special "mv /etc/passwd.engr /etc/passwd"
                special "rm /etc/shadow";
                special "/bin/pwconv";
rhosts.engr:
/common/.rhosts.engr -> ${ENGR}
                install /.rhosts.engr;
                special "mv /.rhosts.engr /.rhosts";
# Define the files to be maintained on the MARKETing hosts.
passwd.market:
/common/etc/passwd.market -> ( ${MARKET} )
                install /etc/passwd.market;
                special "mv /etc/passwd.market /etc/passwd"
                special "rm /etc/shadow";
                special "/bin/pwconv";
rhosts.market:
/common/.rhosts.market -> ${MARKET}
                install /.rhosts.market;
                special "mv /.rhosts.market /.rhosts";
# Define the files to be maintained on other subsets of hosts.
sendmail.cf.clients:
/common/etc/sendmail.cf.clients -> ${CLIENTS}
                install /etc/mail/sendmail.cf.clients;
                special "mv /etc/mail/sendmail.cf.clients /etc/mail/sendmail.cf";
                special "ln -s /etc/mail/sendmail.cf /etc/sendmail.cf";
sendmail.cf.server:
/common/etc/sendmail.cf.server -> mail.astro.com
                install /etc/mail/sendmail.cf.server;
                install /etc/mail/sendmail.cw;
                special "mv /etc/mail/sendmail.cf.server /etc/mail/sendmail.cf";
                special "ln -s /etc/mail/sendmail.cf /etc/sendmail.cf";
```

Running rdist

When the setup is complete, it is time to run the **rdist** command. Using the debugging mode of rdist is recommended when a new *Distfile* is used for the first time. To do so, you invoke rdist as follows:

```
# rdist -n -f Distfile
```

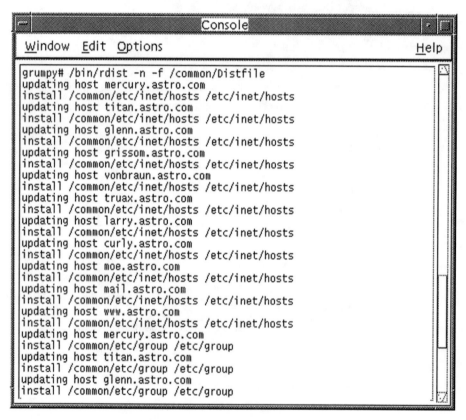

```
┌──┬────────────────────────Console──────────────────────┬─┬─┐
│  └────────────────────────────────────────────────────┘ └─┘│
│  Window   Edit   Options                            Help    │
│ ┌────────────────────────────────────────────────────────┐▲│
│ │grumpy# /bin/rdist -n -f /common/Distfile              │▨│
│ │updating host mercury.astro.com                        │ │
│ │install /common/etc/inet/hosts /etc/inet/hosts         │ │
│ │updating host titan.astro.com                          │ │
│ │install /common/etc/inet/hosts /etc/inet/hosts         │ │
│ │updating host glenn.astro.com                          │ │
│ │install /common/etc/inet/hosts /etc/inet/hosts         │ │
│ │updating host grissom.astro.com                        │ │
│ │install /common/etc/inet/hosts /etc/inet/hosts         │ │
│ │updating host vonbraun.astro.com                       │ │
│ │install /common/etc/inet/hosts /etc/inet/hosts         │ │
│ │updating host truax.astro.com                          │ │
│ │install /common/etc/inet/hosts /etc/inet/hosts         │ │
│ │updating host larry.astro.com                          │ │
│ │install /common/etc/inet/hosts /etc/inet/hosts         │ │
│ │updating host curly.astro.com                          │ │
│ │install /common/etc/inet/hosts /etc/inet/hosts         │ │
│ │updating host moe.astro.com                            │ │
│ │install /common/etc/inet/hosts /etc/inet/hosts         │ │
│ │updating host mail.astro.com                           │ │
│ │install /common/etc/inet/hosts /etc/inet/hosts         │▨│
│ │updating host www.astro.com                            │ │
│ │install /common/etc/inet/hosts /etc/inet/hosts         │ │
│ │updating host mercury.astro.com                        │ │
│ │install /common/etc/group /etc/group                   │ │
│ │updating host titan.astro.com                          │ │
│ │install /common/etc/group /etc/group                   │ │
│ │updating host glenn.astro.com                          │ │
│ │install /common/etc/group /etc/group                   │▨│
│ └────────────────────────────────────────────────────────┘▼│
└─────────────────────────────────────────────────────────────┘
```

Typical output of /bin/rdist -n to test rdist rules.

The administrator should study the resulting output to determine if rdist is will perform the desired steps. Several changes to the *Distfile* may be necessary before the process works perfectly. Once the *Distfile* is correct, run rdist manually with the following command to ensure that everything works the first time.

```
# rdist -f Distfile
```

Subsequent sections of this chapter will discuss ways to make the system automatically run the rdist (and other) commands at specified times using the **cron** facility.

✓ **TIP:** *The* rdist *command may be used to install local packages and configuration files on new machines. Once the administrator installs the operating system and has a new host running on the network,* rdist *may be used to automatically install all of the "localisms" on the new system.*

•→ **NOTE:** *The* rdist *command may also be used to run commands on a remote machine. However, because* rdist *was not intended to perform remote execution of commands, a few extra steps are required. A simple way of accomplishing this is to distribute an empty file to the target machine, then use the "special" directive to cause the machine to remove the empty file, and then execute the desired commands. For instance, in the example* Distfile, *the ENGR machines are commanded to run the* prtconf *command every time* rdist *runs. The output of the command may be e-mailed to the administrator along with the rest of the* rdist *report. This allows the administrator to keep track of the hardware/software configuration of the systems within the engineering department.*

Finding and Removing Unused Files

An old system administration adage states that "Disk usage will expand to fill all available disk space." What steps are necessary to ensure that enough disk space is available for users? One way of policing free disk space might be to locate and remove all files that had not been accessed (used) during a predetermined period of time. Although a valid management trick, this becomes a very time-consuming process if it is done manually. Fortunately, UNIX provides a utility called **find** which permits automation of the process.

find Command

The **find** utility allows the user to search for specific information on system disks. The find utility may be used to search for files, directories, or strings of characters on the disks. When an item is found, the find command provides many additional options.

You could have find execute other commands such as using **rm** to remove files, or printing the location of the item on the screen. You could cascade several functions together and use find to perform more complex operations. The syntax for using the find command follows:

```
find pathlist expression
```

The **pathlist** argument is a list of files to be searched by find. The file list may consist of file names, directory names, or file system names. If the root (/) directory is specified as the pathlist, find will traverse all directories on the system.

The **expression** argument includes the string of characters that find will search for and what to do when items are found. A few commonly used expressions for the find utility appear in the following table.

Commonly used find command expressions

-name	Search for files called "name."
-atime N	Search for files accessed within the past N days.
-ls	Print current path name and related information.
-mtime N	Search for files modified N days ago.
-exec command	Execute "command" if file is found.
-print	If file is found, print path name to file.
-type C	Search for files of type "C."
Valid values for C include:	
b	Search for block special files.
c	Search for character special files.
d	Search for directories.
l	Search for symbolic links.
p	Search for FIFO files.
s	Search for sockets.
f	Search for plain files.
-fstype type	Search for file systems of "type."

The following sections provide a few examples of how the find utility might be used in daily system administration activities.

Eliminating Old Files

In a previous chapter the **find** command was used to locate and remove all *a.out* and *.o* files in a user's home directory that were more than seven days old. The find statement to perform this task follows:

```
% find $HOME \(-name a.out -o -name '*.o'\) -atime +7 -exec rm {} \;
```

> ↔ **NOTE:** *The use of parentheses, quotation marks, backslashes, and curly braces is explained in detail later in this chapter. These symbols are required to cause the shell to interpret the command in a particular manner.*

In the previous example, the parentheses are used to inform find to search for multiple file names. The curly braces following the **rm** command tell find to execute the rm command, and remove any files which match the search specifications. The backslashes are used so that the shell does not interpret the parentheses and semicolon characters, but rather passes them to the find command for interpretation.

How could you modify this command to find the same files in all system directories? Simply change the pathlist argument from *$HOME* to root (/) as follows:

```
# find / \(-name a.out -o -name '*.o'\) -atime +7 -exec rm {} \;
```

> ↔ **NOTE:** *The above example should be run as root because root is the only user that can access all files on the system. Note the root prompt in the example.*

Other Handy Uses for find

Because **find** can locate so many different things on disks, there are other uses for this tool. Experienced system administrators use find on a daily basis. The following sections present a sampling of tasks which can be automated with the find command.

Locating Files

Assume that you wish to locate all instances of a particular file so that information in the file(s) can be updated. With the find utility, this process becomes a simple operation, as shown in the next illustration.

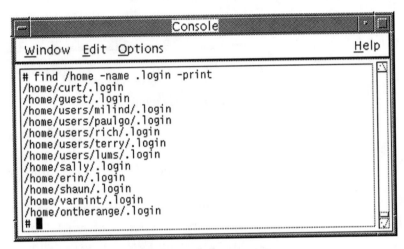

```
# find /home -name .login -print
/home/curt/.login
/home/guest/.login
/home/users/milind/.login
/home/users/paulgo/.login
/home/users/rich/.login
/home/users/terry/.login
/home/users/lums/.login
/home/sally/.login
/home/erin/.login
/home/shaun/.login
/home/varmint/.login
/home/ontherange/.login
#
```

Using find to locate all instances files named .login.

Locating Text Strings

The find command can also be used to locate a file that contains a particular string of characters. Why is this useful? One popular method of breaking into systems is to replace system binaries with new versions that contain backdoor entry points for use by people who wish to break in. Quite often these programs contain a unique string of characters (such as a compiled-in password) that make them easy to search for. The find command makes it easy to search the disk for a specific string of characters.

```
# find / -type f -exec grep -l "floobydust" {} \;
/opt/tmp/.xxsh
/opt/tmp/login
/usr/bin/ .login
/usr/tmp/.xxsh
/var/mail/.xxsh
/tmp/()
/var/tmp/.xxsh
#
```

Using find to locate all instances of a particular string.

Checking File Permissions

Another common use of the find command is to locate files with particular access permissions. For instance, files that are writeable by users other than the file owner could be considered security risks. Changing the access permission of these files so that only the owner has write permission is recommended. Rather than spending hours looking for such files manually, find can be used to automate this operation.

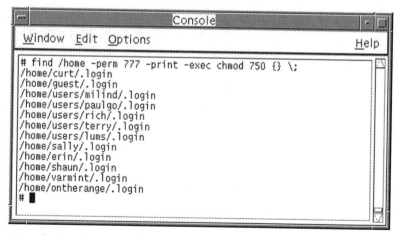

```
# find /home -perm 777 -print -exec chmod 750 {} \;
/home/curt/.login
/home/guest/.login
/home/users/milind/.login
/home/users/paulgo/.login
/home/users/rich/.login
/home/users/terry/.login
/home/users/lums/.login
/home/sally/.login
/home/erin/.login
/home/shaun/.login
/home/varmint/.login
/home/ontherange/.login
#
```

Using find to locate and change permissions on files with mode 777.

In certain situations it may be desirable to change the ownership of files owned by a particular user. For instance, you may wish to change a user's log-in name or user-id for some reason. In order for the user to be able to access his/her files after the *uid/user-name* change, the files would have to be "chowned" and "chmoded" to the new *uid/username*. The find command to locate all files owned by the user and to change ownership to the new user name is shown in the next illustration.

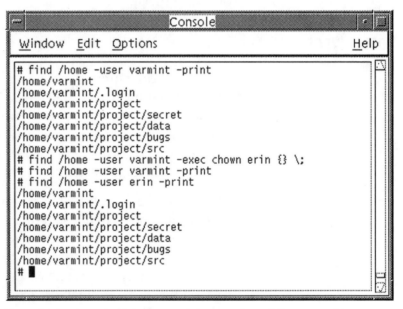

```
# find /home -user varmint -print
/home/varmint
/home/varmint/.login
/home/varmint/project
/home/varmint/project/secret
/home/varmint/project/data
/home/varmint/project/bugs
/home/varmint/project/src
# find /home -user varmint -exec chown erin {} \;
# find /home -user varmint -print
# find /home -user erin -print
/home/varmint
/home/varmint/.login
/home/varmint/project
/home/varmint/project/secret
/home/varmint/project/data
/home/varmint/project/bugs
/home/varmint/project/src
#
```

Using find to locate and change the ownership of files owned by a particular user.

You may wish to build script files which automatically run the find commands necessary to perform normal housekeeping, such as cleaning temporary file systems. The following section examines one way to automate the execution of these scripts.

Automating Commands with cron

The **cron** command is a chronometer (clock) based function which executes commands at specified dates and times. One of the more common uses of cron is to run batch jobs at off-peak hours. Another common use of cron is to run housecleaning and administrative functions at regular intervals. Because cron runs under the control of a system daemon program, understanding the types of processes available under Solaris is crucial.

Process Types

The Solaris operating system is a multi-user/multi-tasking operating system. This means that many users can be active on the system simultaneously, and many jobs (processes) can be active simultaneously. While only one job is active (per processor) at any given time, many jobs can be in the run queue.

The run queue is a list of jobs awaiting respective slices of run time from the scheduler. Because the slice of run time allotted to each process is so short, many jobs appear to be active simultaneously. In many cases, these jobs are interactive processes started by users sitting at a terminal or using a window on a workstation. These processes are generically referred to as *foreground processes.*

Foreground Processes

When users sit at the terminal typing commands to the shell, they are typically executing foreground processes. Foreground processes maintain control of the terminal (window) until the execution is complete, and are therefore referred to as "interactive" processes. Foreground processes typically read input from the terminal and write output to the terminal. Some typical foreground processes include system commands such as **ls**, **cat**, **vi**, and **sh**.

Background Processes

Some of the processes running on the system are not interactive processes. A process that is not run interactively is referred to as a background process. When a process is run in the background, it is disassociated from the terminal. This means that the terminal is free for other uses while the background process executes. Because the background process is not associated with a terminal, the user must arrange for the background process to get input from a file or other source. Likewise, the user must arrange for the background process output to be sent to a file (or another process).

cron Processes

Processes executed by **cron** are background jobs. Many cron jobs are a special flavor of background processes known as *batch jobs*. A batch job reads its input from one place, processes the input, and then writes its output (if it produces any output) to another place.

In order to run a cron job, the user must create a **crontab** entry. A crontab entry tells the cron program which command to run, and more importantly, when to run it. The crontab (cron tables) command allows users to create and modify crontab entries. Once the user has created a crontab entry, the crontab program stores that entry in the */var/spool/cron/crontabs* directory. The cron software scans this directory to identify the commands to run at particular times.

By default, the crontab command is not limited to root user usage. Other users can create their own cron jobs. System administrators using the root user account can control access to cron by editing the */etc/cron.d/cron.allow* and */etc/cron.d/cron.deny* files. As the file names imply, *cron.allow* contains a list of all users permitted cron access while *cron.deny* contains a list of all users denied cron access. By default, both files are empty.

➝ **NOTE:** *Many system administrators limit* cron *access to system staff members. Otherwise, system users could use* cron *to hide (i.e., move) illegal files prior to security sweeps designed to detect such files.*

crontab Entries

While the **crontab** program is used to create cron table entries, it is the cron daemon (a background process) which examines the entries to determine when cron jobs should be executed. In order for cron to understand the crontab entries, they must conform to a specific template. Every entry in a crontab file consists of the fields listed below.

❏ Minute of execution (0-59, asterisk, or list).

❏ Hour of execution (0-23, asterisk, or list).

❏ Day of the month (1-31, asterisk, or list).

❏ Month of the year (1-12, asterisk, or list).

❏ Day of the week (0-6 [Sunday = 0], asterisk, or list).

❏ Command string to be executed.

As noted above, the first five fields may contain a numeric argument. For example, to execute a command on the first day of the month, place a one (1) in the third field of the crontab file. This command would be executed on the first day of the month at the time set by the hour, minute, month, and day fields.

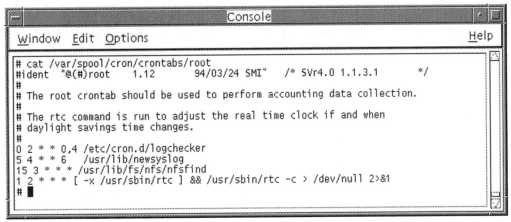

```
# cat /var/spool/cron/crontabs/root
#ident   "@(#)root    1.12      94/03/24 SMI"    /* SVr4.0 1.1.3.1      */
#
# The root crontab should be used to perform accounting data collection.
#
# The rtc command is run to adjust the real time clock if and when
# daylight savings time changes.
#
0 2 * * 0,4 /etc/cron.d/logchecker
5 4 * * 6   /usr/lib/newsyslog
15 3 * * * /usr/lib/fs/nfs/nfsfind
1 2 * * * [ -x /usr/sbin/rtc ] && /usr/sbin/rtc -c > /dev/null 2>&1
# 
```

Example root crontab file.

Alternately, the first five fields may contain a list of numeric values separated by commas. This allows scheduling commands to be run multiple times a day with a single crontab entry. For example, to execute a command on the hour, and 15, 30, and 45 minutes after the hour, place the 0,15,30,45 string in the first field of the crontab file. The command would be executed four times an hour as specified by the other four fields.

The asterisk acts as a wildcard entry in the crontab file. When cron finds an asterisk in a field, it automatically substitutes every valid value for that field. For example, a crontab entry with an asterisk in the first field would instruct cron to execute the command once a minute as long as the conditions set in the other fields were true.

crontab Options

The crontab application allows the options listed below.

☐ **crontab -e** edits a copy of the current user's crontab file, or creates an empty file to edit if crontab does not exist.

☐ **crontab -l** lists the user's current crontab file.

☐ **crontab -r** removes the user's crontab file from the crontab directory.

Example crontab Entries

The use of the **find** command to locate and remove unused files from the system was covered earlier in this chapter. You can automate system housecleaning tasks by cascading the find and cron utilities. The following crontab entries execute find commands to clear out specific files. These files may be system generated temporary files, or files that are otherwise easily recreated.

⟿ ***NOTE:*** *Exercise caution when creating such* crontab *entries to avoid accidentally removing critical files. Manual development and testing of* crontab *entries is recommended prior to having the system perform these tasks automatically.*

The crontab entry appearing below will run at 02:15AM every day of every month. The find command will search every file system for *.o* files (object files created by the compilers). If such files are found and are over seven days old, find will remove them. Note that this command will not remove such files from NFS mounted file systems due to the **-fstype nfs -prune** directives.

```
15 02 * * * find / -name *.o -mtime +7 -exec rm {} \; -o -fstype nfs -prune >>/dev/
null 2>>&1
```

The crontab entry below will run at 03:05AM every day of every month. The **/bin/rdist** command will update the files on every machine listed in the /common/Distfile control file. Because the output of this command will be mailed to the root account, the administrator can check file distribution results.

```
05 03 * * *     /bin/rdist -f /common/Distfile
```

Finally, the next crontab entry will run at 03:10AM every day of every month. The find command will search through the /tmp file system for directories over one (1) day old. If such directories are found, and they are not the /tmp directory or the lost+found directory, they will be removed.

```
10 03 * * *     cd /tmp ; find . ! -name . ! -name lost+found -type d -mtime +1 -
exec rmdir {} \; /dev/null 2&1
```

✓ **TIP:** *The* cron *command may be used to automate many functions. A few simple examples include automated distribution of new system software, file system backups, collection of network usage statistics, and creating system accounting summaries. If you can develop a shell script file to perform these tasks, it is also possible to develop a* crontab *entry which will allow the system to automatically perform the task.*

System Automation with Shell Scripts

The UNIX shell is an interactive programming language, as well as a command interpreter. The shell executes commands received directly from the user sitting at the terminal. Alternately, the shell can execute commands received from a file. A file containing shell programming commands is called a shell file, or *shell script.*

Many of the operations performed by a system administrator are accomplished by typing commands at the terminal. Due to file access permissions, these commands must often be issued by root. Under Solaris 2, the standard shell for root is **/sbin/sh**, a statically linked version of the Bourne shell. On the other hand, the program in **/usr/bin/sh** is a dynamically linked version of the Bourne shell. What is the difference between these two versions of the **sh** program, and why are both needed?

⤙ *NOTE: While shell scripts are powerful system administration tools, they do not fulfill every need.* Perl, awk, *and* Tcl/Tk, *among other utilities, are often found in administrator's toolboxes. For more information on these utilities, consult the on-line manual pages or the Answerbook. There are also several good reference books available on these utilities.*

Review of Dynamic versus Static Linking

A statically linked program is one which links (or includes) all library functions at compile time. If changes are made to a library routine due to compiler or other system software upgrades, the program must be recompiled to take advantage of those changes. A statically linked program will usually be quite large because it includes all library functions.

A dynamically linked program loads the library functions at run time, which allows the program to include the latest version of all library functions upon invocation. No recompilation is required to enjoy the advantages of updated library routines. Dynamically linked programs are usually smaller than statically linked counterparts because the library functions are not included in the on-disk binary image.

> ☞ **WARNING:** *Many of the programs in the* /sbin *directory are statically linked. While it may be tempting to replace these programs with dynamically linked versions to save disk space, this practice is not recommended. The programs in* /sbin *are statically linked for a very good reason. In the event that the dynamic libraries (typically stored in the* /usr *file system) are not available due to a partial system failure, the programs in* /sbin *(such as the root log-in shell) will still function. Consequently, the system administrator can still log in and attempt to recover from the failure.*

The Shell Game

The Bourne shell implements a language commonly referred to as shell programming language, or shell programming. However, the Bourne shell is just one of many shells available for Solaris. Another popular shell program is **/usr/bin/csh** or the C shell. The C shell offers user features that are not available in the Bourne shell, such as advanced job control.

Other available shells include the Bourne Again shell (**bash**), Korn shell (**ksh**), **tcsh** (a variant of **csh**), and the **zsh** shell. Most of these variations share a common command structure with the Bourne and C shells.

Selecting the Right Shell for the Job

Due to the fact that the shell is the primary user interface to the system, its selection is typically intensely personal. Users select the shell that they are most comfortable with. Every shell has strengths and weaknesses which the administrator can exploit. One of the secrets of system administration is knowing which shell will make a particular task easier to perform. In some cases, a system administrator may choose one shell over another based on familiarity alone.

Although the Bourne shell and C shell differ, they share a few concepts. Both shells allow the user to declare local variables; provide and understand variables which customize the execution environment for users; and interact with the operating system to provide error handling and other basic features.

In order to create shell scripts to aid in system administration, you must understand the language used by each shell. Unfortunately, this topic is beyond the scope of a single chapter in a systems administration book. Entire books have been dedicated to the description of the Bourne, C, and Korn shells. The following section examines a few of the most frequently used capabilities of the Bourne shell program. The reader is encouraged to seek out more comprehensive coverage of these programs in other publications.

Basic Shell Features

Most shells offer basically the same capabilities to the user, and can successfully interpret standard Solaris commands. However, the shell programming language recognized by the shell may differ from one program to another. The programming language of most shells conforms to either the Bourne or C shell language. For purposes of illustration, the examples in the following sections use the Bourne shell syntax, because the Bourne shell is the default root shell.

Environment and Local Variables

One of the similarities between the Bourne and C shell programs is the use of variables. Shell programs support local and environment categories of variables.

Local variables consist of variable names chosen by the user, and the variables contain data defined by the user. System binaries have no access to the variables. These variables are much like the local variables that a user would employ when writing a program in a high-level programming language.

Environment variables can be viewed as a combination of reserved words and global variables in a high-level programming language. The names of several of these variables cannot be altered by the user. The user can set many of these variables to a value that can be checked by system binaries. Users can modify some of these variables in order to customize the behavior of the shell to suit their preferences and needs. Some environment variables are reserved, and therefore cannot be modified by users.

Users can create their own environment variables using the **setenv** command. This command allows software developers to create environment variables used by the software packages they distribute. The discussion of environment variables in the following sections is limited to well-known system environment variables used in Solaris.

Environment Variables

Environment variables allow users to customize the operating environment. Some of the more familiar

Bourne shell environment variables are described in the next table.

Popular Bourne shell environment variables

Variable name	Set by	Description
CDPATH	Not set by default	Directory search path
COLUMNS	Set by shell	Width of window
EDITOR	Not set by default	Command editor
ENV	Set by user	Setup script invoked for each command
ERRNO	Set by shell	Last error code from a system call
HOME	Set by log-in	Default argument for cd
LOGNAME	Set by log-in	User's log-in name
LPDEST	Not set by default	Sets the default printer destination
MAIL	Set by log-in	File that contains user's incoming mail
OLDPWD	Set by shell	Previous working directory
OPTARG	Set by shell	Current argument after getopts
OPTIND	Set by shell	Number of arguments remaining after getopts
PATH	Set by log-in	Search path for commands
PPID	Set by shell	Process ID of parent process
PS1	Set by shell	Primary shell prompt
PS2	Set by shell	Secondary shell prompt
PS3	Set by shell	Select shell prompt
PS4	Set by shell	Debugging shell prompt
PWD	Set by shell	Current working directory
RANDOM	Set by shell	A random number between 0 and 32,767
SECONDS	Set by shell	Number of seconds since login
SHELL	Set by login	Path to desired shell
TERM	Not set by default	Sets the terminal type
TMOUT	Not set by default	Idle time-out delay
VISUAL	Not set by default	Command exit mode for editors

The environment variables in the previous table are available for every log-in session under the Bourne shell. For instance, if the user opens two window sessions on a single system, the same list of environment variables is available in each window. If the user changes the value of an environment variable in one window, however, the new value affects only the shell in that window.

➤ **NOTE:** *Environment variable settings propagate from the shell where they are set (the parent shell) to any commands or shells invoked from the parent shell. Local variables do not propagate to commands or shells invoked by the parent shell.*

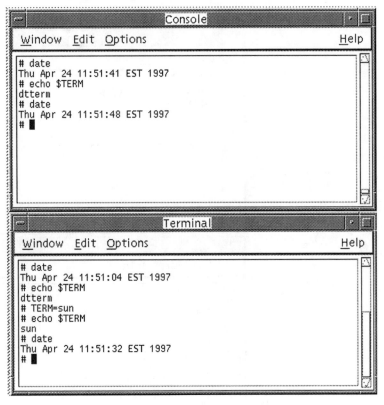

```
┌─────────────────── Console ───────────────────┐
│ Window  Edit  Options                    Help  │
│ # date                                         │
│ Thu Apr 24 11:51:41 EST 1997                   │
│ # echo $TERM                                   │
│ dtterm                                         │
│ # date                                         │
│ Thu Apr 24 11:51:48 EST 1997                   │
│ # █                                            │
└────────────────────────────────────────────────┘
```

```
┌─────────────────── Terminal ──────────────────┐
│ Window  Edit  Options                    Help  │
│ # date                                         │
│ Thu Apr 24 11:51:04 EST 1997                   │
│ # echo $TERM                                   │
│ dtterm                                         │
│ # TERM=sun                                     │
│ # echo $TERM                                   │
│ sun                                            │
│ # date                                         │
│ Thu Apr 24 11:51:32 EST 1997                   │
│ # █                                            │
└────────────────────────────────────────────────┘
```

Setting an environment variable in one window does not set it for all windows.

Local Variables

A local variable is defined for the user's personal use. Users may find local variables such as loop counters and character strings to be very handy. For example, a shell variable that contains the operating system version could be helpful. The following code segment sets a local version variable:

```
% set version = `uname -s -r`
```

To determine the value of the version variable, type *echo $version*. Similar to the environment variable, a local variable is available only under the log-in shell in which it was created.

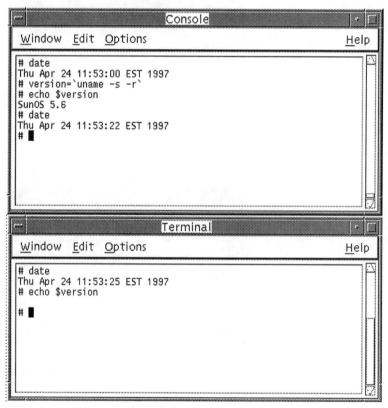

A local variable defined in one log-in session is not available in another.

To set a local variable for all log-in sessions, place the shell commands that perform this operation in the *.login* file in the *$HOME* directory. This file is executed for each log-in session initiated.

✓ **TIP:** *Local variables may also be used in conjunction with other shell features.*

Setting Variables from .login

In order to set an environment variable for all login sessions, users must place the shell commands that perform this operation in the *.login* file in their respective home directories. This file is executed for each log-in session initiated. The following paragraphs present an example of using code in the *.login* file to set an environment variable.

A system administrator at a large site may need to log in on several systems during the workday. The standard Bourne shell dollar sign ($) prompt complicates identifying the system the shell is operating on. It would be useful for the shell's command prompt to provide the name of the system that the user is logged in on.

Setting an environment variable makes possible customization of the log-in prompt such that it informs the user of the system she is logged in on. Under the Bourne shell, this variable is called *PS1 (Prompt String 1)*. To see what the prompt is currently set to, issue the command **echo $PS1**.

➥ **NOTE:** *To cause the shell to print the value of a variable, preface the variable name with a dollar sign. Otherwise, the shell will echo the variable name string to the terminal. For example, typing* echo PS1 *to the Bourne shell would result in the system echoing "PS1" to the terminal.*

Solaris provides a utility called **uname** which prints information about the system. If called with the **-n** flag, uname will print the host name of the system. The following command will capture the output of uname -n and assign it to the PS1 variable:

```
% PS1=`uname -n`
```

Some systems use fully qualified domain names instead of a simple name. For example, the response to uname -n on a system the users call *mercury* might be *mercury.astro.com.* In order to set the prompt to *mercury* alone, issue a command such as the following:

```
PS1=`uname -n| awk -F\. '{print $1}'`
```

The **awk** command is a pattern matching utility which contains its own programming language. In the example, the awk command scans the output of the uname command. The **-F\.** directive to awk tells it to use the period character as a word separator. The directive **{print $1}** tells awk to print out the first word of the input string (everything up to the first period in this case). This output is used as the input for the set directive. Therefore, the prompt variable gets set to the value returned by the awk command.

Using Quotation Marks in Scripts

Several of the previous examples contained backslashes and quotation marks. What do these characters mean to the shell? Quotes are required because the shell interprets some characters differently than might be expected. The following *metacharacters* have a special meaning to shell programs: $ (dollar sign), ∧ (caret), ; (semicolon), : (colon), # (pound sign), ~ (tilde), ? (question mark), & (ampersand), { } (curly brackets), [ ] (square brackets), ' (single quote), ` (back quote),* (asterisk), ( ) (parentheses), | (pipe), < (left angle bracket), > (right angle bracket), newline space, and <Tab>. In order to pass these characters on to commands, they must be quoted or escaped when used as part of a command.

Quoting and Escaping Metacharacters

Quoting or escaping a metacharacter causes the shell to alter its interpretation of the character. There are several ways to quote or escape a metacharacter. The easiest way to tell the shell not to use special interpretations of characters is to precede the character by a backslash (\ ). This is referred to as *escaping* the character.

Using a backslash to escape a character.

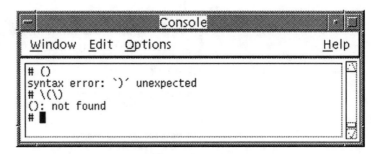

The single quote mark (') will protect text from *any* substitution. Backslashes, wildcards, double quotes, variables, and command substitution are ignored when enclosed in single quotes. Rules for using single quote marks appear below.

❏ Only a single quote can end a single quoted string.

❏ Single quoted strings can span multiple lines.

❏ It is not possible to write a single quote in single-quoted text.

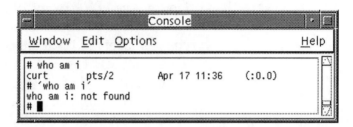

Using single quotes in the shell.

When the shell encounters commands surrounded by back quotes (` ), it performs command substitution. The commands inside the back quotes are executed, and the result is substituted in place of the backquoted command. For example, **set prompt=`uname -n`** causes the shell to execute the uname -n command. The result of this command is a string that provides the system name. This string is used as the input for the set command, resulting in the prompt variable being set to the string which gives the name of the machine.

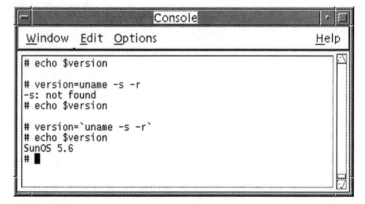

Using back quotes in the shell.

Yet another method of quoting in shell scripts is to use double quotes ("). The following rules apply to the use of double quotes.

❑ Use double quotes to protect text from wildcard substitution.

❑ The shell removes the quotation marks and passes the string to subsequent commands.

❑ Variable replacement, command substitution, and backslashes are effective inside the double quotes.

❑ To write a double quote (") inside double quotes, the double quote is escaped by preceding it with a backslash (e.g., "this \"escapes\" the double quotes").

❑ Single quotes will be treated as text inside the double quotes.

❑ Quoted text can span multiple lines.

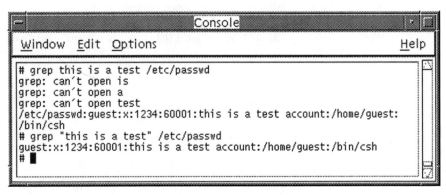

Using double quotes in the shell.

Redirection Metacharacters

The shell programs allow the user to redirect the input and output streams to programs. The following metacharacters are used to invoke this input/output redirection.

Known as *input redirection*, the left angle bracket (<) character causes the shell to take the following word as the name of the input file. The following rules apply when using input redirection.

❏ **< file** causes the shell to read its input from the file.

❏ **<&n** causes the shell to read its input from the file descriptor *n*.

❏ **<&-** causes the shell to close the standard input file.

❏ **<Tab** tag causes the shell to read up to the line starting with tag.

❏ **<~** tag causes the shell to read up to the line starting with tag, while discarding leading white space.

Known as output redirection, the right angle bracket (>) causes the shell to take the following word as the name of the output file. The following rules apply when using output redirection.

❏ **> file** causes the shell to write the output to file.

❏ **>&n** causes the shell to write the output to the file descriptor *n*.

❏ **>&-** causes the shell to close the standard output file.

❏ **>>** file causes the shell to append the output to file.

The pipe character (|) informs the shell to take the output of one command and pipe it into the input of the next command.

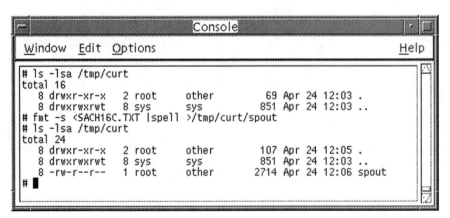

Using input and output redirection in the shell.

Other Metacharacters

The other metacharacters do not fit nicely into a particular group. Nonetheless, they are important characters and should be examined.

❏ The space character is considered white space. White space is used as a delimiter for commands and strings in shell programming.

❏ The <Tab> character is another white space character.

❏ The newline character signifies the end of a line of input. It is typically used to signify the end of a command.

❏ The tilde (~) is a C shell metacharacter, and refers to the user's home directory. For example, **ls ~shaun** would list the contents of Shaun's home directory.

❏ The caret (^) is used as the delimiter in C shell command substitutions.

❏ The dollar sign ($) is used as a prefix for local and environment variables under the shells.

❏ The ampersand character (&) tells the shell to run this job in the background. Commands are normally run in the foreground. This means that the terminal (or window) is not available for other uses until the command completes its execution. When a job is running in the background, the terminal is free for other activity.

❏ The asterisk (*) is a wildcard character. The asterisk matches any string.

❏ The question mark (?) is another wildcard character. It matches any single character.

❏ The [...] construct is another form of wildcard substitution. The pattern will match one character in the input stream, but the matched character can be any of the characters enclosed in the brackets. For instance, *rm file.[ot]* will remove any *file.o* and *file.t* files in the current directory.

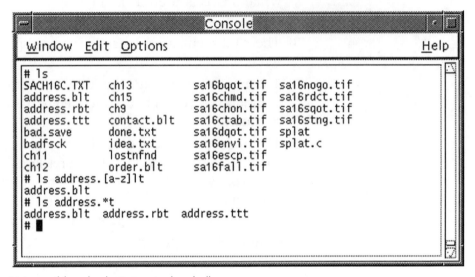

Using wildcard substitution in the shell.

❏ The curly bracket characters ({ }) are operators when appearing by themselves. When embedded in a word, or when used as a command argument, they are treated as text. When used to enclose other text, the brackets are used to preserve operator priority rules.

❏ The parentheses characters are used to group commands much like the braces. Differences between the curly brackets and parentheses are summarized below.

❏ Parentheses do not require leading white space.

❐ Commands enclosed in parentheses are executed by a subshell, and hence cannot alter the execution environment of the current shell.

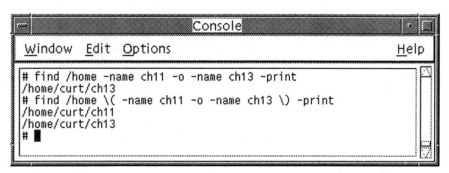

```
# find /home -name ch11 -o -name ch13 -print
/home/curt/ch13
# find /home \( -name ch11 -o -name ch13 \) -print
/home/curt/ch11
/home/curt/ch13
#
```

Using grouping metacharacters in the shell.

❐ The semicolon (;) is a command separator in the shell.

❐ The colon character (:) is a null command to the shell.

❐ The pound sign (#) is recognized as a comment when found at the beginning of a word.

Shell Programming

In addition to its function as the user interface, the shell allows the user to write shell programs. These programs can make use of variables, metacharacters, built-in shell features, and other UNIX system functions. The following sections touch briefly on some of the more useful shell programming constructs available in the Bourne shell. In-depth coverage of these constructs may be found in several commercially available reference books, the manual pages, and the Answerbook.

which Statement

The which statement determines which of multiple programs with the same name will be executed when a user types the name of the program. For instance, if a user has a program named *hello*, and a system program is also called *hello*, which program would be executed when the user types the following?

```
% hello
```

The which command consults the user's *.profile* file and searches the path environment variable to determine which binary occurs first in the search path. Due to the shell's traversal of the search path, the command that is executed will always be the first occurrence of that command in the search path. If a user is experiencing problems with a particular utility from his account, it is often useful to have the user log in and type the following to determine the path to the binary:

```
% which utilityname
```

In many cases, the problem is caused by the user aliasing the utility to another function, or possessing a personal program with the same name as the utility.

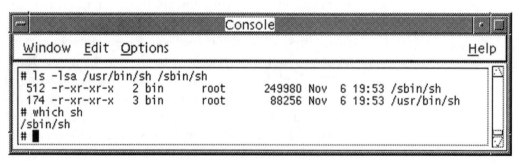

Using the which statement to identify the version of sh that is running.

Loop Control Statements

Shells provides several statements which implement program loop capabilities. Two of the most frequently used constructs are the **for** and **while** statements. Subsequent sections explore statement usage.

for Statement

The **for** statement, a loop control construct in the shell, tells the shell to perform a list of instructions for each element in the argument list. The syntax of the for statement follows.

```
for variable
in argument_list
do
          list of functions to perform
done
```

How could the for statement be used to simplify system administration tasks? Assume that an administrator would like to identify the operational status of several machines on the network. One way would be to make a list of all machine names, and then log in to each machine to ensure that it is up and running. A simpler method to accomplish the same goal would be to use a for loop to check the machines automatically, as shown in the next illustration.

```
Console                                          Window  Edit  Options                                      Help
# for i
> in grumpy wizard lsc xterm4
> do
> rup $i
> done
        grumpy    up 23 days,  4:18,    load average: 1.00, 1.00, 1.00
        wizard    up 89 days,  7:29,    load average: 3.50, 2.09, 1.34
        lsc       up  3 days, 10:12,    load average: 0.03, 0.01, 0.00
        xterm4    up 41 days, 21:46,    load average: 0.20, 0.18, 0.18
#
```

Using a for loop to determine machine status.

while Statement

The **while** statement is another loop control construct offered by the shell. The syntax of the while statement follows.

```
# while ( condition )
do
        list of actions to be performed
done
```

The while statement tests the *condition*, and if the condition evaluates as "true," the actions are performed until the condition is no longer true. The while function can be very useful when constructing shell scripts that require the occurrence of a condition prior to performance of a given action.

For example, some organizations limit outside connectivity to the use of a dial-up link in order to provide better protection from intruders. These sites typically have a single dial-out modem connected to a machine. When several users need access to this modem, competition problems may arise. Sometimes the administrator requires access to the modem to perform functions such as downloading security patches or new versions of software. In order to automate the process of determining whether the modem is free, a simple script could be developed to check the status of the modem port. The next illustration provides an example of such a script.

```
# cat -v modemcheck
#!/bin/sh
#
# check if modem is busy at one minute intervals
#

while [ -f /var/spool/locks/LK* ]
do
    dte=`date`
    echo "Modem busy at $dte"
    sleep 60
done

#
# Modem is free!  Inform operator!
#

echo "^G^G^G^G^G"
echo "Modem is free at $dte"
# ./modemcheck
Modem busy at Thu Apr 24 13:08:57 EST 1997
Modem busy at Thu Apr 24 13:09:57 EST 1997

Modem is free at Thu Apr 24 13:09:57 EST 1997
#
```

Using a while loop to detect when a dial-up modem is in use.

The above script examines */var/spool/locks* to determine whether a lock file exists. If the lock file exists, the script prints the date, and goes to sleep for 60 seconds. As soon as the modem is free (the lock file is gone), the script echoes several <Ctrl>+Gs to the terminal. The <Ctrl>+G character causes the terminal bell to sound, alerting the administrator that the modem is available.

Conditional Statements

The shell programs also allow the use of conditional execution statements. These statements allow modification of the shell script actions based on the results of a test on an argument or variable. Conditional statements test the value of an operation, and if the result of the test is true, a particular action is taken. If the result of the test is not true, another action may be taken.

if Conditional

One of the simple conditional statements available in the shell is the if statement. As seen below, this statement operates in the shell similarly to the if statement in high level languages such as C or Fortran.

```
if [ expression ]
then
    list of actions to perform ;
else
    list of actions to perform ;
fi
```

The expression may use flags or operators to direct the test of the condition. Some of the more useful flags appear in the following table.

Most frequently used expression flags

Flag	Result
-o	Logical OR operation (if [a or b]).
-a	Logical AND operation (if [a and b]).
!	Logical negation operation (if [a is not 0]).
=	Equality test (if [a+1 = b]).
!=	Inequality test (if [a != b]).
-r filename	True if user has read permission on filename.
-w filename	True if user has write permission on filename.
-x filename	True if user has execute permission on filename.
-e filename	True if file exists.
-o filename	True if user owns file.
-z filename	True if file is empty.
-f filename	True if file is a plain file.

The easiest way to explain the use of these expressions is to examine sample shell code. Because UNIX systems use different directories to store the system binaries, users may become confused when they try to set up *.login* and *.profile* files on their accounts. For example, the **uname** utility resides in */bin/uname* on Solaris, but other UNIX dialects may place it in */usr/5bin/uname* or */usr/bin/uname*. Consequently, the user must search for the utility in order to use it.

The following code segment tests to determine where an executable version of the uname utility is installed. Once the utility is located, the variable *version* is set to the operating system version name as returned by uname.

```
if [ -x /usr/5bin/uname ] then
        set version = `/usr/5bin/uname -s -r` ;
else if [ -x /usr/bin/uname] then
        set version = `/usr/bin/uname -s -r` ;
else
        set version = `/bin/uname -s -r` ;
fi
```

case Conditional

Another useful conditional statement available in the shell is the **case** statement. The case statement can be used to replace an if-then-else conditional in a shell script. The case statement allows more flexibility in the value matching process than is afforded by the if statement. Therefore, the case statement is a very popular conditional statement. The syntax of the case statement appears below.

```
case value
in
value_1)
```

```
           list of actions to perform if value is equal to value_1
;;
value_2)
           list of actions to perform if value is equal to value_2
;;
 . . .
esac
```

The shell checks the value of *value* and compares it to *value_1*. If the values are equal, the list of actions between *value_1)* and *;;* are performed. Once those statements are completed, the shell exits the switch and resumes execution at the statement following *esac*.

If the two values are not equivalent, the value of the variable is compared for equivalence with *value_2)* and so on. The **esac** statement tells the shell that there are no more values to compare against, which causes the shell to exit the case statement and continue execution at the statement following esac.

The following code segment examines the value of the *version* variable set in the previous section. Depending on the value of the version variable, shell statements in the *.login* file customize the environment for the operating system.

```
case $version
in
SunOS 5.?)
# Place code after this comment to customize the Solaris 2 environment.
                path=( $path:/usr/sbin:$HOME/bin/solaris:/usr/bin:/sbin )
                PS1=`uname -n`%
                MANPATH=( /usr/man:/usr/local/man )
                LPDEST=myprinter
                    ;;
```

```
SunOS 4.1.?)
# Place code after this comment to customize the SunOS 4.1.X environment.
                set path=( $path:/bin:/etc:/usr/bin:/usr/etc:$HOME/bin/sunos )
                  PS1=`uname -n`%
                  umask 0077
                  set notify
                  ;;
HP-UX B.10.2)
# Place code after this comment to customize the HP-UX environment.
                set path=( $path:/bin:/etc:/usr/bin:/usr/etc:$HOME/bin/hpux )
                  set cpu = `hostname | awk '{FS = .; print $1}'`
                  PS1=$cpu>>
                  PAGER=/usr/bin/more
                  ;;
esac
```

Summary

This chapter briefly examined methods of automating selected tasks performed on a daily basis by system administrators, including the use of utilities such as /bin/rdist, cron, and find, and selected built-in functions of the shell programs. Examples showing how to cascade these constructs to develop useful administration tools were presented.

Part III

Network Services

Network Configuration and Management

Previous chapters covered how to manage peripherals and resources directly connected to local systems. It is now time to examine how computer networks allow the utilization of resources indirectly connected to local systems. This chapter explores a few of the more common network technologies used today as well as a few that promise to be the network technologies of the future. In addition, network configuration, important network administration files, and basic network management and troubleshooting will be discussed.

The information presented in this chapter could easily fill several chapters (or entire books). Hopefully, the overview information provided in this chapter will prepare you for managing network applications such as network file systems (NFS) and network information services (NIS) presented in subsequent chapters.

Overview of the Internet

The Internet is comprised of many dissimilar network technologies around the world that are interconnected. It originated in the mid-1970s as the ARPANET. Primarily funded by the Defense Advanced Research Projects Agency (DARPA), ARPANET pioneered the use of packet-switched networks using Ethernet, and a network protocol called transmission control protocol/Internet protocol (TCP/IP). The beauty of TCP/IP is that it "hides" network hardware issues from end users, making it appear as though all connected computers are using the same network hardware.

The Internet currently consists of thousands of interconnected computer networks and millions of host computers. The Internet spans the entire Earth as well as providing several links into outer space (e.g., the Mir space station, amateur packet radio relays, and the NASA Advanced Communications Technology Satellite).

Connecting to the Internet

In the international telephone system, every phone line is assigned a unique phone number. The hierarchical addressing scheme of country code, area code, exchange, and line number ensures that any phone in the system can establish a connection with any other phone in the system. Telephone numbers are assigned by the telephone companies.

If hosts on a network wish to communicate, they need an addressing system that identifies the location of each host on the network. In the case of hosts on the Internet, the governing body that grants Internet addresses is the Network Information Center (NIC), affectionately known as "The Nick."

Technically speaking, sites that do not wish to connect to the Internet need not apply to the NIC for a network address. The network/system administrator may assign network addresses at will. However, if the site decides to connect to the Internet at some point in the future, it will need to re-address all hosts to a network address assigned by the NIC. Although reassigning network addresses is not difficult, it is tedious and time-consuming, especially on networks of more than a few dozen hosts. It is therefore recommended that networked sites apply for an Internet address as part of the initial system setup.

Beginning in 1995, management of the Internet became a commercial operation. The commercialization of Internet management led to several changes in the way addresses are assigned. Prior to 1995, sites had to contact the NIC to obtain an address. The new management determined that sites should contact respective network service providers (NSPs) to obtain IP addresses. Alternately, sites may contact the appropriate network registry as outlined in the following table.

Network address registry contact information

Site location	Postal	Telephone	Electronic	Notes
Asia/ Pacific Rim	Asia Pacific Network Information Center, Tokyo Central Post Office, Box 351, Tokyo, 100-91, Japan	+81-3-5500-0481	ip-request@ rs.apnic.net	Obtain the file */apnic/ docs/Contents* via anonymous ftp from *ftp.apnic.net.* Determine which form is required, obtain and complete the form, and then submit it electronically.
Europe	RIPE NCC, Kruislaan 409, 1098 SJ Amsterdam, The Netherlands	+31 20 592 5065; +31 20 592 5090 (fax)	hostmaster@ripe.net	Obtain the files */ripe/ forms/netnum-appl.txt* and */ripe/forms/ netnum-support.txt* via anonymous ftp from *ftp.ripe.net.*
United States, Americas, and other locations	Network Solutions, InterNIC Registration Services, 505 Huntmar Park Drive, Herndon, VA 22070 USA	(703) 742-4777; (703) 742-4811 (fax)	hostmaster@internic. net	Obtain the file */ templates/internet-number-template.txt* via anonymous ftp from *ds.internic.net.* Complete the form, and submit it electronically.

> ◆ **NOTE:** *Canada will soon have a network registry as well. At present, Canadian sites should obtain the form /templates/canadian-ip-template.txt via anonymous ftp from rs.internic.net. Submit the form to the U.S. registry according to the instructions for U.S. sites. (Do not follow the submission instructions on the Canadian form.)*

Internet Protocol

The process of connecting two computer networks together is called *internetworking*. The networks may or may not be using the same network technology, such as Ethernet or token ring. In order for an inter-network connection to function, a transfer device that forwards datagrams from one network to the other is required. This transfer device is called a *router* or, in some cases, a *gateway*.

In order for internetworked computers to communicate, they must "speak the same language." The language supported by most computers is the transmission control protocol/Internet protocol (TCP/IP). The TCP/IP protocols are actually a collection of many protocols. This suite of protocols defines every aspect of network communications including the "language" spoken by the systems, the way the systems address each other, how data are routed through the network, and how the data will be delivered. The Internet is currently using version 4 of the Internet protocol (IPv4).

Network Addresses

In order for internetworked computers to communicate, there must be some way to uniquely identify the address of the computer where data are to be delivered. This identification scheme must be much like a postal mail address. It should provide enough information so that the systems can send information long distances through the network, yet have some assurance that it will be delivered to the desired destination. As with postal addresses, there must be some sanctioned authority to assign addresses and administer the network.

The TCP/IP protocol defines one portion of the addressing scheme used in most computer networks. The Internet protocol defines the Internet protocol address (also known as an IP address, or Internet address) scheme. The IP address is a unique number assigned to each host on the network.

The vendors of network hardware also provide a portion of the addressing information used on the network. Each network technology defines a hardware (media) layer addressing scheme which is unique to that network technology. These hardware level addresses are referred to as media access controller (MAC) addresses. The following sections provide details about both IP and MAC addresses and discuss how each are used to facilitate communications between systems.

Internet Protocol Addresses

Hosts connected to the Internet have a unique Internet Protocol (IP) address. IP addresses are comprised of a 32-bit hexadecimal number, but they are typically represented as a set of four integers separated by periods. An example of an Internet address is 154.7.3.1. Each integer in the address must be in the range from 0 to 255. IP addresses in this format are also known as "dotted quad" addresses.

There are five classes of Internet addresses: Class A, Class B, Class C, Class D, and Class E. Classes A, B, and C addresses are used for host addressing. Class D addresses are called multi-cast addresses, and Class E addresses are experimental addresses.

Class A Addresses

If the number in the first field of a host's IP address is in the range 1 to 127, the host is on a Class A network. There are 127 Class A networks. Each Class A network can have up to 16 million hosts. With Class A networks, the number in the first field identifies the network number, while the remaining three fields identify the host address on that network.

☛ **WARNING:** *127.0.0.1 is a reserved IP address called the "loopback address." All hosts on the Internet use this address for their own internal network testing and interprocess communications. Do not make address assignments of the form 127.x.x.x; nor should you remove the loopback address unless instructed otherwise.*

Class B Addresses

If the integer in the first field of a host's IP address is in the range 128 to 191, the host is on a Class B network. There are 16,384 Class B networks with up to 65,000 hosts each. With Class B networks, the integers in the first two fields identify the network number, while the remaining two fields identify the host address on that network.

Address Class	Bit 0 1 2 8 16 24 31	Class Range	# Nets	# Hosts per Net
Class A	0 Network Host	0 – 127	127	16,777,214
Class B	1 0 Network Host	128 – 191	64	65,534
Class C	1 1 0 Network Host	192 – 223	32	254

Example: 129.74.25.98 = 10000001.01001010.00011001.01100010
Network = 129.74   Host = 25.98

Internet protocol address space.

Class C Addresses

If the integer in the first field of a host's IP address is in the range 192 to 223, the host is on a Class C network. There are 2,097,152 Class C networks with up to 254 hosts each. With Class C networks, the integers in the first three fields identify the network address, while the remaining field identifies the host address on that network.

☛ *WARNING: The numbers 0 and 255 are reserved for special use in an IP address. The number 0 refers to "this network." Number 255, the "broadcast address," refers to all hosts on a network. For example, the address 154.7.0.0 refers to the class B network 154.7. The address 154.7.255.255 is the broadcast address for the 154.7.0.0 network and refers to all hosts on that network.*

↝ *NOTE: Due to the immense growth of the Internet, the IPv4 protocol suite is quickly running out of unique addresses. The Internet grew from 6,000 hosts in 1986 to more than 600,000 hosts in 1991. But do not despair: a new version of the IP protocol suite will soon provide more addresses.*

Media Access Controller (MAC) Addresses

In addition to the IP address assigned by the NIC, most networks also employ another form of addressing known as the hardware or media access controller (MAC) address. Each network interface board is assigned a unique MAC address by the manufacturer. In the case of Ethernet interface boards, the MAC address is a 48-bit value. The address is typically written as a series of six two-byte hexadecimal values separated by colons.

Each network interface manufacturer is assigned a range of addresses which it may assign to interface cards. For example, 08:00:20:3f:01:ee might be the address of a Sun Microsystems Ethernet interface. This address would become a permanent part of the hardware for a workstation manufactured by Sun Microsystems.

Ethernet interfaces know nothing of IP addresses. The IP address is used by a "higher" level of the communications software. Instead, when two computers connected to an Ethernet communicate, they do so via MAC addresses. Data transport over the network media is handled by one system's network hardware communicating with another system's network hardware. If the datagram is bound for a foreign network it is sent to the MAC address of the router, which will handle forwarding to the remote network.

Before the datagram is sent to the network interface for delivery, the communications software embeds the IP address within the datagram. The routers along the path to the final destination use the IP address to determine the next "hop" along the path to the final destination. When the datagram arrives at the destination, the communications software extracts the IP address to determine what to do with the data. An example Ethernet address resolution protocol (ARP) and ARP reply packets are shown in the next illustration. The first packet is sent to the network asking "What is the MAC address of banzai.astro.com." The second packet contains the reply from *banzai* stating "My MAC address is 0:60:2f:88:da:63." A more detailed explanation of packet routing and addressing appears later in this chapter.

```
┌─┐                        Console                        ┌─┬─┐
└─┘                                                       └─┴─┘
 Window  Edit  Options                                    Help

# snoop -d hme0 -v arp
Using device /dev/hme (promiscuous mode)
ETHER:  ----- Ether Header -----
ETHER:
ETHER:  Packet 1 arrived at 10:11:3.35
ETHER:  Packet size = 60 bytes
ETHER:  Destination = ff:ff:ff:ff:ff:ff, (broadcast)
ETHER:  Source      = 8:0:20:74:26:c2, Sun
ETHER:  Ethertype = 0806 (ARP)
ETHER:
ARP:    ----- ARP/RARP Frame -----
ARP:
ARP:    Hardware type = 1
ARP:    Protocol type = 0800 (IP)
ARP:    Length of hardware address = 6 bytes
ARP:    Length of protocol address = 4 bytes
ARP:    Opcode 1 (ARP Request)
ARP:    Sender's hardware address = 8:0:20:74:26:c2
ARP:    Sender's protocol address = 192.88.53.112, legvold.astro.com
ARP:    Target hardware address = ?
ARP:    Target protocol address = 129.74.223.81, banzai.astro.com
ARP:

ETHER:  ----- Ether Header -----
ETHER:
ETHER:  Packet 2 arrived at 10:11:3.42
ETHER:  Packet size = 60 bytes
ETHER:  Destination = 8:0:20:74:26:c2, Sun
ETHER:  Source      = 0:60:2f:88:da:63,
ETHER:  Ethertype = 0806 (ARP)
ETHER:
ARP:    ----- ARP/RARP Frame -----
ARP:
ARP:    Hardware type = 1
ARP:    Protocol type = 0800 (IP)
ARP:    Length of hardware address = 6 bytes
ARP:    Length of protocol address = 4 bytes
ARP:    Opcode 2 (ARP Reply)
ARP:    Sender's hardware address = 0:60:2f:88:da:63
ARP:    Sender's protocol address = 129.74.223.81, banzai.astro.com
ARP:    Target hardware address = 8:0:20:74:26:c2
ARP:    Target protocol address = 192.88.53.112, legvold.astro.com
ARP:
#
```

Sample ARP and ARP reply packets show how hosts exchange MAC addresses.

Internet Protocol version 6 (IPv6)

As previously mentioned, Internet Protocol version 4 is running out of address space, result of the enormous expansion of the Internet in recent years. In order to ensure that address space is available in the future, the Internet Engineering Task Force (IETF) is readying Internet protocol version 6 (IPv6) for deployment. Major differences between IPv4 and IPv6 are listed below.

❐ IPv6 addresses will be 128 bits long (as opposed to the IPv4 32-bit addresses). A typical IPv6 address will be comprised of a series of colon-separated hexadecimal digits. For example, *0xFEDC:BA98:7654:3210:0123:4567:89AB:CDEF* might be a valid IPv6 host address. Provisions have been made to allow IPv4 addresses on an IPv6 network. These addresses will have hexadecimal numbers separated by colons, followed by decimal numbers separated by periods. Such an address might look like the following: *0000:0000:0000:0000:0000:FFFF:222.33.44.83*, which can be written in a shortened notation as *::FFFF:222.33.44.83*.

❐ IPv6 will implement a different broadcast addressing scheme. The IPv6 broadcast scheme allows for several types of broadcast addresses (organizational, Internet-wide, domain-wide, and so forth).

❐ IPv6 will not contain address classes. Some IPv6 address ranges will be reserved for specific services, but otherwise the idea of address classes will vanish.

❐ IPv6 will use classless Internet domain routing (CIDR) algorithms. This new routing algorithm allows for more efficient router operation in large network environments.

❐ IPv6 will be able to encrypt data on the transport media. IPv4 has no facility that allows network data to be encrypted.

➥ **NOTE:** *IPv6 will employ new packet formats not available in IPv4. These new formats will allow services such as reservation of bandwidth, guaranteed signal quality, and better multicasting capabilities.*

Network Services

Computers are generally networked in order to take advantage of the information stored on the network's individual systems. By networking many systems, each of which manages a small portion of the data, the corporation has access to all of the data. This is one of the underlying foundations of distributed computing.

In order for the systems on the network to take advantage of distributed computing, the network must provide network services. A few of these services might be IP address to MAC address resolution, hostname to IP address resolution, network file services, and World-Wide Web (WWW) services. The following sections cover the basics of address and name resolution. File services are detailed in Chapter 19, "Network File Systems," and Web services are detailed in Chapter 22, "World Wide Web Administration."

Name Services

Computers are very adept at dealing with numbers. In general, people are not capable of remembering the myriads of numbers associated with computer networking. People prefer to work with alphabetic names. The Internet Protocol provides a way to associate a hostname with an IP address. The "naming" of computers makes it simple for humans to initiate contact with a remote computer simply by referring to the computer by name.

Unfortunately, computers want to refer to remote systems by IP address. Name services provide a mapping between the host name that the humans prefer to use, and the IP addresses that computers prefer to use. These name services require a host (or hosts) connected to the local network to run special name resolution software. These *name server* hosts have a database containing mappings between host names and IP addresses.

Several name services are available under Solaris. Sun's Network Information Service (NIS+) and the Domain Name Service (DNS) are the two name services used most frequently in the Solaris environment. The NIS+ name service is designed to provide local name service within an organization. The DNS name service is designed to provide Internet-wide name services. Chapter 21, "Network Name Services," provides more information about name services.

Name Resolution Example

When a user on one computer wants to contact another computer, a specific sequence of events must take place. First, the user at host *glenn* enters a command to contact the remote computer (*grissom*). The *glenn* machine searches the */etc/hosts* file for the name *grissom*. If the name is found, *glenn* has the IP address of *grissom*, and name resolution is complete.

If *glenn* does not have an address for *grissom* in the */etc/hosts* file, it issues a request on the network asking "Does anyone know the IP address of *grissom?*" A name server on the network will receive this request. The name server searches a master database of host names and IP addresses to determine if it knows how to contact the *grissom* machine. If it has an entry for *grissom*, it returns this information to *glenn,* and the name resolution is complete.

If the local name server does not contain an address for the *grissom* machine, it must query other name servers to determine if any of them have an address for the *grissom* machine. The mechanics of name resolution, illustrated in the next figure, will be covered more comprehensively in Chapter 21, "Network Name Services."

Sample domain name look-up process.

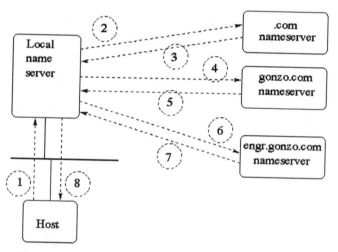

Step	Action
1	Host requests address of remote host dilbert. engr. gonzo. com
2	Local nameserver polls nameserver in .com domain
3	.com nameserver refers local nameserver to gonzo. com
4	Local nameserver polls gonzo. com
5	gonzo. com refers local nameserver to engr. gonzo. com
6	Local nameserver polls engr. gonzo. com
7	engr. gonzo. com replies with address of dilbert. engr. gonzo. com
8	Local nameserver passes address of dilbert. engr. gonzo. com to host

Address Resolution Protocol (ARP)

As mentioned previously, systems connected to an Ethernet prefer to use MAC level addresses for communications. But how do the hosts determine the MAC level address of other computers on the network once they know the IP address? The address resolution protocol (ARP) provides a method for hosts to exchange MAC addresses, thereby facilitating communications.

ARP Example

Suppose two computers, Host A and Host C, are connected to a network as shown in the next illustration. Before these two computers can exchange data, they must know one another's MAC addresses. The host software provides a method for the two machines to determine the address information required in order to communicate.

A simple address resolution scenario.

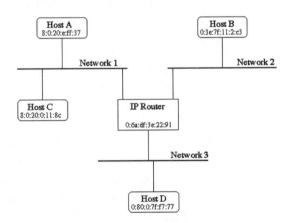

Host Requesting Connection	Destination Requested	Source MAC Address	First Hop MAC Address	Next Hop MAC Address
A	B	8:0:20:e:ff:37	0:6a:df:3e:22:91	0:3e:7f:11:2:c3
C	A	8:0:20:0:11:8c	8:0:20:e:ff:37	
A	D	8:0:20:e:ff:37	0:6a:df:3e:22:91	0:80:0:7f:f7:77

The ARP software runs on every host on the network. This software examines every packet received by the host, and extracts the MAC address and IP address from each packet. This information is stored in a memory-resident cache which is commonly referred to as the *ARP cache*.

If Host A needs to communicate with Host C, it first checks its own ARP cache to determine if it knows the address for the Host C machine. If Host A has the MAC address in its cache, it places the datagram on the media with the MAC address of Host C in the destination address field.

But what happens if the MAC address for Host C is not in the ARP cache? Somehow Host A must determine the address before it can send the pending datagram. The Host A machine resolves the Host C MAC address by sending a broadcast ARP packet to the network. Every machine on the network receives the broadcast packet and examines its contents. The packet contains computerese asking "Does anyone know the MAC address for the Host C machine at IP address W.X.Y.Z?"

Every machine on the network (including Host C) examines the broadcast ARP packet it just received. If the Host C machine is functioning, it should reply with an ARP reply datagram which (in computerese) says "I am Host C, my IP address is W.X.Y.Z, and my MAC address is 8:0:20:0:11:8c." When the Host A machine receives this datagram it adds the MAC address to its ARP cache, and then sends the pending datagram to Host C.

Network Design Considerations

Very few system administrators have the opportunity to design a corporate network. This function is typically managed by a network administrator. Unfortunately, many corporations require the system administrator to perform both functions for the company. In cases where the system administrator also has knowledge of network design and management practice, this situation may not present a problem.

All too often, however, the system administrator does not know how to design a network, or how to go about managing the network that gets implemented. The following sections outline some of the topics that the network and/or system administrator must be concerned with when designing a corporate network.

Network Topologies

Computer networks may be classified by many methods, but geographic, technological, and infrastructural references are three of the most common classifications. The terms network geography, network technologies, and network infrastructure are used in this book to refer to common classifications. Each term is defined below.

❏ *Network geography*—Physical (geographic) span of the network. Network geography may dictate the technologies used when the network is implemented.

❏ *Network technology*—Type of network hardware used to implement the network. The network hardware will (in many cases) dictate the infrastructure required to implement the network.

❏ *Network infrastructure*—Nitty gritty details of the corporate network wiring/interconnection scheme.

Network Geography

Computer networks can be divided into three general categories: local area networks (LANs), metropolitan area networks (MANs), and wide area networks (WANs). Subsequent sections discuss a few of the characteristics of each of these network geographies.

➥ **NOTE:** *Network throughput measurements are generally listed in terms of bits per second instead of bytes per second as is customary with disk subsystems. The difference is attributed to the fact that many network technologies are serial communications channels, while disk subsystems are parallel communications channels. The abbreviations Kbps, Mbps, and Gbps refer to kilobits (thousands of bits) per second, megabits (millions of bits) per second, and gigabits (billions of bits) per second.*

Local Area Networks (LANs)

Local area networks are, as the name implies, confined to a small area. These networks typically operate at bandwidths ranging from 4 to 622 megabits per second. LAN networks are typically used to connect systems within a single office building or university campus. Because LAN networks are typically confined to a single building, the underlying cable plant is often owned by the corporation instead of a public communications carrier.

Metropolitan Area Networks (MANs)

MANs are slower-speed connections that operate at 9.6Kbps to 45Mbps but cover larger areas than a LAN.

As the name implies, MANs typically provide network services for an area the size of a large city. Many corporations use MANs to connect geographically dispersed offices within a city or state. MANs typically involve a single communications carrier, which makes them much simpler to implement than WAN networks.

Wide Area Networks (WANs)

WANs are generally the most complex network geography because they span vast geographic areas. A WAN typically involves the interconnection of circuits through several commercial communications carriers. Because each carrier's equipment and technology is unique, the complexity of providing an end-to-end connection can be extreme. Most WANs operate at data rates of 56Kbps to 155Mbps, but cover large areas such as a state or the entire United States. WANs are used to interconnect metropolitan networks and local area networks. Many corporations use wide area networks to connect worldwide offices into a single corporate network.

Network Technologies

The variety of network technologies available today often leaves new (as well as veteran) system administrators in a fog of confusion. Each technology has advantages and disadvantages that must be considered carefully when selecting equipment for the corporate network. Subsequent sections provide a brief background on several of the more popular network technologies currently available.

○⇥ **NOTE:** *While several common LAN technologies used on Solaris systems are introduced in this section, the discussion is not exhaustive. A few recommended references to help broaden the reader's background on the subject of computer networking are listed in the references section at the end of this chapter.*

Ethernet

Ethernet is one of the dominant (if not the most dominant) local area network technologies in use today. A "bus" network developed in the late 1970s by Xerox Corporation, Ethernet is based on a network signaling method called carrier sense multiple access with collision detection (CSMA/CD) that was originally designed to run on coaxial cable. Each host connected to an Ethernet network constantly monitors the network for the presence of the carrier signal (CS). The carrier is a network control signal that lets hosts know that the network is in use. A host may transmit data onto the network only when the network is idle.

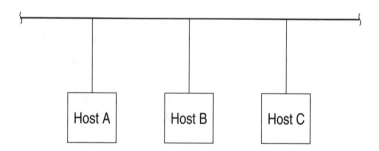

Example of an Ethernet bus infrastructure.

Ethernet has no prearranged order in which hosts transmit data. Any host can transmit onto the network whenever the network is idle. Thus Ethernet is said to be "multiple access" (MA). If two hosts transmit simultaneously, however, a data collision occurs. Both hosts will detect the collision (CD) and stop transmitting. The hosts will then wait a random period of time before attempting to transmit again.

It is important to note that Ethernet is a best-effort delivery LAN. In other words, it does not perform error correction or retransmission of lost data when network errors occur due to events such as collisions. Any datagrams which result in a collision or arrive at the destination with errors must be retransmitted. On a highly loaded network, retransmission could result in a bottleneck.

Ethernet networks typically transmit data at 10 Mbps (megabits per second). However, in 1993 Ethernet hardware operating at 100 Mbps became available. Ethernet is best suited for LANs whose traffic patterns are data bursts. NFS and NIS are examples of network applications that tend to generate "bursty" network traffic.

Integrated Services Digital Network

Through the 1950s, the public telephone network was wholly analog; the switching and transmission of voice signals were handled by analog circuitry. In the early 1960s the telephone industry began implementing digital transmission systems. In the 1970s, the industry began deploying digital switching systems. The Integrated Services Digital Network (ISDN) is a multiplexed digital networking scheme for use over existing telephone facilities. ISDN is a CCITT (Consultative Committee on International Telephony and Telegraphy) standard.

The major advantage of ISDN is that it can be operated over most existing telephone lines. An ISDN connection is typically capable of transmission rates of 64Kbps (kilobits per second) per channel. The 64Kbps circuits are referred to as basic rate interfaces or BRIs. A BRI is comprised of two 64Kbps bearer (B) channels and a single delta (D) channel. The B channels are used for voice and data, while the D channel is used for signaling and X.25 packet networking.

↝ **NOTE:** *In telephone terminology, 64Kbps refers to 64,000 bits per second. In computer terminology, 64Kbps typically refers to 65,536 bits per second.*

ISDN transmission rates can be increased to 128 Kbps by bonding two ISDN B channels together, which is still considered "slow" for digital data. It is, however, significantly faster than the typical 9600 baud (9.6 Kbps) or the newer 56Kbps modems used to connect computer systems via analog telephone lines. The 128Kbps ISDN circuits are referred to as "2B" circuits. At the customer end of an ISDN link, the connection is a single pair of wires. The maximum length of this link is 5,500 meters (18,000 feet). Due to the length constraints ISDN is typically used in LANs and MANs.

Token Ring

Token ring networks are another widely used local area network technology. A token ring network utilizes a special data structure called a token which circulates around the ring of connected hosts. Unlike the multiple access scheme of Ethernet, a host on a token ring can transmit data only when it possesses the token. Token ring networks operate in receive and transmit modes.

In transmit mode, the interface breaks the ring open. The host sends its data on the output side of the ring, and then waits until it receives the information back on its input. Once the system receives the information it just transmitted, the token is placed back on the ring to allow other systems permission to transmit information, and the ring is again closed.

In receive mode, a system copies the data from the input side of the ring to the output side of the ring. If the host is down and unable to forward the information to the next host on the ring, then the network is down. For this reason many token ring interfaces employ a dropout mechanism that, when enabled, connects the ring input to the ring output. This dropout mechanism is disabled by the network driver software when the system is up and running. But if the system is not running, the dropout engages and data can get through the interface to the next system on the ring.

Fiber Distributed Data Interconnect

Fiber distributed data interconnect (FDDI) is a token ring LAN based on fiber optics. FDDI networks typically operate at 100 Mbps. Unlike Ethernet, FDDI is well-suited for LANs whose traffic patterns include sustained high loads, such as relational database transfers and network tape backups. The FDDI standards define two types of topologies: single attachment stations (SAS) and dual attachment stations (DAS).

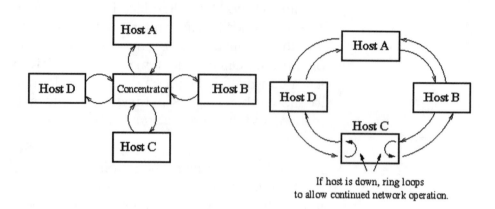

Example FDDI infrastructures.

An SAS FDDI ring allows a star infrastructure network. A concentrator is placed in the center of the star. A single pair of fibers is connected from the concentrator to each host. If a host needs to send information to another host, the concentrator must forward the information. Single attachment FDDI rings require less fiber than dual attachment FDDI rings, but they contain a serious disadvantage: if a connection between the host and the concentrator is broken, the host is not available to the network. Another disadvantage of a single attachment FDDI ring is the cost of the concentrators.

The DAS FDDI ring employs two counter-rotating fiber rings. In a counter-rotating token ring network, one ring sends information in a clockwise direction, and the other ring sends information in a counterclockwise direction. In normal operation, one FDDI ring is used for all communications, while the second ring sits idle. If the primary data ring is broken, the interface reconfigures itself to use the second ring as a loopback to form a single ring between the remaining hosts. This feature provides some fault tolerance to the network at the cost of extra fiber.

Asynchronous Transfer Mode

Asynchronous transfer mode (ATM) networks are rapidly becoming a popular networking technology. Because ATM networks are based upon public telephone carrier standards, the technology may be used for local area, metropolitan area, and wide area networks. ATM networks can operate over fiber optic cables at speeds of up to 622 Mbps. Recent additions to the specifications will allow ATM networks to operate at rates over 1 Gbps (gigabits/second), and 2.4Gbps standards are currently under discussion. Connections over twisted-pair copper media are currently possible at speeds up to 155Mbps.

ATM, like today's telephone network, is a hierarchical standard which employs a connection-oriented protocol. In order for two hosts to communicate, one must "place a call" to the other. When the conversation is over, the connection is broken. Most ATM networks are implemented over fiber optic media, although recent standards also define connections over twisted-pair copper cable plants.

ATM is a "cell" networking technology. While Ethernet allows 1,500 byte data packets, and FDDI allows 4,500 byte data packets, ATM operates on 53-byte data cells. A 53-byte cell allows a 5-byte header (which contains control information), and 48 bytes of data. The control information in the header tells how this data cell is to be used in relation to the other data cells which are received. The use of small cells reduces wasted bandwidth on the transmission media, and reduces transmission delays on the media.

Network Infrastructures

One of the most difficult (and expensive) problems in network design is defining the network infrastructure. The infrastructure of a network includes the cable plant required to tie the hosts together, the physical topology of the wire within the buildings, the type of electronics employed in the network, and the bandwidth requirements of the hosts connected to the network.

These topics could (and do) fill several books. A few such books are listed in the references section at the end of this chapter. The points listed in this section are meant to be general guidelines which should be considered before a network design is implemented.

Bandwidth Requirements

A good rule of thumb for network design is to provide a hierarchical network structure. At the top (root) of the hierarchy is the corporate backbone. Systems connected to this network typically provide corporation-wide services such as routing, name service, and firewall protection. These services are typically in high demand, and therefore require the most network bandwidth. Connections at this level of the network should be the fastest available, typically 100 Mbits/ second or faster.

One layer down from the root of the network is the portion of the network which provides connectivity to the other hosts on the network, and other network services. A few examples of the services at this level include network file system servers, e-mail servers, application servers, and license servers. These services, while still in high demand, are typically less bandwidth intensive than the top level network services. Connections at this level are typically moderate speed connections (100 Mbits/second).

At the bottom layer of the network are individual workstations. These machines typically require the least bandwidth, and therefore get the slowest links to the network, typically 10 Mbits/second.

⇥ **NOTE:** *These guidelines are meant to be generic suggestions. It is possible, and indeed probable, that some desktop workstations may require more bandwidth than the corporate services backbone. The network designer must determine the typical use for each host connected to the network, and "size" the connection appropriately.*

Typical bandwidth requirements of a corporate network.

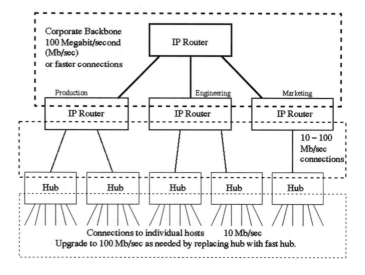

Cable Plant

Once the bandwidth requirements have been determined, the next task is to determine the physical implementation of the network cable plant. In many cases, the layout of the cable plant is determined by the architecture of the buildings which house the corporation.

Due to the high bandwidth requirements of the top level services, the cable plant for this portion of the network should be as compact as possible. If the top level services can all be located in a single room within the building, then the high speed portion of the network can be built using very short cable segments. These segments could be implemented with fiber optic cabling at a reasonable cost.

The middle layer of the network is responsible for connecting the remote reaches of the network into the top level of the network. Due to the need for this layer to span large distances (100 meters to several kilometers) fiber optic cabling is the media of choice. This layer of the network is typically implemented with a "star" cabling topology. In brief, the cables from the high speed portion of the network to the low speed portion of the network are point to point links emanating from a central location.

The bottom layer of the network is responsible for connecting individual hosts to the middle layer of the network. These connections typically cover fairly short distances (100 meters). Again, these connections are typically implemented with a star cable topology. Cables emanate from a central point on the floor of a building to individual offices on that floor.

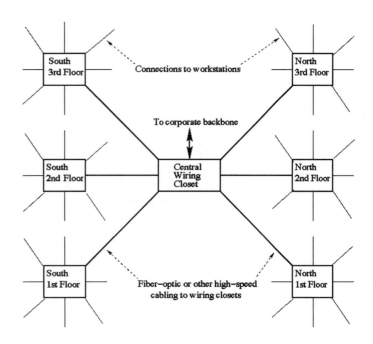

*A simple
star cable topology.*

Network Electronics

With the cable plant complete, the next step in the network design is to determine the type of electronics required to implement the network. For example, should the low speed connections be implemented via network hubs, network switches, or routers? Should the high speed connections be implemented via FDDI, ATM, or Fast Ethernet? Should hosts be used as routers, or should dedicated routers be used to handle the routing of network traffic?

Network Implementation Issues

Once the network is physically implemented, the task is not yet complete. Several other issues require the attention of the network designer. For example, how does the manager determine if the network media is overloaded? What can be done to minimize media overload conditions? How should the packets be routed from one host to another? The following sections touch upon some of these issues.

Netmasks

Netmasks provide a method for segmenting large networks into smaller networks. For example, a Class A network could be segmented into several Class B and Class C networks. The "boundaries" of each network can be defined by providing a "mask" value which is logically ANDed with the IP address for a host.

Each bit in the netmask with a value of one represents a bit of the network address. Netmask bits with zero values represent bits used for the host address. For example, a netmask of 255.255.255.224 provides for 29 bits of network address and 5 bits of host address. Such a netmask would provide a network of 30 hosts.

A Class B network number uses 16 bits for the network address, and 16 bits for the host address on that network. What would happen if a Class B network was treated as a group of Class C networks instead? This would allow the site to have 254 individual networks instead of a single network. The implications of this segmentation are summarized below.

❏ External routers still route packets to the host listed as the gateway for the Class B network. This means that the gateway machine will be required to forward all incoming traffic to other hosts on the internal network.

❏ Internal routers will be required to keep track of how to forward packets to internal network segments. This implies the existence of a name server, and the use of an internal routing protocol within the internal network.

Subnets

All IP address classes allow more than 250 hosts to be connected to a network. In the real world, this is a problem. Ethernet is the dominant network technology in use in private sector and campus networks. Due to the shared bandwidth design of Ethernet, more hosts on a network means less bandwidth is available to each host. How can this bandwidth problem be overcome?

One method of partitioning the network traffic is through a process called "subnetting" the network. Some organizations place each department on a separate subnet. Others use the geographic location of a host (such as the floor of an office building) as the subnet boundary. By segmenting the network media into logical entities, the network traffic is also segmented over many networks.

Typical network segmented into subnets.

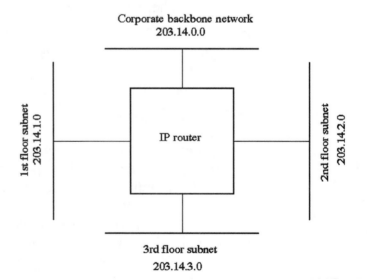

For example, the machines on the third floor no longer have to contend with the machines on the second floor for network access. Because the "floor" subnets are tied together by a router (or gateway), the machines may still communicate when required, but otherwise the third floor machines do not "see" traffic bound for the second floor machines. But should subnets be incorporated into a network, and how are subnets implemented?

When to Subnet

There are no absolute formulas for determining when or how to subnet a network. The network topology, the LAN technology being implemented, network bandwidth, and host applications all affect a network's performance. However, subnetting should be considered if one or more of the following conditions exist:

❐ There are more than 20 hosts on the network.

❐ Network applications slow down as users begin accessing the network.

❐ A high percentage of packets on the network are involved in collisions.

Calculating Network Load

Obtaining accurate network load statistics requires sophisticated (and expensive) network analysis equipment. However, network load can be estimated by calculating the network interface collision rate on each host. This can be done by using the **netstat -i** command. Use the command's output, divide the total collisions (*Collis*) by the output packets (*Opkts*), and then multiply the result by 100. For example, on *grissom*, the total collisions are 2,553 and output packets total 242,220. The collision rate on *grissom* is

1.05% (2553/242220 * 100 = 1.05%). To obtain a rough idea of the network collision rate, collect netstat -i statistics for all hosts on the network and average them.

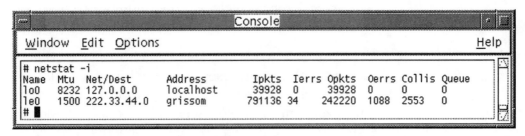

Using netstat -i to estimate network load.

Collision rates under 2% are generally considered acceptable. Rates of 2% to 8% indicate a loaded network. Rates over 8% indicate a heavily loaded network that should be considered for subnetting.

✓ **TIP:** *From a troubleshooting perspective, subnetting allows administrators to isolate (disconnect) pieces (subnets) of a network when trying to resolve network problems. By isolating a subnet, it is often easier to determine the source of a network problem. Most routers, gateways, hubs, and multi-port transceivers have a switch allowing them to operate in standalone mode. If the switch is set to local mode, the subnet is disconnected from the rest of the network. With the switch set to remote mode, the subnet is connected to the rest of the network.*

Subnet Example

Suppose an organization's offices occupy a six-story building. There is a large server room on the first floor that provides services to clients on all six floors. The IP address for the organization is 222.33.44.0. The corporation would like to subdivide the address space into seven networks (one per floor, plus one for the server room) with an equal number of hosts on each network.

A Class C network may serve 255 hosts. One address on each network is reserved as the broadcast address, and another address on each network is reserved as the "zero" host. Packets sent to the broadcast address are monitored by all hosts on the network. The "zero" address is used by routing protocols to determine how to get information to a particular network.

Dividing 224 by seven (the number of subnets) yields 36. This means seven subnets of 36 hosts each. To simplify matters (and to provide some future expansion capabilities), the administrator decides that each subnet will be limited to 32 addresses. Five bits are required in order to address 32 entities. From this the administrator determines that the corporation will have eight networks, to be configured as shown in the following table.

Network subnet example

Network	Zero host	Broadcast	Netmask	# hosts	Serves
One	222.33.44.0	222.33.44.31	255.255.255.224	30	1st Floor
Two	222.33.44.32	222.33.44.63	255.255.255.224	30	2nd Floor
Three	222.33.44.64	222.33.44.95	255.255.255.224	30	3rd Floor
Four	222.33.44.96	222.33.44.127	255.255.255.224	30	4th Floor
Five	22.33.44.128	222.33.44.159	255.255.255.224	30	5th Floor
Six	222.33.44.160	222.33.44.191	255.255.255.224	30	6th Floor
Seven	222.33.44.192	222.33.44.223	255.255.255.224	30	Spare
Eight	222.33.44.224	222.33.44.255	255.255.255.0	29	Servers

With this information in hand, the administrator can install and configure the corporate hosts, resulting in a network topology shown in the next illustration.

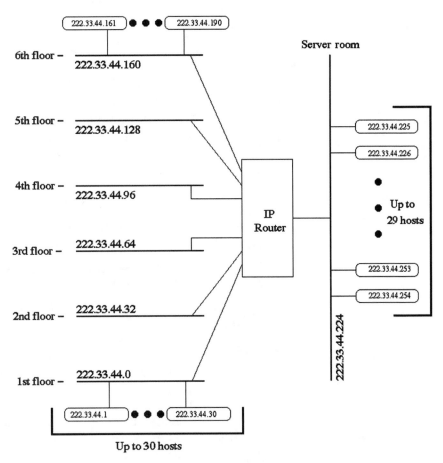

An example of a subnetworked corporate network.

Routing Concerns

All networks require some form of packet routing to ensure that packets get from one host to another. On a simple (single wire) network, the routing is also simple: a source host sends a packet out onto the media, and the destination host picks it up. On a multi-tiered, or subnetted network, routing becomes much more difficult. On the Internet, routing can become a major nightmare.

Routing Overview

Network routing is an iterative process. Consider the problem a visitor to a large city would have trying to locate an office in a building in a city that she is unfamiliar with. Generally this problem is broken down into a sequence of smaller problems as summarized below.

❐ Determine how to get to the destination city.

❐ Determine which quadrant of the city contains the desired street.

❐ Determine where the desired building is located on the street.

❐ Determine where the office is located in the building.

The first two steps in this process are roughly equivalent to external routing. The last two steps are roughly equivalent to internal routing.

In general, an organization should advertise a single routing address. Generally, this address is the network portion of the IP address registered to the organization. Routers on the Internet search the network portion of the IP address, and then forward packets toward the destination based on this information. This process is known as external routing.

Once the packet gets to the corporation, the internal routers search the host portion of the address (and any applicable network mask) to determine how to deliver the packet to its final destination. This portion of the routing process is known as internal routing.

The following discussion of routing is organized into three areas. First, *internal routing* refers to the routing algorithm used within a corporation's network. *External routing* refers to the routing algorithm used to allow the corporate network hosts to communicate with the outside world (e.g., other hosts on the Internet). Finally, *supernets*, or classless inter-domain routing (CIDR) blocks, are discussed.

External Routing

External routing allows packets from the outside world to be delivered to the local network, and vice versa. Several external routing protocols are available for use on the Internet. Instead of providing details on each of these protocols, this section presents some of the basics of how these protocols operate. For more information of external routing protocols, consult the publications listed in the "References" section of this chapter.

External routing typically involves determining the next hop along the path to the final destination. A distant router that wants to deliver a datagram to a host on the local network does not need to know the exact set of gateways required to get the packet to the final destination. Instead, the external router just needs to know the address of another router which is one step closer to the final destination.

External routers periodically exchange routing information with their "peers." Each router knows the specifics of packet delivery within their local network. By exchanging information with other routers they build a routing table which contains information about how to get datagrams to other networks. When the routers exchange routing information they are "advertising" routes to destinations. In other words, the information exchanged between routers includes an implied promise that the router knows how to get datagrams delivered to specific destinations.

External Routing Example

The following example presents a simplified view of external routing based on the next illustration. Assume that Host A wants to communicate with Host B. Host A sends the datagram to the network. Router 1 determines that the destination address is not a local network. Router 1 examines its routing table in an attempt to find a route to the destination. Router 1 determines that Router 2 has advertised a route to the destination, so the datagram is forwarded to Router 2.

Router 2 receives the datagram and goes through the same process that Router 1 followed. Router 2 determines that Router 3 has a route to the destination, so the datagram is forwarded to Router 3. Router 3 is directly connected to the destination network. Router 3 examines the host portion of the IP address, and forwards the datagram on to host B.

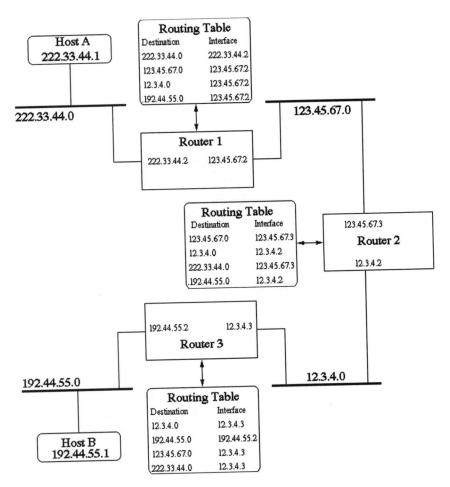

Simplified external routing.

Configuring External Routing

The process of configuring an external routing protocol can be very involved. First, the administrator
needs to determine which device on the network will
provide external routing services. The decision might
be to use a host on the local network, or a dedicated
router which is connected to the local network. The
router/host which handles external routing may have

immense routing tables stored in RAM memory and on the system disks. In general, it is preferable to use a dedicated router to provide this service, thereby freeing the local hosts to provide other services for the corporation.

Once the external routing hardware has been chosen, the administrator must determine which external routing protocol to use. This is generally dictated by the routing protocol(s) used by the Internet service provider which provides connectivity to the administrator's site. The administrator must set up configuration files specific to the chosen protocol. If multiple protocols will be used, the administrator may need to set up extra configuration information such that the protocols can share routing information with each other.

In order to enable external routing, the administrator must complete several additional tasks. First, the site must have a host configured to provide name service for the local network. Next, the site's name server must be registered with the Network Information Center (NIC). The NIC adds this host to a master routing database. Any host on the Internet that wants to communicate with a host at this site must contact the site name server to determine how to get data to the final destination.

Internal Routing

One host or router on the local network must be configured to run one of the external routing protocols. This router/host will receive all (externally sourced) datagrams bound for hosts on the local network. The router/host will have to make internal routing decisions in order to facilitate the delivery of the data to the local destination host. Likewise, internal network traffic must be routed to destinations on the local network. Therefore, the router which provides external routing must also know how to perform internal routing.

Internal routing may be handled in two ways. First, the network manager may tell each host to send all packets to a central router (or host) for delivery (static routing). Alternatively, the individual hosts may be allowed to determine how to get a packet from point A to point B (dynamic routing). Each routing algorithm has strengths and weaknesses.

Static Routing

Static routing requires that every host on the network send all network traffic to a single central routing authority for data delivery. This routing algorithm is somewhat similar to the U.S. Postal Service delivery method. A host "drops" a packet on the network media destined for the router. Once the host has sent the packet to the router, the host forgets about it, and allows the router to determine how to deliver the data to the final destination. Hopefully, the packet will eventually arrive at the destination.

Static routing is typically used on internal networks where there is little change in the network structure. The administrator creates a */etc/defaultrouter* file on each host on the network. This file contains the IP address of the router that will provide the datagram delivery services. The administrator builds static routing tables in the router's memory to tell the router how to deliver data to any location on the internal network.

Static routing requires minimal host processing support, but it does require intelligent routers and considerable human intervention to build the routing tables. Every time the network configuration changes, the administrator must update the routing tables in each router on the network. Otherwise, if the routing information is allowed to become outdated, a certain subset(s) of hosts may not be able to communicate.

Static routing is typically very efficient. Unfortunately, it is often plagued by a "single-point-of-failure" problem. If the network's static router fails, the entire network is down because all hosts send their data to the router for delivery.

Dynamic Routing

Dynamic routing is typically used in situations where the network configuration is changing rapidly. Dynamic routing is much more processor intensive that static routing, but requires little human intervention once it is configured and working. Dynamic routing requires each host to maintain a local routing table. This table tells the host how to get data from one point to another on the network.

In order to create and maintain these routing tables, each host runs a program that builds the routing table. When one host wants to communicate with another host on the network it must determine how to get the data to the final destination. If a dynamic routing algorithm does not know how to get data to the requested destination, it will ask its neighbors if they know a route it can use.

Dynamic routing is typically more robust than static routing. If a network contains multiple paths between points A and B, the dynamic routing algorithms can adjust the routing tables to survive network failures.

Supernets

The opposite of subnetting is referred to as "supernetting." Because the number of available network addresses is dwindling under IPv4, some entities end up with multiple Class C network numbers when they really need a class B network number. The process of supernetting allows the organization to maintain a single gateway to the outside world by combining the Class C network addresses into a single addressable supernetwork. These supernetworks are also known as classless inter-domain routing (CIDR) blocks.

Supernet Example

Assume that an organization with 400 hosts wished to connect to the Internet, and applied for an IP address. The Internet service provider may not have a free block of contiguous addresses to assign to the organization. The ISP could assign two class C network numbers to the organization, and offer to provide the external routing for the organization by advertising a single supernet address to the rest of the network.

For example, suppose the organization was assigned the network numbers 203.14.4.0 and 203.14.7.0. Expressing these two addresses as binary numbers shows that they share a common 22 bit prefix. The ISP could advertise a route to the network 203.14.4/22. This tells the rest of the world that the first 22 bits of the address are used as the network address, while the remaining 10 bits are used as the host address. The external routers know to forward all datagrams bound for hosts on the corporation's two networks to the ISP's router based on this information. The ISP's router forwards the datagrams to the corporate router for final delivery.

Example of supernet routing.

203.14.7.0 = 1 1 0 0 1 0 1 1 0 0 0 0 1 1 1 0 0 0 0 0 0 0 1 1 1 0 0 0 0 0 0 0

203.14.4.0 = 1 1 0 0 1 0 1 1 0 0 0 0 1 1 1 0 0 0 0 0 0 1 0 0 0 0 0 0 0 0 0 0

First 22 bits of both networks are the same

Netmask = 1 1 1 1 1 1 1 1 1 1 1 1 1 1 1 1 1 1 1 1 1 1 0 0 0 0 0 0 0 0 0 0

Advertise 1 1 0 0 1 0 1 1 0 0 0 0 1 1 1 0 0 0 0 0 0 1 0 0 0 0 0 0 0 0 0 0

203   .   14   .   4   /22

Host Configuration

The following sections outline the steps required to configure a system for use on a network. While these sections cover a lot of material, they are not an exhaustive guide to configuring a Solaris system for use on a network. Consult the Solaris manual pages and the System Administration AnswerBook for more information about configuring systems for use on the network. Above all, consult the local network manager before plugging a new system into the network.

General Setup Procedure

In order to connect a host to the network, the administrator requires the information summarized below.

❏ Host name for the system

❏ Domain name for the system

❏ IP address for the system

❏ Netmask for the network (if applicable)

❏ Name service used on the network

❏ Name or address of the machine which provides name service

Once you have the above information, you can configure the machine and "plug it in" to the network. The next section explains how the information above is used to complete this task.

Configuration Files

Network configuration information is stored in several files on the system. Some of these files specify information about the host's address and host name, or unique setup parameters. Other files specify which network services the host will allow, and which other hosts on the network provide services that the host may require.

/etc/inet/hosts File

One of the most frequently used network administration files is */etc/inet/hosts*. The file is a registry of IP addresses and associated host names known to a system. At a minimum, it must contain the loopback address (127.0.0.1) and the IP address for the host. Many networking commands consult the host file in order to resolve the IP address when communications are requested. The format of host file entries follows:

```
IP address<Tab>Fully.Qualified.Name<space>host_alias
```

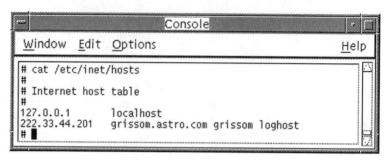

Minimum required entries in the hosts file.

All hosts connected to the corporate network need not be listed in the *hosts* file. Typically, only those hosts that are routinely contacted are listed. For security reasons, *root* is the only user who can edit */etc/inet/hosts*. If a host routinely communicates with many hosts (e.g., more than a dozen) on the network, maintaining a large *hosts* file can be cumbersome. If this is the case, using an on-line distributed name service database system, such as the Domain Name System (DNS) or Network Information Service (NIS+), should be considered. These name services are discussed in Chapter 21, "Network Name Services."

☛ **WARNING:** *Trailing blank characters on a line in the* hosts *file will cause that entry to fail.*

/etc/hostname.le0 File

Solaris uses */etc/hostname.device/instance* to simplify system configuration at boot time. For a host connected to a single network, the */etc/hostname.le0* file would contain the host name of the machine. Machines connected to multiple networks would have multiple */etc/hostname.xxx* files. Each network interface will probably have a unique host name assigned to it. For example, a dual-homed machine with an *hme0* and *le0* interface on separate networks might be called *thishost* on one network interface, and *thishost-that_network* on the other. The */etc/inet/hosts* file should list both addresses for the host in order to facilitate mapping the unique interface names to an IP address.

/etc/nodename File

Solaris employs another file to maintain an "alias" for the host on the network. The */etc/nodename* file contains the "alias" name from */etc/hosts*. For a multi-homed host, this allows the host to respond to service requests from all connected networks by a single host name. This allows users to remember only one name for the host, no matter which network interface they use to contact the host.

/etc/services File

The */etc/services* file contains a list of network ports and the services which correspond to those ports. For example, port 25 is defined as the *sendmail* port, while port 80 is reserved as the hypertext transport protocol daemon (*httpd*) port. To add a new service to a host, the administrator must add a port number and service name pair to the */etc/services* file.

In general, ports numbered zero through 512 are reserved Internet protocol ports. These port numbers are assigned to particular services by the NIC. Port numbers between 512 and 1024 are generally recognized UNIX network service ports. These ports are not reserved, but are generally treated as though they were reserved port numbers. Port numbers above 1024 are unassigned, and may be used for local network services as required by the site.

➥ **NOTE:** *The use of ports which are not assigned by the NIC may vary from site to site. Some sites may reserve a group of ports above 1024 for internal use. Checking with the local network administrator before assigning local ports and/or services is recommended.*

/etc/inetd.conf File

The inet daemon (**inetd**) monitors incoming network packets for service requests. When a request is received, inetd examines the network packet to determine which service is being requested. It then consults the */etc/inetd.conf* file to determine which application to invoke to service the request. You may wish to limit the service requests that a machine answers. This is done by removing the service from the */etc/inetd.conf* file and restarting inetd. Conversely, to add a new service to a host, add information to the */etc/inetd.conf* file and then restart inetd.

✓ **TIP:** *It is sometimes possible to send a signal to the* inetd *process to cause it to re-read the* /etc/inetd.conf *file. This is achieved by sending a* kill -HUP *signal to the process ID (PID) of the running* inetd *process. For example, if the* inetd *process is running with a PID of 73, the command* kill -HUP 73 *should cause the* inetd *process to re-read the configuration file.*

Configuration Commands

Solaris provides several ways for the administrator to enter system network configuration information. During the installation of the operating system, the SunInstall procedure prompts you for network information and builds the files as required for the installation. Once the system is up and running there are other methods available to update network information.

/usr/sbin/sys-unconfig

The **/usr/sbin/sys-unconfig** command allows the administrator to change the network information of the host. This command removes all network-specific information files, and halts the machine. When the machine is rebooted, a portion of the SunInstall program is invoked to prompt the administrator for the new network connection information.

ifconfig

The **/usr/sbin/ifconfig** command is used to bind the IP address, system name, netmask, broadcast address, and other network configuration parameters to a particular network interface. The ifconfig command is run automatically at boot time by the **init** process. You can use the ifconfig command to examine and/or modify interface parameters while the system is up and running. Consult the ifconfig manual page for more information on this command.

Multi-homed Hosts

Hosts with more than one network interface are referred to as *multi-homed hosts*. Such hosts may be configured to act as routers on the internal network. Because these hosts are connected to multiple networks, it may be desirable to have them forward packets from one network to another to facilitate communications. This process is sometimes referred to as *packet forwarding*.

⊸ **NOTE:** *Packet forwarding will require considerable processor overhead, and is usually best left to routers designed to perform this function.*

/etc/ndd

If you wish to use a host as a packet forwarding system, the kernel must be configured to provide this service. The appropriate kernel configuration is accomplished via the **/etc/ndd** command. In particular, you should use ndd to turn on packet forwarding at the IP protocol stack level. To accomplish this task, use the following command:

```
# /etc/ndd -set /dev/ip ip_forwarding 1
```

Conversely, the following command will disable packet forwarding:

```
# /etc/ndd -set /dev/ip ip_forwarding 0
```

Network Metrics

Hosts with multiple network connections have another unique setup situation. These hosts have multiple paths to the rest of the network. The routing algorithms may decide that one path is better than another without paying attention to underlying network technology. For example, a network which offers a 10 Mbit connection from host A to host B might be considered "better" than a 100 Mbit connection between the same hosts.

In such circumstances it may be desirable for the administrator to be able to force the hosts to use the high speed network link in lieu of the low speed link the routing algorithms discovered. This may be accomplished by setting the network "metric" value on each network link.

Many routing algorithms determine the "best" route by counting the number of hops (or gateways) the packet has to pass through between points A and B. The route with the fewest hops is considered the "best" route. The network metric value is a way to "add" hops to the route such that the routing algorithm will choose another route to deliver the data. You can set the metric for each network interface with the **/usr/sbin/ifconfig** command.

Network Monitoring and Troubleshooting

As with most computer operations, networks are bound to have problems. Troubleshooting networks can be a tedious process. One method of troubleshooting is to use monitoring tools that determine how the network is being used. In some cases, it may not be possible to monitor a network because physical connections may be damaged or gateways may be down. Another method of monitoring the network is to watch the console error messages generated by the machines connected to the network.

Console Error Messages

The error messages which Solaris sends to the system console may provide you with a lot of information about the health of the system and the network. Unfortunately, many administrators do not pay attention to the console messages (or worse yet, leave the console window iconized).

Fortunately, Solaris provides the administrator with a simple facility to capture all console messages in a single location. The Solaris kernel uses the **syslog** facility to log messages to the system console. However, syslog also allows these messages to be sent to remote hosts.

/etc/syslog.conf File

The */etc/syslog.conf* file controls the action of the syslog facility. The Solaris kernel defines several levels of message severity. Entries in the *syslog.conf* file configure the syslog facility to handle these different message categories. For example, a simple informational message may be ignored, while a system hardware error may be reported to the console with much fanfare. A typical */etc/syslog.conf* file is shown in the next illustration.

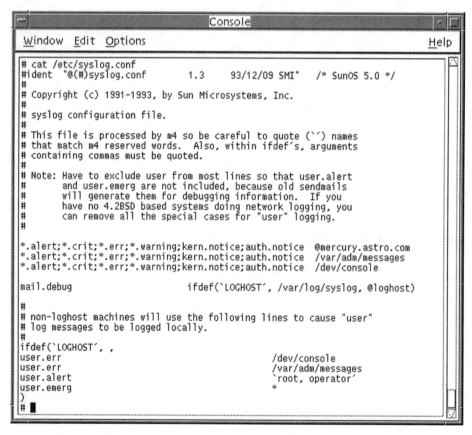

```
# cat /etc/syslog.conf
#ident  "@(#)syslog.conf     1.3     93/12/09 SMI"   /* SunOS 5.0 */
#
# Copyright (c) 1991-1993, by Sun Microsystems, Inc.
#
# syslog configuration file.
#
# This file is processed by m4 so be careful to quote (`') names
# that match m4 reserved words.  Also, within ifdef's, arguments
# containing commas must be quoted.
#
# Note: Have to exclude user from most lines so that user.alert
#       and user.emerg are not included, because old sendmails
#       will generate them for debugging information.  If you
#       have no 4.2BSD based systems doing network logging, you
#       can remove all the special cases for "user" logging.
#

*.alert;*.crit;*.err;*.warning;kern.notice;auth.notice  @mercury.astro.com
*.alert;*.crit;*.err;*.warning;kern.notice;auth.notice  /var/adm/messages
*.alert;*.crit;*.err;*.warning;kern.notice;auth.notice  /dev/console

mail.debug                      ifdef(`LOGHOST', /var/log/syslog, @loghost)

#
# non-loghost machines will use the following lines to cause "user"
# log messages to be logged locally.
#
ifdef(`LOGHOST', ,
user.err                                        /dev/console
user.err                                        /var/adm/messages
user.alert                                      `root, operator'
user.emerg                                      *
)
#
```

Typical contents of the /etc/syslog.conf file.

Logging Messages to a Remote Host

In addition to controlling the local disposition of messages, the */etc/syslog.conf* file also allows the administrator to send error messages to a central log host on the network. This procedure enables you to monitor one log file which will contain messages from all machines on the network. In order to enable remote syslog capabilities, you should add a line similar to the following to the */etc/syslog.conf* file.

```
*.alert;*.crit;*.err;*.warning;kern.notice;auth.notice @mercury.astro.com
```

Once the previous directive has been added to the */etc/syslog.conf* file, you can restart the *syslogd* process on the client machine by sending a **kill -HUP** signal to the process ID of the syslogd process. The kill -HUP signal will cause the syslog process to re-read the configuration file, and restart with the new parameters. Once you have set up all hosts on the network to log errors to a single location, monitoring the error log is as simple as using the **tail -f /var/adm/messages** command on the log host.

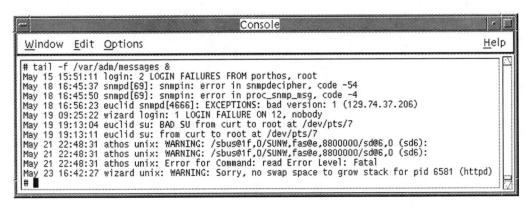

```
# tail -f /var/adm/messages &
May 15 15:51:11 login: 2 LOGIN FAILURES FROM porthos, root
May 18 16:45:37 snmpd[69]: snmpin: error in snmpdecipher, code -54
May 18 16:45:50 snmpd[69]: snmpin: error in proc_snmp_msg, code -4
May 18 16:56:23 euclid snmpd[4666]: EXCEPTIONS: bad version: 1 (129.74.37.206)
May 19 09:25:22 wizard login: 1 LOGIN FAILURE ON 12, nobody
May 19 19:13:04 euclid su: BAD SU from curt to root at /dev/pts/7
May 19 19:13:11 euclid su: from curt to root at /dev/pts/7
May 21 22:48:31 athos unix: WARNING: /sbus@1f,0/SUNW,fas@e,8800000/sd@6,0 (sd6):
May 21 22:48:31 athos unix: WARNING: /sbus@1f,0/SUNW,fas@e,8800000/sd@6,0 (sd6):
May 21 22:48:31 athos unix: Error for Command: read Error Level: Fatal
May 23 16:42:27 wizard unix: WARNING: Sorry, no swap space to grow stack for pid 6581 (httpd)
#
```

Monitoring syslog messages using tail -f /var/adm/messages.

Network Monitoring

While the kernel syslog messages may provide some information about the health of the network, it is sometimes necessary to more closely examine the network. The following sections cover selected utilities to inspect network status, and tips for using these utilities to troubleshoot network connectivity problems.

Simple Network Management Protocol (SNMP)

A more sophisticated method of monitoring networks is to use a network management tool based on the simple network management protocol (SNMP). The SNMP package allows a network administration host to constantly monitor the network. Information available to the SNMP software includes network utilization, host error counts, host packet counts, and routing information. SNMP allows you to determine normal usage of the network and implement alarms to warn of impending problems.

SNMP operates in a client/server mode. One machine on the network is designated as the SNMP network monitor station. It is configured to poll hosts on the local network in order to collect data into a central repository. The data may be made available to other packages in order to implement alarms, generate graphs of network utilization, and other off-line processing.

The other hosts on the network are configured as SNMP clients. These hosts run a process that watches for polling requests from the network management station. When such a request is received, the SNMP agent code collects the data from the workstation's kernel and forwards it to the management station.

Solaris provides the SNMP server and agent utilities as optional software packages in the SUNWconn software cluster.

Remote MONitor (RMON)

Extensions to the SNMP software packages have recently permitted more advanced monitoring of network usage. The Remote MONitor (RMON) package allows the administrator to monitor which applications are utilizing the network and which users are running these applications, giving you the ability to perform bandwidth utilization scheduling for the corporate network. RMON also provides you with the capability of monitoring disk space usage, processor utilization, memory usage, and other system monitoring functions. RMON is not bundled with the Solaris kernel, but may be obtained from several commercial third party sources.

Solstice Domain Manager

Tools such as Sun Microsystem's Solstice Domain Manager package use the data collected by the SNMP software to perform higher-level network monitoring. These packages typically include utilities that have the capacity to "discover" systems connected to a network. The discovered systems are then used to construct an interactive schematic of the network. Alarms (thresholds) on numerous network "health" statistics can be set. If a system on the network fails, an alarm for that condition would be tripped, alerting the network manager of the failure by flashing the icon of the failed system on the schematic.

Sun's Solstice Domain Manager is not bundled with Solaris 2. It is a separate product marketed by Sun-Connect, a division of Sun Microsystems. Hewlett Packard's OpenView product provides capabilities similar to Domain Manager, and is available for a wide range of platforms including UNIX, Windows NT, and Windows 95.

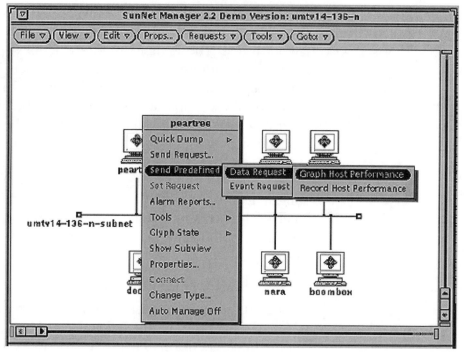

Monitoring network traffic using Solstice Domain Manager.

Network Troubleshooting

Tracking down network problems is similar to detective work. In time, administrators develop an instinct for where to seek clues. The following discussion is by no means an exhaustive troubleshooting scenario. It will, however, provide some basic tools and techniques that all seasoned network managers use.

snoop Utility

There are several ways to monitor network activity and usage. One basic method is to use the **/usr/sbin/snoop** command. The output from snoop is generally voluminous and somewhat cryptic. However, it is a good command to use when trying to determine whether two hosts on a network are communicating, or to identify the type of network traffic being generated. Some of the more common uses of snoop are summarized below.

❏ Monitoring traffic between two hosts. For example, snoop can be very useful when troubleshooting two systems that cannot communicate. For example, a diskless client might issue **tftpboot** requests that are never answered. You could use snoop to capture packets from the diskless client to determine if the client is issuing "proper" requests. Next, you could monitor the boot server network traffic to determine if the server is replying to the boot requests as follows:

```
snoop -d hme0 between client_name server_name
```

❏ Monitoring traffic using a particular protocol. For example, a host on the network may not be able to resolve IP to MAC addresses. This typically means that the address resolution protocol (ARP) process is failing. You could use the snoop command to watch the network for ARP requests and the ensuing replies from the name server as follows: **snoop -d hme0 arp**.

❏ Monitoring traffic containing specific data.

❏ Monitoring a network to determine which host is creating the most network traffic. The snoop command even provides a "cheap and dirty" way to determine which hosts are generating the most traffic on the network. You could start snoop and visually monitor which hosts are generating traffic (**snoop -d hme0**). Alternately, you could generate some shell scripts to invoke snoop to capture a specified number of packets, and then analyze which hosts were sending and receiving packets during the sample window. For example, **snoop -d hme0 -o /tmp/snapshot -c 10000** will collect 10,000 packets and save them in the */tmp/snapshot* file. You can then use **snoop -i /tmp/snapshot | grep hostname | wc -l** to determine how many of the 10,000 packets *hostname* contributed.

☞ **WARNING:** *By default, snoop is a root restricted command because the network interface device special file is created with root read permission. Some application packages (such as Macintosh Application Environment, MAE) require that the network interface run in promiscuous mode. This changes the read/write modes of the device special file such that any user on the host can use the* snoop *command to capture packets. You may wish to have the* root *crontab invoke a utility such as* ifstatus *periodically to determine the mode of the network interface(s) on a host.*

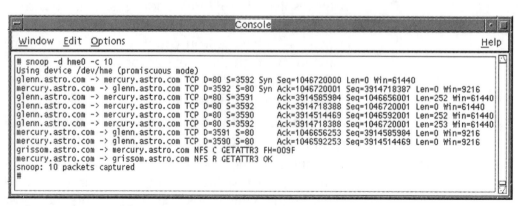

Monitoring network traffic using snoop.

ping Utility

If a host cannot be contacted on a network, the first question to answer is whether the problem is specific to the host or the network. A very useful command to make this determination is **ping**. The command name stands for "packet Internet groper," and is used to test reachability of hosts by sending an ICMP echo request and waiting for a reply. The command name is often used as a verb, such as in "Ping *cooper* to see if it's up."

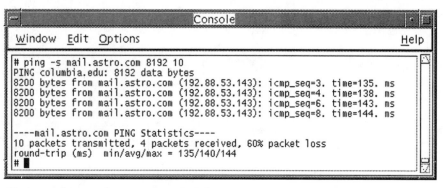

```
# ping -s mail.astro.com 8192 10
PING columbia.edu: 8192 data bytes
8200 bytes from mail.astro.com (192.88.53.143): icmp_seq=3. time=135. ms
8200 bytes from mail.astro.com (192.88.53.143): icmp_seq=4. time=138. ms
8200 bytes from mail.astro.com (192.88.53.143): icmp_seq=6. time=143. ms
8200 bytes from mail.astro.com (192.88.53.143): icmp_seq=8. time=144. ms

----mail.astro.com PING Statistics----
10 packets transmitted, 4 packets received, 60% packet loss
round-trip (ms)  min/avg/max = 135/140/144
#
```

Pinging a host to determine reachability.

By logging on to several hosts on a network and "pinging" other hosts, you can quickly determine whether the problem is network or host related. For example, if *mercury* is able to "ping" all hosts in its network except *cooper,* you can then assume that the network is functioning and the problem centers around *cooper.* The lack of a reply from *cooper* could be the result of a variety of difficulties. Some of the more common problems are listed below.

❏ Faulty physical connection

❏ Faulty host interface

❏ Corrupted */etc/inet/hosts* file

❏ Missing or corrupted */etc/hostname.le0* file

❏ Interface is turned off

❏ Host is down

Rebooting the non-communicating hosts will frequently solve the problem. If rebooting does not work, examining the files and physical components listed above would be the next step. In the absence of network diagnostic equipment, replacing components may be the only way to track down the failure.

A simple network.

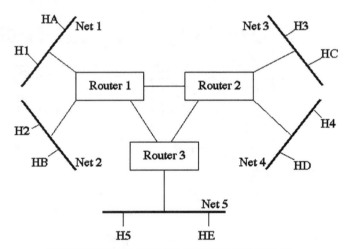

If, on the other hand, no hosts reply to pings, then the problem is specific to the network. Some of the more common network faults are listed below.

❏ Missing or damaged terminators

❏ Broken or damaged cable (may not be visible)

❏ Faulty connectors

❏ Damaged transceiver (may be jamming the network)

Once again, in the absence of network diagnostic equipment, replacing components may be the only way to precisely identify the cause of failure.

If a host on one subnet is unable to contact a host located on a different subnet, and there is no difficulty "pinging" hosts on the local subnet, then the problem could be one of the items listed below.

❏ Malfunction of the router connected to the subnet

❏ Malfunction of the network on the other side of the router

❏ Malfunction of a distant router not directly connected to the subnet

❏ Malfunction of the remote host

❏ Failure of a component or connection between the local host and the remote hosts

To locate the problem, use the techniques discussed earlier in this section for each subnet in question. With subnetted networks, ping is often all that is required to locate (or at least narrow down) the source of a network failure.

traceroute Utility

The **traceroute** utility does exactly what its name implies: it traces the route from point A to point B.

This utility can be very useful when trying to determine why two hosts cannot communicate. For example, to trace the route from a specific local host to a distant host, the administrator could invoke the following command:

```
traceroute fully_qualified_hostname
```

The traceroute command prints an informational message about each "hop" along the route from the local host to the remote host. Routing loops and inoperative hosts are very easy to spot in the traceroute output.

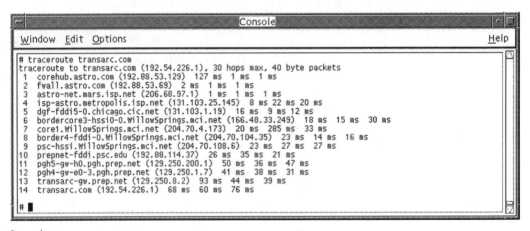

```
# traceroute transarc.com
traceroute to transarc.com (192.54.226.1), 30 hops max, 40 byte packets
 1  corehub.astro.com (192.88.53.129)  127 ms  1 ms  1 ms
 2  fwall.astro.com (192.88.53.69)  2 ms  1 ms  1 ms
 3  astro-net.mars.isp.net (206.68.97.1)  1 ms  1 ms  1 ms
 4  isp-astro.metropolis.isp.net (131.103.25.145)  8 ms 22 ms 20 ms
 5  dgf-fddi5-0.chicago.cic.net (131.103.1.19)  16 ms  9 ms 12 ms
 6  bordercore3-hssi0-0.WillowSprings.mci.net (166.48.33.249)  18 ms  15 ms  30 ms
 7  core1.WillowSprings.mci.net (204.70.4.173)  20 ms  285 ms  33 ms
 8  border4-fddi-0.WillowSprings.mci.net (204.70.104.35)  23 ms  14 ms  16 ms
 9  psc-hssi.WillowSprings.mci.net (204.70.108.6)  23 ms  27 ms  27 ms
10  prepnet-fddi.psc.edu (192.88.114.37)  26 ms  35 ms  21 ms
11  pgh5-gw-h0.pgh.prep.net (129.250.200.1)  50 ms  36 ms  47 ms
12  pgh4-gw-e0-3.pgh.prep.net (129.250.1.7)  41 ms  38 ms  31 ms
13  transarc-gw.prep.net (129.250.8.2)  93 ms  44 ms  39 ms
14  transarc.com (192.54.226.1)  68 ms  60 ms  76 ms
#
```

Sample traceroute output.

References

The following references are listed in an attempt to provide the interested reader with more information on the topics covered in this chapter. The list is not intended to be exhaustive, but should provide the reader with a knowledge base on the topic of computer networking.

❏ Douglas Comer, *Internetworking with TCP/IP, Volume 1.* Prentice Hall, 1991 (ISBN 0-13-216987-8).

❏ Craig Partridge, *Gigabit Networking.* Addison-Wesley, 1994 (ISBN 0-201-56333-9).

❏ Craig Hunt, *TCP/IP Network Administration.* O'Reilly & Associates, Inc., 1992, 1993 (ISBN 0-937175-82-X).

❏ Scott Ballew, *Managing IP Networks with Cisco Routers.* O'Reilly & Associates, Inc., 1997 (ISBN 1-56592-320-0).

❏ William Stallings, *SNMP, SNMP-v2, and RMO, 2nd ed.* Addison Wesley, 1996 (ISBN 0-201-63479-1).

Summary

This chapter examined selected aspects of the terminology associated with computer networking. Topics included the different classes of Internet addresses, how a corporation can obtain a registered Internet address, network routing, network configuration commands, and network database files. Finally, the chapter explored ways to monitor and troubleshoot local area networks. The next few chapters focus on how the network is used for network file system (NFS) file sharing, as well as a few of the popular network name services.

Network Security

Connecting Solaris systems on a network or to the Internet brings with it a new level of security concerns. Primary among these concerns is the trust placed in the management of each of the interconnected systems and the control of access to each system. Securing systems on the Internet is particularly difficult. In some instances, you may wish to make certain services widely available to systems that may not be trusted while preventing access to other services. New security flaws in existing operating systems and application softwares are continually being uncovered. Meanwhile, tools to exploit these flaws are distributed.

In this chapter, the basics of securing a Solaris system from network security problems are discussed. The types of potential security problems are outlined along with some typical solutions. System administrators who intend to connect systems to the Internet should consider reading further from the list of sources at the end of the chapter.

Local versus Network Security

The principal difference between network and local security discussed earlier in Chapter 7, "Managing System Security," is the focus on access methods, and control and monitoring of access. Network services provide a wide variety of access methods. Each type of service has different methods for access control and monitoring of usage.

Once a user has access to a system, the security of the system depends on local security. Therefore, the first job in securing a computer for use on a network is to deal with local system security, the quality of user passwords, and the protection given to files stored on the system. Tools such as ASET, password quality, and aging controls previously described in Chapter 7 can be employed to make the local system more secure.

Access Methods

A standalone system locked in a room is usually secure, but few systems are configured this way. For most Solaris systems some form of network access is part of normal operations. Additional access considerations pertain to connections via modem.

Connection via Modem

There are several ways in which a modem can be configured and used. From a security standpoint, a connection via a modem is a method of remote access that requires control. Dial-in services include both terminal access and network connections via the point-to-point protocol (PPP).

Dial-out service is secure in the sense that your machine is initiating the connection. Users making the dial-out connection should be familiar with the system to which they connect. Users also need to examine any downloaded software or data for potential security problems (as would be the case when downloading such items to a PC or Macintosh). Although viruses are very rare in the Solaris environment, programs that masquerade as useful tools but execute undesirable functions such as deleting files have been distributed.

Dial-in service is a more significant problem. Dial-in modems will respond for anyone who knows or accidentally discovers the modem's phone number. Merely making the number unlisted is not always sufficient because determined individuals will use techniques such as automated dialers or reverse phone directories to identify unlisted numbers. The modem connection will provide a log-in prompt to the person connecting to it, just like a local terminal. At this point, the security of each account depends on its password. Consider the password aging and inactivity controls presented in Chapter 6, "Creating, Deleting and Managing User Accounts."

Using Dial-in Passwords

Solaris provides a dial-in password, an additional aid for defending dial-in terminal access. The dial-in password is controlled by two files, */etc/dialups* and */etc/d_passwd*. As seen in the following steps, creating a dial-in password is not as easy as it should be.

1. Create */etc/dialups*. This file contains a single entry per line with the path of each dial-in modem line (e.g., */dev/term/a*).

2. Create */etc/d_passwd*. This file contains a single entry for each possible shell that a dial-in modem user might use. A sample *d_passwd* file appears below.

```
/usr/lib/uucp/uucico: passwd-goes-here:
/usr/bin/csh: passwd-goes-here:
/usr/bin/ksh: passwd-goes-here:
/usr/bin/sh: passwd-goes-here:
```

➤ **NOTE:** *The* passwd-goes-here *components will be filled in during a later step.*

3. Set the owner, group, and permission bits for the files using the following commands:

```
chown root /etc/dialups /etc/d_passwd
chgrp root /etc/dialups /etc/d_passwd
chmod 600 /etc/dialups /etc/d_passwd
```

4. Insert an encrypted password for each shell in the place of the *passwd-goes-here* strings shown above. There are several ways to accomplish this. One way is to temporarily set the root password to each of the shell passwords, save the results, and then set the root password back to what it was when you started. This procedure is inconvenient, but not difficult. An example of the sequence for the *uucico* shell follows. Note that the password is typed after each of the password prompts but does not appear on the screen.

```
# passwd
New password:
Re-enter new password:
grep root /etc/shadow | awk -F: '{print $2}' > /tmp/foo
```

```
# passwd
New password:
Re-enter new password:
#
```

The */tmp/foo* file now contains the encrypted password string which will be placed in the */etc/d_passwd* file. You can now use **vi** or your favorite editor to insert this string in between the colons after the shell you want to protect.

PPP Security

Network connections over modems using PPP also provide additional security measures in the form of the password access protocol (PAP) and the challenge-handshake authentication protocol (CHAP). PAP adds a password check on top of any password required to log in and begin the PPP session. CHAP provides a more secure authentication method by periodically verifying the identity of the connected systems. CHAP works as follows: a challenge message is sent from one PPP node to the other; a response is returned; and the response is checked against a secret key that is not exchanged.

Configuring PPP to add one or both of these authentication schemes requires editing the PPP configuration file, */etc/asppp.cf*. Only the security option keywords in */etc/asppp.cf* are discussed here. See Chapter 13, "Adding Terminals and Modems," for a complete description of the PPP configuration file and PPP setup.

PAP and CHAP Setup

Both the server and client *asppp.cf* files must be modified to enable authentication and include the PAP and CHAP identification strings and passwords. Starting with the server, the lines added include *require_ authentication, will_do_authentication, pap_id, pap_ password, chap_name,* and *chap_secret.* The *require_ authentication,* and *will_do_authentication* directives control the authentication protocol to be used. The system administrator can choose to use PAP, CHAP, both PAP and CHAP, or to leave these lines out and use no authentication. An example of a PPP server's *asppp.cf* file using both PAP and CHAP follows.

```
# PPP server grissom, serving client shuttle
ifconfig tpdptp0 plumb shuttle down
path
        interface ipdptp0
        peer_system_name shuttle
        require_authentication pap chap
        pap_peer_id atlantis
        pap_peer_password bigboote
        chap_peer_name abilene
        chap_peer_secret 23$56%afbf(-+=P
```

NOTE: *While the PAP name and password are exchanged in the clear, the CHAP secret is never transmitted over the modem link. To preserve the security of the CHAP secret, it should be provided to the remote host by talking to the remote system administrator directly rather than by using e-mail.*

The client's *asppp.cf* is almost exactly a mirror of the server's except for the use of the *will_do_authentication* directive in place of the *require_authentication* directive. Appearing below is an example for *shuttle,* the remote host listed in the server example.

```
# PPP client shuttle, connecting via server grissom
ifconfig ipdptp0 plumb shuttle grissom up
```

```
path
        interface ipdptp0
        peer_system_name grissom
        will_do_authentication pap chap
        pap_peer_id atlantis
        pap_peer_password bigboote
        chap_peer_name abilene
        chap_peer_secret 23$56%afbf(-+=P
```

Once the PPP link has been established, security between the remote host and server depends on the same factors as for hosts directly connected to the network. The remainder of this chapter is focused on systems connected via network connections.

Network Services

Solaris, like other systems using TCP/IP networking, provides access to server processes from the network by using connection points known as ports. A *port* is an addressable location that a remote machine can refer to in order to obtain a service from a local machine. Ports are sometimes called *sockets*, based on the name given to the programming library interface made popular by Berkeley UNIX.

To understand ports, think about a large building. The name of the building is similar to the *hostname* of a machine. The building's street address is similar to the IP address of the machine. The loading dock and the mail room are similar to ports on a machine. Ports with agreed upon names and numbers are "well-known," used for providing services such as e-mail, Web service, telnet, and ftp. Other ports are defined as needed and agreed upon by cooperating machines. A typical server process accepts a connection on its well-known port and then starts a new process to handle the service request and informs the connecting client to switch to another port to conduct the service. This procedure frees the well-known port for another connection.

A typical network service request.

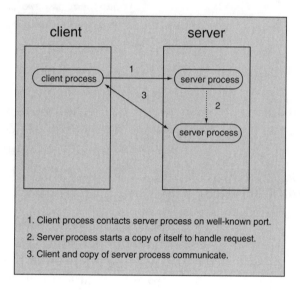

1. Client process contacts server process on well-known port.
2. Server process starts a copy of itself to handle request.
3. Client and copy of server process communicate.

/etc/inet/services

The collection of well-known ports is listed in the */etc/inetd/services* file as discussed in Chapter 17, "Network Configuration and Management." This list is used by client and server processes to determine the port number of a given service from its name.

For example, the mail transport agent, **sendmail**, uses the simple mail transport protocol (**smtp**) to move electronic mail between systems. The smtp port, number 25, is a well-known port where a sendmail client or mail agent can connect to a remote sendmail process to send mail and where a server sendmail listens to accept incoming mail.

Daemons

There are three basic methods of providing a network service, each with its own security implications. In the first method, server processes such as **sendmail** are started when a Solaris system boots, and run continuously listening for and processing service requests arriving on a particular port. These processes are also known as *daemons* to denote their continual background activity.

New services are added by installing new server daemons. An example of this might be a software license server as is used by some software packages to control the number of concurrent users. To add the service, a line would be added to */etc/inet/services* on all the machines using the service. Then the server program itself would be installed and started running as per directions provided by the server's manufacturer.

inetd

Infrequently required services may be started on demand by a special daemon known as **inetd**. The inetd daemon listens to a collection of ports as listed in its configuration file, */etc/inet/inetd.conf,* as described in Chapter 17, "Network Configuration and Management." When inetd detects a connection on a port, it starts the server process for that port as listed in the configuration file.

rpcbind and RPC services

An NFS (network file system) is built on a remote access technique known as remote procedure call (RPC). Access to RPC services is made via sockets assigned dynamically by a service called **rpcbind**. To access an RPC service, the client first checks with the server's rpcbind daemon to determine the port number being used by the required service and then connects to the service using the port number returned by the rpcbind daemon.

r Services

The commonly used remote terminal access and remote process execution commands, **rlogin** and **rsh**, provide access similar to a directly connected terminal or window on a workstation. The server processes for these services are started by **inetd**. However, the **r** commands present additional concerns because they have access controls on a per host and per user basis as well as the usual log-in password protection. Individual users can place pairs of system names and user names in a file in respective home directories known as *.rhosts*.

A secure .rhosts file has the fewest possible entries.

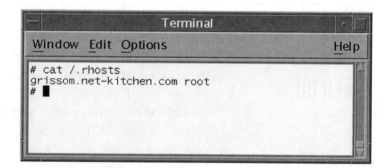

Placing a machine and user pair in a *.rhosts* file confers a high level of trust on the listed machines and users by *removing the need for passwords when establishing remote connections.* If user A and machine Y are listed in user B's *.rhosts* file, then user A can connect to a machine where user B has an account from machine Y without being asked for a password.

While it is very convenient to allow such easy access between machines, the security implications of such access are considerable, especially if the root account is supplied with a *.rhosts* file. Allowing access between machines without requiring passwords makes the collection of machines as secure as the least secure machine in the group. Furthermore, unless the system administrator is certain of the accessibility to each machine, allowing access via *.rhosts* may open a system up to much more access than is intended. In general, the *.rhosts* file should contain the fewest number of entries possible.

Access Control Methods

As discussed above, a networked Solaris system has numerous daemons listening on numerous ports for connections. Each port is an access point to the system. Securing a networked Solaris system is a matter of controlling access to each service in turn. This is far from simple because access control methods differ according to type of service.

Turn It Off

The simplest and most complete access control is to turn off a given service. If no service daemon is running, no entry for a port or RPC service is made in the **inetd** configuration file. Consequently, no connection can be made on the port. Turning a service off is analogous to walling up the loading dock in the building example used earlier. In general, if the service is not used and the environment demands additional security, turn the service off. In the following example, the *finger* service is turned off by inserting a pound sign (#) ahead of its entry in */etc/inet/ inetd.conf*, and then rebooting the system.

```
# fingerd has been turned off --ddm
# finger stream  tcp     nowait  nobody  /usr/sbin/in.fingerd   in.fingerd
```

This approach is of limited value because the most significant benefit of a distributed system is accessing remote resources over the network. In most situations, controlled access to services allowing a specific set of machines or a specific type of service is required, such as e-mail, NFS, or WWW. Due to the wide variety of service methods, securing every service is challenging. The major services, service methods, and access controls are discussed in subsequent sections.

r Services

In addition to the user level *.rhosts* file, the commonly used remote terminal access and remote process execution commands, **rlogin** and **rsh**, provide access controls on a per host basis. Per host access is based on a host being listed in */etc/hosts.equiv*. For systems running NIS+, this file may contain the plus symbol (+) which allows the addition of hosts from the NIS hosts database. Consequently, the list of hosts may be much larger than desirable. Individual users can place pairs of system names and user names in a file in respective home directories known as *.rhosts*.

Although allowing access between machines without requiring passwords is very convenient, the security implications of such access are considerable. In general, the */etc/hosts.equiv* file should be empty or simply absent (if at all possible).

Absence of the /etc/hosts.equiv file, the most effective security measure.

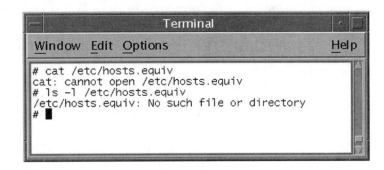

```
# cat /etc/hosts.equiv
cat: cannot open /etc/hosts.equiv
# ls -1 /etc/hosts.equiv
/etc/hosts.equiv: No such file or directory
#
```

Secure NFS and RPC

Chapter 19, "Accessing Network Resources with NFS," describes how to set up a secure NFS by enabling a secure form of the underlying RPC services. This provides a reasonable amount of security for the data being transported over the network by NFS. The security depends on the site using NIS+ to provide a mechanism to share the cryptographic keys among the systems sharing files.

Wrapper Software (tcpd)

What can be done for the other services besides the **r** commands and NFS? There are three basic approaches. First, current secure versions of the programs used to provide the service should be installed and carefully configured. This is particularly important with services that are usually made widely available, such as e-mail and Web page services. (See Chapter 22, "WWW Administration," for a discussion of Web server security.) Old versions of software with known security problems and misconfigured software are among the most common network security problems.

Sites connected to the Internet are encouraged to track security updates provided by Sun and other software vendors and should consider installing alternative versions of services that provide additional security features. Suggestions are listed at the end of the chapter.

A second approach is to control access to the service by adding software to evaluate every access before the service is invoked as shown in the next illustration. This approach is aimed at services started by **inetd** and has been successfully used for several years. The software, known as a *wrapper,* is available from the CERT Web site listed at the end of the chapter.

TCP based service protected by the tcpd wrapper.

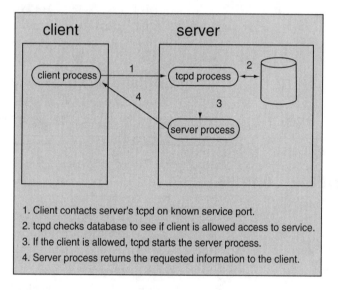

1. Client contacts server's tcpd on known service port.
2. tcpd checks database to see if client is allowed access to service.
3. If the client is allowed, tcpd starts the server process.
4. Server process returns the requested information to the client.

Once the wrapper program is installed, a given service is protected by modifying its entry in */etc/inetd.conf* to invoke the wrapper program which validates the connection according to access rules and then starts the service. An example of this configuration for the *finger* service appears in the following illustration.

```
# grep finger /etc/inetd.conf
#finger stream  tcp      nowait  nobody  /usr/sbin/in.fingerd      in.fingerd
finger  stream  tcp      nowait  nobody  /usr/local/etc/tcpd      in.fingerd
# cat /etc/hosts.allow
ALL: LOCAL grissom.net-kitchen.com apollo.net-kitchen.com
#
```

Protecting the finger service using tcpd.

The two lines extracted from */etc/inet/inetd.conf* using **grep** show the original configuration line for *finger* beginning with a pound sign (#) to disable it. The second line is used in its place. The sixth field in this line shows **tcpd** as the program to be run with the seventh field showing the program name as *finger.* The tcpd consults the */etc/hosts.allow* file, shown next in the example. If the connecting host is one of those listed, tcpd will start the *finger* process and allow the request to be completed. If the connecting host is not listed, tcpd will simply close the connection. (For more information on configuring tcpd, see the software documentation.)

While a wrapper program works well, it is not a comprehensive answer to access security. Wrappers involve an extra process invocation resulting in a performance loss. Building and installing tcpd requires a compiler and other development tools because the program is provided in source code form. The tcpd wrapper program is also only usable for TCP based services; it does not handle UDP or RPC based services. Wrapper programs also require updating the access rules as necessary.

Firewalls and Filtering Routers

A more general approach is to pass all network traffic between a secured network and other networks or the Internet through a filtering router and/or a firewall. This approach constitutes an attempt to deal with all possible avenues of access by carefully controlling the network traffic as it passes through. An advantage is the securing of all services, including those added later or created by individual users.

All network traffic filtered by a firewall.

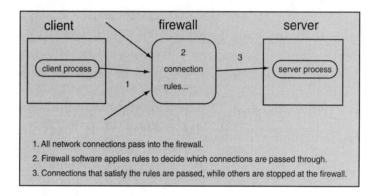

1. All network connections pass into the firewall.
2. Firewall software applies rules to decide which connections are passed through.
3. Connections that satisfy the rules are passed, while others are stopped at the firewall.

A filtering router is a router which conditionally passes network traffic between networks. Depending on the particular router it may pass traffic based on the source or destination IP address, port number, or some combination of the same. Access to a service on a machine in a network protected by the filtering router is controlled by allowing or disallowing the network traffic to enter or leave the network.

However, there are situations in which it is desirable to exercise finer control over the network traffic. A firewall provides additional control over the network traffic passing through it. Depending on design, firewalls may provide the services of a filtering router as well as hiding the IP addresses of machines behind the firewall and controlling specific content from passing through via specific services such as WWW.

Setting up an effective firewall requires careful consideration of a large number of factors. If you are working at a site requiring this form of access control, consider referring to the source on firewalls listed at the end of the chapter.

Monitoring Access

No matter how fine tuned the control over access to a given machine, breakdowns in access control may still occur. Software bugs and configuration errors in any of the previously mentioned control files are two common ways in which access control can break down. Tools are available to probe systems for purposes of uncovering known bugs and configuration errors. To cope with such problems and to evaluate the quality of the access control, monitoring is needed.

Log Files

The security conscious system administrator should routinely monitor system log files. These files can give you clues to potential problems in other areas such as failing hardware. Some of these files are written by the syslog service daemon, **syslogd**.

su Command and Log File

The **su** command provides a method for gaining the authority of another user by supplying the user's password when prompted. Employed without a user name, the su command offers access to the root user. Because of the authority given to the root user, the su command maintains its own log file.

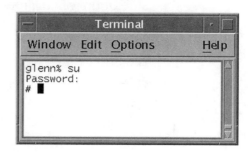

The su command in action.

The log file produced by the su command is one of the first places a system administrator should check on a routine basis. The command logs both successful and unsuccessful attempts to obtain root privileges. Appearing below is a sample su log from the machine named *glenn.* The successful su attempts have a plus symbol (+) listed after the date and time; unsuccessful attempts show a hyphen or minus sign (-) after the date and time. Unsuccessful attempts should be investigated, especially by persons not authorized to perform system administration tasks.

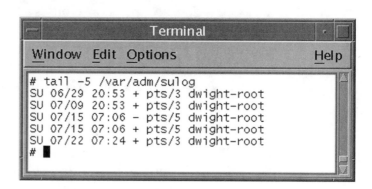

Last five lines of the su command log file.

Log files that should be routinely checked are listed in the next table.

File	Contents
/var/adm/sulog	Log of the su command. Entries containing a minus sign (-) after the date and time indicate a failed password entry.
/var/adm/log/asppp.log	PPP software logs errors and connections here. Watch for failed connections as indicators of problems.
/var/log/syslog	By default, the syslog service writes error messages produced by sendmail here. Watch for errors of all kinds.
/var/adm/messages	By default, the syslog service writes error messages from a number of different daemons to this file. Watch for errors of any kind.

⟶ **NOTE:** *The* syslog *daemon is controlled by a configuration file,* /etc/syslog.conf, *which can be modified to increase the detail of recorded log file messages, and optionally, to display messages on the system console or broadcast to the terminals of specific users.*

Be aware that a determined individual who has gained unauthorized access to a machine you manage will most likely know about these log files and will modify them to remove any indication of her actions. Use log file monitoring needs in concert with tools like **ASET**, **COPS**, or **tripwire** to provide a high level of assurance that unauthorized access is discovered.

Checking for Unauthorized Access

As described in Chapter 7, "Managing System Security," Solaris provides a local security evaluation tool called **ASET**. This tool can be used not only to alter system security, but also to monitor a system for changes that may indicate the system has been accessed in an unauthorized manner. Consider running ASET and checking the log files it produces on a routine basis. As an additional precaution, ASET logs should be saved off line on tape or diskette.

A still more advanced approach is to use an additional local security checking tool such as **COPS** or **tripwire**, available from the CERT Web site listed in the "References" section at the end of the chapter. These tools permit the detection of unauthorized access.

Checking for Potential Security Problems

Systems can be checked for potential network security problems by employing tools to examine available network services. **SATAN**, the most well-known of these tools, will automatically search the network services of a selected system for known problems and produce a report on what it finds. The program can be found in the "References" section below.

References

The discussion of network security in this chapter has been brief. If you require more detailed information, consider the sources listed in the following tables.

Books

Source	Comments
Simson Garfinkle and Gene Spafford, *Practical UNIX and Internet Security*, 2nd edition. O'Reilly & Associates, April 1996 (ISBN 1-56592-148-8).	A detailed and complete guide to security for UNIX systems connected to the Internet.
Simson Garfinkle and Gene Spafford, *Web Security and Commerce*. O'Reilly & Associates, June 1997 (ISBN 1-56592-269-7).	Targets Web server security and securing transactions executed over the Web.
Brent Chapman and Elizabeth D. Zwicky, *Building Internet Firewalls*. O'Reilly & Associates, September 1995 (ISBN 1-56592-124-0).	Covers the setup and maintenance of firewalls.

Web sites and software

Contact information	Comments
Computer Emergency Response Team (CERT); http://www.cert.org/; cert@cert.org; 1-412-268-7090	The national contact for computer security. This Web site contains a wealth of information on security matters and pointers to the security software mentioned in this chapter.
Computer Incident Advisory Capability (CIAC); http://ciac.llnl.gov/; ciac@llnl.gov; 1-510-422-8193	Government and education contact for computer security. Provides an excellent mailing list with updates on the latest security problems and their solutions.
http://www.cs.purdue.edu/homes/spaf/hotlists/csec-top.html	Professor Gene Spafford's well-maintained list of security related Web sites.

Summary

Securing networked systems is a challenge, especially for systems connected to the Internet. Solaris provides numerous network services, and all have the potential of offering unauthorized access. Careful attention to access methods and controls, and continual monitoring for unauthorized access can help to make any networked system more secure.

Accessing Network Resources with NFS

The network file system (NFS) is an integral component of many Solaris installations. Developed by Sun Microsystems in the mid-1980s, NFS is a published standard, or protocol, which allows file sharing among hosts over a network. It is a standard feature on most UNIX systems and is available for most operating systems, including VMS, MS-DOS/Windows, and Macintosh (MacOS). Solaris provides a secure version of NFS called Secure NFS, which provides enhanced security features to protect NFS operations from network security breaches. A discussion of secure network communications is covered later in the chapter.

NFS usefulness derives from its ability to transparently access files located on remote hosts via a network connection. For example, NFS allows hosts on a network to access data or programs from a central

location. This reduces system maintenance requirements by limiting the number of hosts, or locations,
where programs and data files must be installed. Furthermore, in environments where users regularly
work on different hosts, NFS can automatically mount
their centrally located home directories to any host
they log on to. To users, it appears as though respective home directories follow them from host to host.
NFS can also be used to share network devices such
as CD-ROMs, or files from different operating systems
such as the aforementioned VMS, UNIX, Macintosh,
or MS-DOS.

NFS Function and Terminology

The most basic NFS configuration consists of a single
host operating as a server and another operating as a
client. The server *shares* or makes available some or
all of its files by running a special set of processes.
These processes are run in the background and are
often called *daemons*.

The **share** command is used to instruct the daemons
about what to share with other hosts. By placing configuration information in a special file, */etc/dfs/dfstab*,
the share commands can be run automatically when
the server is booted.

The client attaches or *mounts* one or more of the file
system sections made available by the server. The
mount command incorporates the identified file system section into the local host's file tree in the same
manner that local disk-based file systems are
mounted. The client also runs a special set of daemon
processes and has information concerning the file systems to mount from the server in its configuration files.

Servers, Clients, and Sharing

A host that shares file systems via NFS is called a *server.* A host that mounts NFS file systems from other hosts is called a *client.* A host can be both client and server. Networks on which there are servers and clients are called *client-server networks.* Networks where hosts are both client and server to each other are often called *peer-to-peer networks.* Most large networks have a mix of clients, servers, and peers.

A file system which is offered for mounting by a server via NFS is said to be *shared.* This term comes from the **share** command used to offer file systems for mounting. Similar to other file system types, a remote file system is mounted on a client using the **mount** command. The shared portion of the server's file systems can be mounted in the same or in a different location on the client.

Remote Procedure Call (RPC)

A key to NFS operation is its use of a protocol called *remote procedure call* (RPC). The RPC protocol was developed by Sun Microsystems in the mid-1980s along with NFS. A *procedure call* is a method by which a program makes use of a subroutine. The RPC mechanism extends the subroutine call to span across the network, allowing the procedure call to be performed on a remote machine. In NFS, this is used to access files on remote machines by extending basic file operations such as read and write across the network.

To understand the purpose of RPC, first consider how the computer system interacts with the files stored on its local disk. When a Solaris program is read into memory from disk or a file is read or written, system calls are made to the read or write procedures in the SunOS kernel. As a result data are read from or written to the file system.

When a file is read from or written to an NFS file system on a network, the read or write calls are made over the network via RPC. The RPC mechanism performs the read or write operation needed on the server's file system on behalf of the client. If the NFS client or server is a different type of host or operating system, such as VMS or a PC running Windows, the RPC mechanism also handles data conversion. The result is that the information returned by the read or write call appears to be the same as if the read or write had occurred on a local disk. While this appears complex, it all occurs at the file system level and requires no changes to normal operations on files or internal programs. It is similar to reading and writing a floppy disk versus a hard disk or a CD-ROM. Regarding programs, NFS is just another file system. The transparent nature of NFS is one of its best features.

Typical NFS operation.

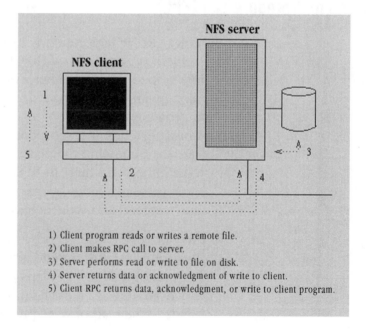

1) Client program reads or writes a remote file.
2) Client makes RPC call to server.
3) Server performs read or write to file on disk.
4) Server returns data or acknowledgment of write to client.
5) Client RPC returns data, acknowledgment, or write to client program.

NFS File System Differences

The major differences between an NFS file system and other file system types derives from the network connection between NFS clients and servers. NFS file systems can be used by several clients simultaneously. For example, it is possible to share the */usr/share/man* portion of a server's file system among many clients. Sharing a single file system among many clients allows the administrator to reduce the amount of disk space required for every system on a network. It also reduces the effort required to install and maintain software because a single copy of a program is shared among many clients.

While NFS is quite handy and versatile, it does have certain drawbacks. NFS is not as fast as local disks on Ethernet, especially some of the faster SCSI disks currently available. NFS can cause problems on clients when the server of a mounted NFS file system becomes unavailable or network problems occur. NFS is only as good as the underlying network connection. These problems and some preventive measures are discussed later in this chapter and in Chapter 20, "Automating NFS with automount."

Setting Up an NFS Server

The setup of an NFS server involves dealing with the administrative issue of UID and GID numbering, editing the *dfstab* file, and verifying that the required daemon processes are started. Once an NFS server has been set up, the necessary **share** commands and daemons will be automatically started when the server is booted.

Match UID and GID Numbers for All Users

Setting up a host as an NFS server is fairly easy. First, the administrator must ensure that the clients and servers use identical UIDs and GIDs for all users and groups. In other words, a user must have the same UID number on all hosts on which he or she will be sharing files via NFS. Mismatches can result in files being inadvertently accessible to users who might not normally have access to them. If the site is using NIS or NIS+ to handle the */etc/passwd* and */etc/group* files, the UID and GID numbers should already match up across the hosts that share the same NIS or NIS+ server. NFS uses the UID and GID numbers when identifying file access permissions on remote file systems. See Chapter 21, "Network Name Services," for information on setting up name services.

Insert share Commands in /etc/dfs/dfstab

With uniform UID and GID numbering in place, the next step is to add lines to the */etc/dfs/dfstab* file. Each line is a **share** command specifying the portion of the server's file tree to be shared, which hosts are allowed to mount it, and how they are allowed to mount it. It is also possible to issue the share command on the command line to allow a file system to be shared. Putting the command in */etc/dfs/dfstab* causes the file system to be shared every time the host is booted. A simple example appears in the next illustration.

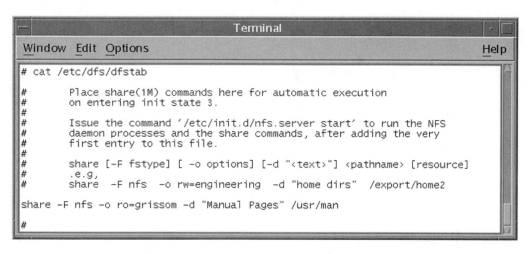

```
# cat /etc/dfs/dfstab

#       Place share(1M) commands here for automatic execution
#       on entering init state 3.
#
#       Issue the command '/etc/init.d/nfs.server start' to run the NFS
#       daemon processes and the share commands, after adding the very
#       first entry to this file.
#
#       share [-F fstype] [ -o options] [-d "<text>"] <pathname> [resource]
#       .e.g,
#       share  -F nfs  -o rw=engineering  -d "home dirs"  /export/home2

share -F nfs -o ro=grissom -d "Manual Pages" /usr/man

#
```

A simple /etc/dfs/dfstab file.

The final line of the *dfstab* file is the actual share command. The lines preceding it begin with a pound sign (#) and are treated as comments. The first option to share, **-F nfs**, specifies that the portion of the file system is to be shared via NFS.

The next flag reading from left to right is the **-o** flag, which specifies how the portion of the file tree is to be shared and by whom. Several options are available. The **ro=** and **rw=** options specify a list of hosts allowed to access the file system on a read-only or read-write basis, respectively. Other options can be used to control secure access to a file system and are discussed later in this chapter.

In the previous example, the **ro=grissom** example allows the host named *grissom* to mount the shared resource, */usr/man*, as read-only. Additional hosts can be added by separating their names with a colon (e.g., **ro=grissom:glenn**). This procedure can become unwieldy for large numbers of hosts. Creating a net group using NIS+ allows the administrator to group collections of hosts together and refer to them by a single name. See Chapter 21, "Network Name Services," for information on setting up NIS+ and creating a net group.

In the example, the **-d** option appears in the **share** command past the **-o** option flag and its arguments. This option allows a description of the shared item to be displayed when the server is queried concerning available shared items. The final item on the line is the path for the portion of the file system to be shared.

Start the Daemons

The next step is to start the required daemon programs. The daemons will start automatically during the boot sequence only if **share** commands are found in */etc/dfs/dfstab*. When setting up an NFS server for the first time, it will be necessary to start them manually after configuring the */etc/dfs/dfstab* file. A simple way to start these daemons is to run the shell script that would have been run during the boot sequence, as shown in the next illustration.

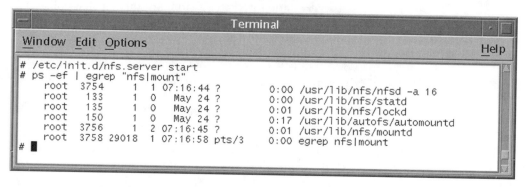

Manual start of nfs server daemons.

A Look at the nfs.server Script

In the previous example, the **nfs.server** script has been run, the **ps** command was used, and then its output was filtered with the **egrep** command to produce a list of the NFS daemons. The **statd** and **lockd** daemons provide for server crash recovery and file locking across the network. If the host will be running only as an NFS client, then these are the only daemons started.

When a server crashes and comes back into service, the statd daemon handles the resolution of any locks pending at the time of the crash. The statd on the client informs the system that a server has crashed and rebooted pending locks; information about file locking and status is not maintained across server crashes. The lockd enforces file locking on the shared file system by communicating with the lockd process on clients which mount and use a shared NFS file system.

Next, the **mountd** and **nfsd** daemons are started by the **nfs.server** script. The mountd daemon handles requests from clients to mount shared items. It uses the information provided by the **share** command to help it determine permission for mounting and the access allowed to a particular shared item. The nfsd daemon sets up the NFS mechanism inside the SunOS kernel and provides its interface to the network.

↝ **NOTE:** *The* nfsd *process does not directly handle RPC requests, but instead passes them on to the SunOS kernel. As a result, the* nfsd *process may not accumulate CPU time while NFS operations are occurring. Avoid using CPU time accumulated by* nfsd *as a measure of NFS operations.*

The number 16 as provided to the nfsd is used to allocate threads in the SunOS kernel. Each thread can handle one RPC operation at a time. A thread is a separate line of execution or control, allowing each request to be handled asynchronously without the overhead of separate processes handling each request. The number of threads set depends on the amount of NFS traffic the server is expected to handle. The amount of traffic is a function of the number of clients and how frequently RPC requests will be made. More threads will improve service under heavy loads up to the point at which the server's resources are fully utilized. The proper setting for a specific environment can be determined only by experimentation, and depends on factors such as the server's CPU type, CPU speed, disk system performance, network interconnections, and so forth. The default of 16 works well for most small to moderately sized sites.

Checking the NFS Server

In addition to using the **ps** command, it is possible to check on an NFS server by using the **showmount** and **share** commands. With showmount's **-e** option, it is possible to specify the name of a remote host and view the list of file systems it is currently sharing. Alternatively, the share command, when used without arguments, will list the shared file systems on the local system.

Checking the NFS server with the showmount and share commands.

NFS server setup is a one-time activity. Once a server is configured, the share commands and daemon processes will automatically be started as part of the system boot procedure. Adding shared directories is a matter of inserting them in the *dfstab* file and running the **share** command.

Setting Up an NFS Client

Configuring an NFS client is even easier. All the administrator need do is to specify remote file systems to mount in */etc/vfstab* and then issue the requisite **mount** commands. An NFS file system entry looks similar to entries for other file system types. Like other file systems listed in */etc/vfstab*, NFS file systems are automatically mounted at system boot. An example is shown in the next illustration.

```
—                               Terminal                        · ▢
 Window  Edit  Options                                        Help
# cat /etc/vfstab
#device         device          mount         FS    fsck    mount   mount
#to mount       to fsck         point         type  pass    at boot options
#
#/dev/dsk/c1d0s2 /dev/rdsk/c1d0s2 /usr         ufs   1       yes     -
fd        -        /dev/fd fd    -        no    -
/proc     -        /proc   proc  -        no    -
/dev/dsk/c0t1d0s1     -       -         swap          no      -
/dev/dsk/c0t1d0s0     /dev/rdsk/c0t1d0s0  /      ufs   1       no
-
/dev/dsk/c0t1d0s6     /dev/rdsk/c0t1d0s6  /usr   ufs   1       no
-
swap      -        /tmp    tmpfs  -        yes   -
glenn:/usr/share/man  -       /mnt/man      nfs   -       yes     ro
# /etc/init.d/nfs.client start
# df
/                (/dev/dsk/c0t1d0s0 ):   287898 blocks    96825 files
/usr             (/dev/dsk/c0t1d0s6 ):   605666 blocks   324084 files
/proc            (/proc           ):        0 blocks      418 files
/dev/fd          (fd              ):        0 blocks        0 files
/tmp             (swap            ):   105760 blocks     5194 files
/mnt/man         (glenn:/usr/share/man):  605664 blocks   324084 files
# ▮
```

Adding an NFS file system to /etc/vfstab.

In the upper part of the window is the */etc/vfstab* file. The last line mounts */usr/share/man* from the host named *glenn*. It is mounted read-only on */mnt/man*. Running the **/etc/init.d/nfs.client** shell script executes the **mountall** command to mount the remote file systems using the same procedure used when the system starts up. The output of the **df** command shows the mounted file system. Note that for NFS file systems the remote host name and shared path take the place of the device name shown for local file systems, and the number of files is shown as *-1*.

NFS File System Mount Options

Like other special file system types, NFS file systems have a variety of specialized **mount** options. Most of these options pertain to the networked nature of an NFS file system. Some options are paired as alternative choices separated by slashes.

mount Command Options

Option	Description
rw/ro	Read/write and read only. NFS file systems which are mounted *read/write* can block activity on the client when the server providing the file system becomes unavailable. See intr and bg below.
hard/soft	The hard option mounts an NFS file system in such a way as to ensure that data are written to the remote file system. If the file server becomes unavailable, a file system mounted with the hard option will stop all remote file operations until the file server becomes available again. All file systems mounted with the rw option should also use the hard option to ensure the integrity of data written to the file system. The soft option does not provide assurance of data writes to the remote file system, but does not stop remote file operations in the case of a file server becoming unavailable. This option is useful for file systems that are mounted read-only.
suid/nosuid	The nosuid option negates the effect of programs on the remote file system for which respective *setuid* bits are set. *Setuid* programs run from NFS file systems mounted with the nosuid option are executed with the normal permissions of the user executing the program, *not* those conferred by the *setuid* bit. This option is used to increase the security of the client by preventing setuid programs on remote file systems from being used on the client system.
bg/fg	This option pair controls how to handle a failed mount of an NFS file system. Mounts with the bg option are retried in the background, freeing the shell which issued the mount command. Use this option when mounting file systems in */etc/vfstab* to prevent a workstation from stopping during the boot sequence because a file server is down.
intr/nointr	The nointr option prevents program interrupts when programs cause an NFS operation to occur. This can result in programs being uninterruptible when an NFS file server becomes unavailable. The default is to allow interrupts so that programs can be aborted in the event of server failures.
retry=n	Number of times to retry a failed mount. The default of 10,000 is usually sufficient.

Option	Description
timeo=n	Time-out value for retrying NFS operations. Increase this value to permit very slow systems, such as near-line file stores, more time to complete basic operations.
retrans=n	Number of retransmissions of a given NFS operation. The setting depends on the network and type of server being used. Some networks where packet loss is high benefit from an increase in the number of retransmissions.
rsize=n	Read buffer size. Some servers and clients (e.g., those with slower or less reliable network connections) perform better when the buffer used for NFS operations is a different size than the default.
wsize=n	Write buffer size. Similar to rsize in usage.
vers=n	NFS protocol version number (2 or 3). By default, the mount command will attempt to use the highest version number protocol available from the server. This option allows the protocol version to be fixed for use with servers that support only the older version 2 protocol.
proto=?	Controls the network protocol used to transport NFS data. NFS uses IP datagrams by default. By setting *proto=tcp*, NFS will use tcp, thereby improving performance when moving data over wide area networks and the Internet.

Using mount Command Options

Choosing the correct mix of **mount** command options for various circumstances is a bit tricky. General guidelines to help select the necessary options for a particular situation are discussed below.

hard Option

In general, NFS file systems mounted with the **rw** option should also use the **hard** option. When a problem occurs in writing to an NFS file system that is mounted **soft**, it is reported when the file is closed. Few programs check the return value from the **close( )** system call. The hard option handles this at the file system level, forcing the client to retry failed writes until they succeed (e.g., the server returns to operation). Data loss is thus prevented. However, using the hard option has serious ramifications for overall system functioning as discussed later in this chapter.

bg Option

To prevent a client from hanging before completion of system start-up, use the **bg** option whenever possible. This option allows the boot process to complete by placing any NFS file system mounts that cannot be executed at boot time in the background. The mount will complete automatically once the problem preventing the mount from completing (e.g., network connections or server down) is fixed.

Use rsize and wsize with Care

Exercise great care when changing the values for **rsize** and **wsize**. NFS puts packets of the largest possible size on a network to improve efficiency. Reducing the size of these by adjusting the rsize and wsize parameters should be done only when a constraint requires such a change. For example, some older equipment with limited network hardware can cause network problems that are solved only by changing the NFS buffer sizes.

Use timeo with Equal Care

In a similar vein, the **timeo** value should be adjusted with care. Check to be certain that the underlying network is sound before adjusting this parameter. Flooding a problematic network with extra NFS request retransmissions will only make a bad situation worse.

However, slow NFS servers may require increasing the value of timeo to avoid sending unnecessary retransmission of NFS requests.

In general, take care to examine the network hardware before making adjustments to the various NFS parameters. NFS is very demanding on a network and can bring to light problems not seen under lesser network loads. Network hardware failures can often result in additional network traffic as a result of NFS request retransmissions. See the "Troubleshooting NFS" section below for additional information.

Improving NFS Security

When using NFS to share files the network becomes more like a single large computer system than a group of separate machines. The implications of this for system security are important to note. Access to files shared via NFS depends on the security of each machine involved in the sharing. Because the contents of files are transferred over the network, you must consider access to the network itself as well. As the network grows, the ability of other machines on a network to monitor network communications or even alter them becomes more significant as a security concern.

File and Access Security

NFS security is only as good as the security of the individual hosts sharing file systems. Some things can be done to harden NFS against security problems, but the proper place to begin is by evaluating the security of each client and server. See Chapter 7, "Managing System Security," and Chapter 18, "Network Security," for discussions of the basic tools and areas that should be addressed on each host.

Server Sharing Security Options

With the individual hosts secured, move on to evaluate the permissions given by the file server for mounting its file systems. By default, an NFS file server maps access requests made by the root user on a client to a special UID. This UID does not have root file access privileges, and is normally set to that of the *nobody* account. You can change it by using the **anon=** option to the **share** command. For example, appearing below is an entry that would be placed in */etc/dfs/dfstab* to share */export/progs* with the root user on remote systems mapped to UID 10.

```
share -F nfs -o anon=10,rw=local -d "local programs" /export/progs
```

In some circumstances, you may wish to offer root access to a server's file system, perhaps to facilitate remote software installation. This can be implemented on a host-by-host basis using the **root=** option. This option accepts a list of hosts in much the same way as the **ro=** and **rw=** options to the **share** command. Hosts listed in the **root=** list will have full root access to the file system.

In situations where the administrator may not have complete control over the security of client workstations, it may be prudent to limit access to file systems offered by the server.

Client Mounting Security Options

On the client side, the **mount** commands should use the **nosuid** option to improve security. This option disables the effect of the *setuid* and *setgid* bits when a program from a remote file system is executed. The programs are run using the normal user permissions and *not* those conferred by the *setuid* or *setgid* bits. This helps to prevent non-secure programs offered by a server from being used by users on clients to affect client security.

Security at the Wire Level

Solaris provides for additional security for NFS operations through a secure version of the base remote procedure call (RPC) mechanism. By default, RPC uses the UNIX UID and GID numbering information as the basis of its authentication method. These numbers can be easily seen by network monitoring tools as they go by on the network, and it is impossible for the client and server to determine if the requests they send and receive are genuine.

Secure RPC Uses a Public Key Crypto System

Secure RPC solves the problem of authenticating the client and server by the use of a public key crypto system. A cryptographic system uses a mathematical formula to encrypt or encode information in such a way as to be unreadable except for those who have the needed numerical keys to decrypt or decode the information to a readable form.

A public key crypto system works as follows: two large numbers are generated for each user. One of the numbers (the private key) is kept secret, while the other number (the public key) is published in a public database available to all hosts. Something encrypted using one number as the key can be decrypted using the other number as the key. By using a combination of keys from the client and server, information can be exchanged securely with assurance of the sender's and recipient's identity.

The public and private keys are used in the following ways. First, the client randomly picks a session key to be used to encrypt data being sent to and from a server. The client encrypts the session key using a combination of the current time, the client's private key and the server's public key, and then sends this encrypted information to the server. The server decrypts the message using the current time, the client's public key, and the server's private key. The server now has the session key and is ready to respond.

Flow of information through crypto system.

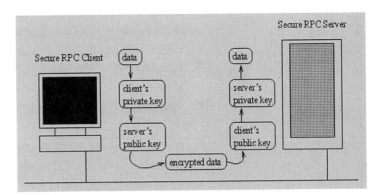

Due to the mathematical properties of the keys and the encryption method, the message the server and client exchange can only be decrypted by the specific server whose keys were used and could only have come from the specific client whose keys were used. This process works to ensure that no other host on the network is masquerading as either the client or the server.

The time stamp and session key are used to avoid "playback attacks." A playback attack is when a host simply records RPC requests and then reuses the encrypted information later in forged RPC requests. A secure RPC server will reject requests containing invalid time information.

⊷ **NOTE:** *Time synchronization between client and server hosts is required for secure RPC to function properly.*

Secure RPC Setup

To set up a secure RPC, the administrator may first need to assign a domain name to the collection of hosts that will share public keys. (This procedure is covered in Chapter 21, "Network Name Services.") Once a domain has been established, the sequence of tasks to perform is listed below.

1. Generate new public and private keys using **newkey**.

2. Start the key server by typing **/usr/sbin/keyserv**.

3. Edit */etc/dfs/dfstab* on the servers to add the **secure** option to the **share** commands and NFS file system entries.

Troubleshooting NFS

NFS depends on the underlying network to function. If systems are encountering problems mounting a remote file system, check the network first. Refer to Chapter 17, "Network Configuration and Management," for specific items to check. For NFS in particular, check to be sure that large data packets can be reliably sent between client and server.

On an Ethernet, check using the **ping** command. Use the following command on the client:

```
/usr/sbin/ping -s server 1500 100
```

The above command will send 100 maximum size packets (1500 bytes) from the client to the server and report statistics for each packet with totals at the end. On a quiet network, packet loss should be minimal. Even on busy networks, losses of more than one or two packets should be examined. While NFS will continue to function even in the face of severe network problems, performance will suffer greatly. Attempting to adjust NFS mount parameters such as **rsize**, **wsize**, and **retrans** without first checking the network can mask network problems.

Server Problems

Once the network connection is proven to be sound, the next items to check are the various NFS daemon processes. These background processes must be present for NFS to function. Use the **ps** command to look for the daemons listed below on the NFS server.

❏ **mountd**—If this daemon is missing, clients will be unable to mount NFS file systems from the server. Restart the daemon by becoming *root* and typing **/usr/lib/nfs/mountd**. Check for the daemon again using **ps**. If it fails to start, check the */etc/dfs/dfstab* file for errors.

❏ **nfsd**—Absence of nfsd means no NFS service. To restart the daemon, become root and type **/usr/lib/nfs/nfsd -a X**, where X is the number of threads used by nfsd. The default is 16. If another number is used for a particular server, use it here when restarting nfsd.

❑ **rpcbind**—NFS is built on RPC. The rpcbind daemon provides an RPC name look-up service. Without this daemon, RPC calls will not work and NFS and other RPC services will stop. Because several services are dependent on rpcbind, the best course of action is to reboot if this daemon is missing.

If the clients still do not seem to be able to get NFS services from the server, try killing the nfsd process and restarting it. This can be accomplished by using the **nfs.server** script found in */etc/init.d*, which is reproduced below.

```
# /etc/init.d/nfs.server stop
# /etc/init.d/nfs.server start
```

If using the script fails to correct the problem, try rebooting the server.

Client Problems

The problems that can occur on clients are usually less a matter of daemons not running and more a matter of errors in configuration. If a particular client cannot mount NFS file systems that other clients can, carefully check the configuration files. Check for errors in the server's */etc/dfs/dfstab* file and the client's */etc/vfstab* file. If the site is using NIS or NIS+, verify that the client and server are receiving current information from those services. Refer to Chapter 21, "Network Name Services," for ways to test and verify that the NIS+ service is functioning properly.

NFS Design Pitfalls

As mentioned earlier, the simple NFS configuration described here has a few drawbacks. To mount a file system read/write and maintain file integrity, file systems must be mounted using the **hard** option. However, if the file server of such a file system becomes unavailable, processes on the client which refer to the remote file system will stop working, and wait for the server to return to service. This can happen even in the case of file systems not directly involved in the processing that the client is executing at the time.

As networks grow, maintaining the */etc/vfstab* files on all clients becomes increasingly laborious. The */etc/vfstab* on each client will usually require modification as each new file server is added or as file servers are reorganized, renamed, or moved. The likelihood of mistakes in the file escalates, and the fact that all clients need not mount all file servers at any given time becomes increasingly evident. If files are shared from a common file server, the */etc/vfstab* method makes no allowance for adding file servers to help balance the load or to provide redundancy in the face of failures.

To deal with the above problems, Solaris uses a special daemon called an **automounter** and a special file system type known as *autofs*. These tools are covered in Chapter 20, "Automating NFS with automount."

Summary

Sun's network file system (NFS) is a widely available mechanism for sharing files between Solaris systems and a large number of other UNIX, VMS, PC, and Macintosh systems. Through the use of specific options to

the share and mount commands, special circumstances arising from the networked nature of NFS can
be handled. However, a simple NFS setup has drawbacks due to problems which occur when file servers
are unavailable. These problems can be addressed by
using the automount service described in the next
chapter.

20

Automating NFS with automount

As mentioned in Chapter 19, "Sharing Resources with NFS," growing networks present the system manager with many challenges. As new machines are added, correctly maintaining */etc/vfstab* files becomes an increasingly daunting chore. In addition, when a file server becomes unavailable, processes on client workstations may stop even if they do not depend on that file server. The automount service was developed to minimize these problems.

General Concept of automount

In most cases, an NFS client does not require continuous access to all NFS file systems that may be available on a network. If you can avoid mounting NFS file systems until they are needed, problems caused by unavailable file servers are greatly reduced. In brief, mechanisms to identify and use alternative file servers when possible are required for this purpose. The combination of the *autofs* file system type and the **automount** daemon provides an automated NFS file system mounting service as a solution to the unavailable file server problem, while also providing an efficient means to manage the mounting of large numbers of shared file systems.

The automount service works by intercepting file access requests, mounting the required NFS file system, and then allowing the request to proceed. Assume that the */usr/man* directory, an unmounted NFS file system under control of the automount service, was accessed. The access request would be caught by the automount daemon, which in turn would mount the file system and then allow the request to continue. This mounting process is transparent; the file access occurs as if the file system had been mounted all along. The automount service also cleans up in that NFS file systems no longer in use are unmounted.

Advantages of automount for the Small Network

While the demands of a large network drove the development of the automount service, it provides benefits for even very small networks. In all cases where an unavailable file server can cause problems, the automount service is useful. As a small network grows, the automount service makes adding and managing additional shared file systems easy. As additional resources become available, critical programs and data files can be offered on duplicate file servers to further improve reliability and performance. Through the use of the *auto_home* feature, user accounts can appear to have a single home directory while the system manager is freed to move the location of the user's files as needed to adapt to changing configurations and disk usage demands.

autofs and automountd

The *autofs* is a special file system type which resides in the SunOS kernel. When the automount daemon, **automountd**, is started, it reads a series of configuration files known as *map files*, and creates *autofs* file systems associated with each of the mount points listed in the maps. Unlike other mount points, which are attachment points for specific data storage devices in the UNIX file tree, the *autofs* mount points attach the automountd process to the UNIX file tree. Imagine that the mount point for a local disk is a shelf or cubbyhole. An autofs mount would be similar to a person who would wheel into place a shelf or cubbyhole as needed based on the request and the rules found in a map file.

A few autofsmount points in a UNIX file tree.

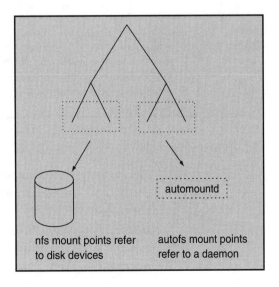

nfs mount points refer to disk devices

autofs mount points refer to a daemon

automountd

When a file or directory access request traverses an *autofs* mount point, the request is held and a message is sent to the automount daemon. The automount daemon consults the map files for entries pertaining to the mount point and mounts the referenced NFS file system. The access request is then allowed to proceed by being directed to the newly mounted NFS file system.

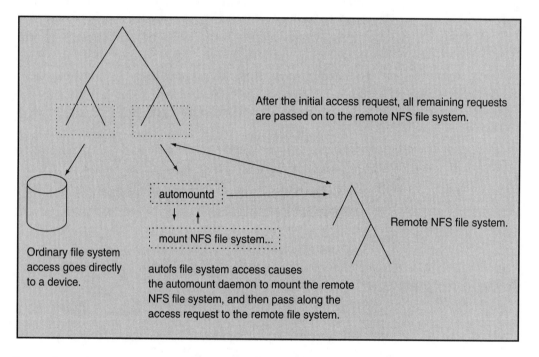

After the initial access request, all remaining requests are passed on to the remote NFS file system.

automountd

Remote NFS file system.

mount NFS file system...

Ordinary file system access goes directly to a device.

autofs file system access causes the automount daemon to mount the remote NFS file system, and then pass along the access request to the remote file system.

Steps taken when automount initially mounts an NFS file system.

All subsequent requests for file access that traverse this particular autofs mount point are directed to the mounted NFS file system. The automount daemon monitors the autofs mount point. If the mount point is not traversed during a predefined time period, the referenced NFS file system is automatically unmounted.

The automatic mounting imposes a brief delay when an unmounted NFS file system is first accessed. Little additional overhead occurs after the initial mount and the automatic unmount occur in the background. Consequently, there is no interference with ongoing processing on the client. Great flexibility and ease of maintenance are the payoff in return for a tiny increment to overhead.

Starting and Stopping the automount Service

Starting the automount service is very simple. It is most easily done by manually running the run control or **rc** shell script used by Solaris to start the automount service during the system boot process. From the root user prompt, type the following:

```
/etc/init.d/autofs start
```

Conversely, to stop the automount service, key in the following:

```
/etc/init.d/autofs stop
```

These commands are automatically performed as part of the system boot and shutdown procedures, respectively.

automount Maps

The heart of the automount service is its map files. These files specify the *autofs* mount points, control the mount options used when mounting specific NFS file systems, and allow for automated selection of NFS file systems based on various parameters. The automatic selection of redundant file servers and the general flexibility of the map files allow for a single collection of map files to be used across an entire network. This process greatly eases the maintenance chore when shared file systems are added, removed, or moved.

Map files are divided into direct and indirect types. Direct maps list the mount points for remote file systems as their keys. Indirect maps list mount points within a specified directory and use symbolic links which point from the key entry to the remote file system which is mounted elsewhere. Every map file is listed in the master map file.

Master automount file and respective map files.

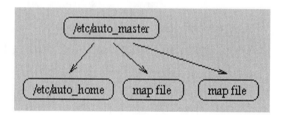

Master Map

The master map (*/etc/auto_master* file) forms the basis of the map system. The format of lines in the file is simple, consisting of a mount point, the name of a map file which specifies NFS file systems to be mounted at that mount point, and any mount options to be applied to the *autofs* file system at the mount point.

Sample auto_master file.

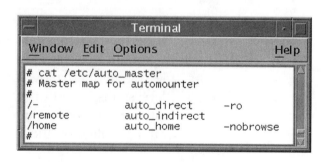

Each line in the sample *auto_master* file uses a different feature of the master map. The first two lines begin with a pound sign (#) and are treated as comments. Adding comment lines helps document changes made to the file or to describe what each of the listed map files does. The third line shows a direct map file. The dash (-) listed for the mount point is a place holder. Direct maps list the path name of each mount point as the key entry in the map file itself. NFS file systems mounted by a direct map are mounted at the mount point listed as the key in the direct map file. This line also specifies the mount point as **-ro** or read-only. A read-only mount point might be used when mounting a file system containing programs or data that do not change (e.g., manual pages).

Direct Maps

Upon reading each line in the master map file, the **automount** daemon reads the listed map file. Map files are assumed to be in the */etc* directory.

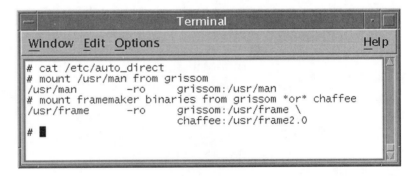

Direct map file.

Each line in the direct map file consists of a key, a list of mount options, and the location of the NFS file system to be mounted. In a direct map the key is the path to the mount point on the client where the NFS file system is to be mounted. The second line of this file shows a simple example where */usr/man* is mounted using the **-ro** or read-only option. The */usr/man* directory is shared by the file server *grissom.*

The fourth and fifth lines show how multiple file servers can be used to redundantly serve applications or data. In this case the */usr/frame* directory contains a popular text processing program. It is available on two different file servers in two different directories. The map file specifies that either of these file servers can be used to obtain */usr/frame.*

To decide which file server to use, the automount daemon consults the client workstation's network address information and tries to use the *closest* file server to reduce network traffic. The closeness measure used is to first try servers on the same network as the client, and then those on other networks. This is done to minimize NFS traffic across routers which join networks. If more than one file server is close, the automount daemon will contact all close file servers and use the file server that responds first. If the chosen file server does not respond, the automount daemon will automatically try the other listed file servers. The result is that the nearest available file server will be used.

Note also that the line has different remote directory names being used to locate */usr/frame* (i.e., */usr/ frame* and */usr/frame2.0*). When the remote file system is mounted by the automount daemon, the name of the key is used regardless of the name of the remote file system. Thus, in this case, if *chaffee* were chosen as the file server, *chaffee:/usr/frame2.0* would be mounted as */usr/frame* on the client. This allows the map file to mask directory naming schemes which might be confusing to users on client workstations. System managers should be aware that this feature can be used to provide easily remembered names for users, while a different naming scheme can be used on the file server for purposes of labeling the software or data to aid in its management.

✓ **TIP:** *When setting up a heterogeneous network of NFS servers and clients from different vendors, consider using the* auto_home *map. Vendors have differing conventions for the location of user home directories. With the use of* auto_home, *these different locations can be consolidated into a single, easy-to-use scheme.*

•• **NOTE:** *The multiple location feature of the* auto-mount *service is the key to building highly reliable NFS file sharing. When designing a network where shared collections of programs and data will be used, consider replicating the shared items and using* automount *to avoid single points of failure.*

Variables Used in Map File Entries

As a further aid to creating maps, the **automount** daemon understands and can substitute special symbols into map entries. These symbols can be used to create map entries which mount different file systems from the same key depending on the machine type or operating system version in use.

Map file variable names

Variable	Meaning	Example
ARCH	CPU architecture type	sun4
CPU	CPU type	SPARC
HOST	Host name	glenn
OSNAME	Operating system name	SunOS
OSREL	Operating system release number	5.6
OSVERS	Operating system version	FCS1.0

For example, consider the following direct map file entry:

```
/usr/man        chaffee:/export/manuals/$OSREL/man
```

Assuming that the file server *chaffee* had several different versions of the manual pages available, this entry would mount the correct version of the manual pages for the client on which it is being used. The ARCH and CPU variables can be used in a similar manner to create maps used on Solaris systems based on both SPARC and Intel CPUs without modification. For example, to mount the correct version (SPARC or Intel) of a local collection of programs with a direct map, you could use the following statement:

```
/usr/local/bin        grissom:/export/$CPU/local/bin
```

Another map feature is the use of the plus symbol (+). Map files can be included inside one another by using the +*mapfile* notation. For example, to include a map file named */etc/more_direct_maps* in the direct map file, the system administrator could add a line like the following:

```
+more_direct_maps
```

↝ **NOTE:** *Where the* automount *daemon will look for plus sign (+) entries depends on the* automount *entry in the* /etc/nsswitch *file. By default the* automount *daemon will look in* /etc *for files first, and then consult the NIS service. See Chapter 17, "Network Name Services," for more information.*

The objective of using these map features is to create general purpose maps that can be used anywhere on a given network. By carefully crafting a general purpose map, the problem of configuring a large number of NFS file systems on many different clients can be reduced to working on a single file.

Indirect Maps

Sample auto_master file.

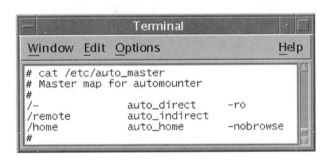

The fourth line of the sample *auto_master* file in the illustration above lists an indirect map. In an indirect map, the NFS file systems referred to by the map are mounted at a different mount point than the one listed in the table and a symbolic link is made from that location to the directory listed in the master map. This allows for some more exotic automount tricks. A diagram of an indirect mount with */home* as its entry in *auto_master* and a key of *frank* in the map file appears in the next illustration.

Mount points and symbolic links created by an indirect map.

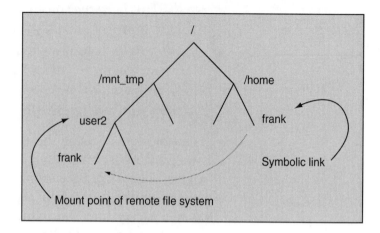

auto_home Map

One of the most common indirect maps is *auto_home*. The *auto_home* map has keys in the */home* directory. It is used to collect users' home directories into a single location, and greatly eases the task of maintaining home directory location information.

The next illustration shows a typical *auto_home* map. Indirect maps have the same items in each entry line as found in a direct map: a key, an optional list of mount options, and the location of a remote file system.

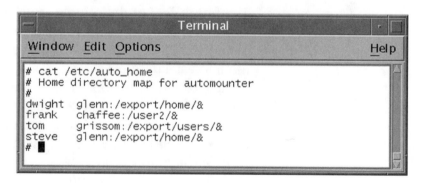

An auto_home indirect map.

Note how the line starting with the key *frank* does not contain a full path. The key in an indirect map specifies the name of the symbolic link to be made in the directory specified in the entry for the map in the *auto_master* file. In this case, all the keys in the *auto_home* map will create symbolic links in the */home* directory.

The ampersand character (&) stands in place of the key. It is a shorthand notation that can be used in any type of map file. For example, it shortens the entry that begins with the key *tom* to *grissom:/export/users/&* instead of *grissom:/export/users/tom*.

Indirection by use of symbolic links is useful for several reasons. First, the actual mount point can be placed so as to avoid unnecessary mounting of remote file systems. Consider what could happen if the command **ls -F** is invoked in a directory that contained numerous *autofs* mount points controlled by direct maps. All remote file systems would be referenced and mounted. The symbolic link used in an indirect map avoids unintended mount point traversal and mounting. Mounting the remote file system and creating the connection occur when the mount point is traversed for access to files or directories beyond the link itself. When a mount occurs, the referenced file system is mounted, and then a connection is made from the key to the file system.

Why Use auto_home?

The major administrative reason for using the *auto_home* map is to allow for easier maintenance of users' home directories. Under this system a user can always refer to his or her home directory as being in */home*. From the above example, Tom's home directory could be referred to as */home/tom*, Frank's as */home/frank*, and so on, even though they are located on different file servers or in different directory trees. Because this path is an automatically created symbolic link, the actual location of the directory (e.g., */export/users/tom* on *grissom*) can easily be changed if necessary.

Moving user home directories is a common occurrence for the system administrator due to system changes such as new disks, existing file systems filling, and so on. With the use of *auto_home*, the only change required when making a move is to adjust the *auto_home* map entry. The user's password file entry and programs or shell scripts written by users that refer to their home directory can remain untouched. This is easier for the system manager and more convenient for users.

➥ **NOTE:** *Do not use the* /home *directory for local file systems. The* automount *service by default will use this directory for the* auto_home *map. By convention, home directories are placed in the* /export *file system.*

Map Maintenance

How maps are maintained depends on whether NIS+ is being used. When using NIS+, the table administration command, **nistbladm**, is used to make changes to the map files. The command updates the NIS+ databases as necessary to make any changes in the automount maps available via NIS+. If changes to the *auto_master* map have been made, the **automount** command must be run on each client to ensure that new mount points are created. Chapter 21, "Network

Name Services," provides a more detailed description of nistbladm and other commands used to manage NIS+.

Maintaining the auto_home Map File

The *auto_home* map is a special maintenance case. The OpenWindows user account administration tool described in Chapter 6, "Creating, Deleting and Managing User Accounts," allows the administrator to easily update the *auto_home* map. If you do not use the account administration tool, you will need to manually edit the map file when moving user home directories around. Be sure to distribute the updated file to all client workstations as soon as the home directory has been moved to avoid mismatches between the map entry and the actual location of the user's home directory. Mismatches would prevent the user from accessing her/his home directory.

When you are not using NIS+, edit the files first. Then, if changes were made to the *auto_master* map, run the **automount** command. A convenient method for editing the files on a large number of clients is the **rdist** tool. As described in Chapter 16, "Automating Routine Administrative Tasks," rdist can use a table of files, commands, and host names to automate file distribution such as the automount map files. Appearing below is a sample rdist file that will distribute the *auto_master, auto_direct* and *auto_home* maps used as examples in this chapter. The rdist file will automatically check each file and update files which are out of date on the listed clients. Whenever the *auto_master* file is updated, rdist will also run the automount command.

```
# Define the clients who will receive the files.
SLAVES = ( grissom shepard )
```

```
# Define the files to be distributed
FILES = (
        /etc/auto_master
        /etc/auto_direct
        /etc/auto_home
        )
# Distribute the files to the clients.
# Specify special handling for the /etc/auto_home file.
${FILES}->> ${SLAVES}
        install;
        special /etc/auto_master /usr/sbin/automount;
```

If the above rdist instructions were placed in a file named *auto.dist*, the rdist command could read and use the file with the following command:

```
rdist -f auto
.dist
```

Given the above command, rdist would compare the date and size of the files on the host with the listed clients. Files on the distribution host would be installed on the clients in instances where differences are encountered. If the */etc/auto_master* file were updated on a client, rdist would also run the */usr/sbin/automount* command on the client.

Troubleshooting automount

When trying to locate and fix problems with the **automount** service, bear in mind that automount is built on top of NFS which in turn is built on top of the basic Solaris TCP/IP networking functions. The first step should be to check that NFS and the network features on which it depends are working. Refer to Chapter 19, "Accessing Network Resources with NFS," and Chapter 17, "Network Configuration and Management," for additional help. Then, if NFS is working, proceed to check the automount map files.

Disappearing File Systems

A common problem when beginning to use **automount** is the placement of **autofs** mount points over the top of existing mount points. This is most commonly seen when you start the automount service and a file system seems to disappear beneath the autofs mount point.

To use features such as the *auto_home* map, the user file systems must be mounted in a location other than */home*. The most common convention is to place local file systems in */export*, and the *autofs* mount points in the final location. For example, it might be desirable to mount a user file system such as */export/home* and then use the *auto_home* map to make the user's home directories appear in */home*. This method allows both the file server and clients to use the same automount maps.

Home directories in /export can be automounted by both clients and servers.

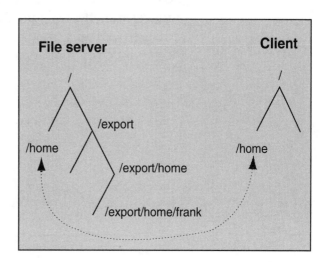

Syntax Errors

Another common problem is syntax mistakes in the map files. When error messages appear or incorrect syntax is suspected, try using the **-v** option flag to the **automount** command to obtain additional messages. The next step you could attempt is using the pound sign comment character (#) to "comment out" lines to help track down the problem.

```
# bob mercury:/export/home/bob
```

Lines which start with a pound sign are ignored by the automount daemon. Inserting a pound sign on a suspect map file line disables it to test whether a problem exists on a particular line.

Summary

Using the automount service can add flexibility and greater reliability to the already useful NFS file service. A single set of map files can be created for use across the entire network. This process greatly reduces the effort required to maintain NFS file access in large networks. By using the auto_home map, user home directory maintenance can be simplified.

Network Name Services

Previous chapters focused on how to manage system operations information (e.g., host and password files) on individual hosts. Maintaining system information stored on individual hosts is cumbersome on networks as small as five or six hosts, and becomes unmanageable on larger networks. For example, the */etc/inet/hosts* file is accessed by a variety of network applications in order to establish communications between hosts. If a new host is added to a network of 100 existing hosts, the */etc/inet/hosts* file on each of the existing hosts would have to be updated with the new host's name and IP address.

Chapter 16, "Automating Administrative Tasks," illustrated how the **/bin/rdist** utility could be used to distribute important files to systems on the network.

Unfortunately, some files change too frequently to be kept current by an overnight rdist run. The problem of keeping system information current is especially troublesome with very dynamic information such as passwords.

The NIS+ (Network Information Service Plus) environment addresses this problem by maintaining system information on a central host known as an NIS+ server. The NIS+ server provides its clients with system information (e.g., host and password) upon request through the NIS+ look-up service. The Domain Name Service (DNS) is also a name look-up service. Because NIS+ does not scale to networks the size of the Internet, DNS is often used in conjunction with NIS+ to provide the name services required for efficient network communications.

This chapter focuses on configuring the NIS+ environment, the setup of DNS services, and how to use NIS+ in conjunction with DNS to provide efficient name services for the corporate network.

What is NIS+?

NIS+ is a network look-up service that provides information about the users, workstations, and network resources to hosts that reside within an NIS+ domain. An NIS+ domain is a collection of NIS+ related hosts. NIS+ is a component of ONC+ (Open Network Computing Plus) in Solaris 2, and replaces its predecessor, NIS, which is used in Solaris 1. Like NIS, NIS+ is not proprietary, but may be licensed to any vendor who wishes to include the service as part of its operating environment.

⇥ **NOTE:** *NIS is also known as YP (Yellow Pages).*

NIS+ Servers and Clients

NIS+ solves the problem of multiple machines sharing the same information by storing the data in databases on NIS+ server hosts. The server hosts make the data available to NIS+ clients on demand. This arrangement is known as the NIS+ Client-Server Model. A NIS+ client process uses remote procedure calls (RPC) when making information look-up requests. A NIS+ server is a process that services an NIS+ client RPC request, searches for requested information in a database, and returns the information to the client process.

Every NIS+ domain is served by one master server, and may also utilize one or more replica servers. A NIS+ master server contains the master set of system information in the form of NIS+ tables. If changes are made to the NIS+ tables on the master NIS+ server, they are automatically pushed to the replica servers. A NIS+ replica server maintains copies of the tables in order to distribute the load in answering client requests and to provide backup sources of information in the event the master server fails.

NIS+ client-server model.

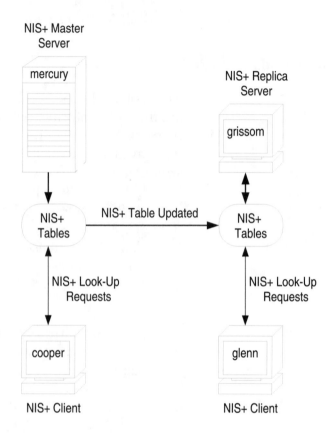

NIS+ Master
Server

mercury

NIS+ Replica
Server

grissom

NIS+
Tables

NIS+ Table Updated

NIS+
Tables

NIS+ Look-Up
Requests

NIS+ Look-Up
Requests

cooper

glenn

NIS+ Client

NIS+ Client

NIS+ Commands

Solaris provides a suite of commands which allow
alteration of NIS+ data. Before proceeding to explore
the inner workings of NIS+, a basic understanding of
these commands is essential. The names of many of
the NIS+ commands are very similar to the names of
standard UNIX commands. The NIS+ commands only
operate on the NIS+ data, but are otherwise very similar in operation to standard UNIX commands.

NIS+ User commands

NIS+ user level commands may be invoked by any user on the system, but only return data if that user is authorized access to the NIS+ data.

- ❐ **nisaddent**—Add /etc files and NIS maps to NIS+ tables.
- ❐ **niscat**—Display NIS+ tables and objects.
- ❐ **nischgrp**—Change group ownership of an NIS+ object.
- ❐ **nischmod**—Change access rights on an NIS+ object.
- ❐ **nischttl**—Change lifetime of NIS+ object.
- ❐ **nisdefaults**—Show NIS+ default values.
- ❐ **niserror**—Show NIS+ error messages.
- ❐ **nisgrep**—Search NIS+ information.
- ❐ **nisgrpadm**—NIS+ group administration command.
- ❐ **nisln**—NIS+ symbolic link utility.
- ❐ **nisls**—List NIS+ directory.
- ❐ **nismatch**—NIS+ search utility.
- ❐ **nismkdir / nisrmdir**—Create / remove NIS+ directory.
- ❐ **nispasswd**—Change NIS+ password.
- ❐ **nisrm**—Remove NIS+ object.
- ❐ **nisshowcache**—Print contents of shared NIS+ cache file.
- ❐ **nistbladmin**—NIS+ table administration utility.
- ❐ **nistest**—Test the state of NIS+ name space.

NIS+ Administrator Commands

In addition to the user level NIS+ commands, Solaris provides a suite of administrator commands for NIS+. The NIS+ administrator commands are limited to use by the system administrator or other authorized NIS+ administrators. These commands are used to initialize and manage the NIS+ databases.

- ❏ **aliasadm**—Manipulate NIS+ aliases.
- ❏ **nis_cachemgr**—NIS+ cache manager utility.
- ❏ **nisaddcred**—Create NIS+ credentials.
- ❏ **nisclient**—Initialize NIS+ credentials for NIS+ clients.
- ❏ **nisd / nisd_resolv**—NIS+ service daemon.
- ❏ **nisinit**—NIS+ client/server initialization utility.
- ❏ **nislog**—Display the NIS+ transaction log.
- ❏ **nisping**—Send ping to NIS+ servers.
- ❏ **nispopulate**—Populate the tables of the NIS+ domain.
- ❏ **nisserver**—Set up an NIS+ server.
- ❏ **nissetup**—Initialize an NIS+ domain.
- ❏ **nisstat**—Report NIS+ server statistics.
- ❏ **nisupdkeys**—Update public keys in an NIS+ directory.
- ❏ **rpc.nisd / rpc.nisd_resolv**—NIS+ service daemon.
- ❏ **sysidnis**—System configuration tool.

NIS+ Domains

A collection of related hosts and the information managed and utilized among the hosts is known as an

NIS+ domain. Every NIS+ host must be associated with a specific NIS+ domain. For example, *astro.com.* might be the domain name issued to a company named Astro Inc. Because Astro is a commercial entity, it would be assigned to the subdomain named *com* (commercial). Similar to IP addresses, domain names are issued by the NIC. The domain name issued by the NIC is often referred to as the root domain or top level domain. A root domain must contain two components ending with a dot, such as *astro.com.*

•• **NOTE:** *NIS+ domain names are not case-sensitive. Thus, Astro.com. and astro.com. are equivalent.*

The name of the NIS+ domain that a host is associated with is declared in the */etc/defaultdomain* file. Domains can be divided into subdomains to accurately reflect the hierarchical structure of an organization. For example, a company named Astro Inc. with hardware, software, marketing, and sales divisions might set up the domain hierarchy shown in the following illustration.

NIS+ hierarchical structure for astro.com.

NIS+ Objects

An NIS+ name space is a hierarchical structure in which NIS+ information is stored. The structure includes the root domain and all domains below it. Every name space has a root master server that serves the root domain at the top of the name space. An NIS+

name space is similar in structure to a UNIX file system hierarchy. Directory, table, and group objects are the most common types of objects in the NIS+ name space. These objects are stored in the *etc/nis* directory on the NIS+ server.

❑ Directory objects are similar to UNIX directories in that they can contain other objects such as table and group objects.

❑ Table objects store information in the NIS+ name space and are analogous to a UNIX plain file. The Solaris 2 environment provides 16 default tables, each of which stores information within a domain about hosts, users, networks, services, and so forth. A set of NIS+ tables stores information for that particular domain only.

❑ Group objects are used for NIS+ security. Like the group ownership within a UNIX file system, an NIS+ group is a collection of users and hosts identified by a single name and is used to facilitate NIS+ security. The default NIS+ group created in every domain is called *admin*. Although additional NIS+ groups are optional, they are a security convenience allowing an NIS+ administrator to assign access rights to a group of users and hosts.

Directory Objects

Directory objects make up the framework of a name space and are created with the **nismkdir** command. Each subdomain in an NIS+ hierarchy is analogous to a subdirectory in a UNIX file system. Two default NIS+ directories reside within every domain: the *org_dir* directory stores NIS+ table objects, and the *groups_dir* directory stores NIS+ security group objects. The directory object at the top of a name space (root domain) is called the "root directory." A name space is said to be "flat" if it consists of a root domain only. A flat domain can become hierarchical by creating subdomains.

Default objects in every NIS+ domain.

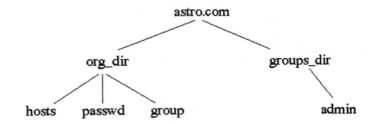

Using NIS+ Object Names

NIS+ object names are formed by appending the root directory name (including the period) to the names, known as a "fully qualified name." For example, *sales.astro.com.* and *software.astro.com.* are both fully qualified names that represent the *sales.astro.com.* and *software.astro.com.* subdomains within the *astro.com.* name space. Like UNIX files and directories, NIS+ objects can be referred to by partially or fully qualified names. Characteristics of those names are summarized below.

❏ Similar to a relative path name in a UNIX file system hierarchy, a partially qualified NIS+ name is simply the name of the component. For example, the hosts table's partially qualified name would be *hosts.*

❏ A fully qualified name resembles an absolute UNIX name in that it is the complete name of the component. Such a name is formed by starting with the name of the component and appending the names of all involved components up to the root domain, and is delimited by dots (.). For example, the fully qualified name of the sales division's hosts table at Astro Inc. is *hosts.org_dir.sales.astro.com.* Likewise, the fully qualified name of the software division's hosts table would be *hosts.org_dir.software.astro.com.*

Configuring an NIS+ Domain

This section provides a step-by-step procedure for setting up a flat domain called *domain.astro.com.* using the NIS+ installation scripts. The first step in the process is to configure the root master server. The second step consists of populating the root master server tables with information. The files in the */etc* directory are frequently used to achieve this step. Finally, clients will be added to the domain.

Creating a Flat NIS+ Domain Using NIS+ Scripts

Solaris 2 includes the following NIS+ administration shell scripts to facilitate the creation of NIS+ servers, tables, and clients.

❑ **/usr/lib/nis/nisserver**—Used to set up a root master or its replica NIS+ servers.

❑ **/usr/lib/nis/nispopulate**—Used to populate NIS+ tables from corresponding files in a specified domain.

❑ **/usr/lib/nis/nisclient**—Used to initialize NIS+ hosts and users.

Configuring a Root Master Server

1. Log in as *root* on the host that is to become the NIS+ root master for the name space *astro.com.*

2. Run the **nisserver** script with the **-r** and **-d** options. The -r option is used to indicate that the server being initialized is a root server, while the -d option specifies the domain name the server is administering.

```
mercury# /usr/lib/nis/nisserver -r -d astro.com.
```

3. The **nisserver** script will ask for setup verification as seen below.

```
This script sets up the machine mercury as an NIS+ Root Master Server for domain
astro.com.
Domainname : astro.com.
NIS+ Group : admin.astro.com.
YP compatibility : OFF
Security level : 2=DES
Is this information correct? (Y or N) y
```

4. The **nisserver** script will then run the **nisinit** and **nissetup** commands and start an NIS server process. You will be prompted to enter an NIS+ network password required by NIS+ for security purposes. Enter either the master server's root password or another password. To keep things simple, most sites use the root password as the NIS+ password.

```
adding credential for mercury.astro.com.
Enter login password: rootpasswd
```

Populating a Root Master Server's Tables

1. On the master server for the domain, run the **nispopulate** script to populate the empty tables created by the **nisserver** script. Typically, the system file located in the */etc* directory is used as the source for NIS+ table population.

```
mercury# cd /etc
mercury# /usr/lib/nis/nispopulate -F
```

2. The **nispopulate** script will ask for setup verification as shown below.

```
NIS+ Domainname: astro.com.
Directory Path: (current directory)
Is this information correct? (Y or N) y
```

3. The source files are listed and you are prompted once again to verify the creation of the destination NIS+ tables.

```
This script will populate the following NIS+ tables for domain your.domain.name.
from the files in current directory:
auto_master auto_home ethers group hosts networks passwd protocols services rpc
netmasks bootparams netgroup aliases shadow
Do you want to continue? (Y or N) y
```

Setting Up the NIS+ Client

Use the **nisclient** script on the root master to create the credentials for the client. In this case *mercury* is the master server for the domain, and *grissom* is the client being initialized.

```
mercury# /usr/lib/nis/nisclient -c grissom
```

1. The user is asked to verify the client.

```
You will be adding DES credentials in domain astro.com. for mercury
**nisclient will not overwrite any existing entries ** in the credential table
Do you want to continue? (Y or N) y
checking astro.com. domain...
checking cred.org_dir.astro.com. permission...
adding DES credential for grissom...
Adding key pair for unix.grissom@astro.com.
Enter grissom.astro.com. root login
password:rootpasswd
Retype password:rootpasswd
```

2. For all new NIS+ clients added, you should run the following statement on the client's machine.

```
nisclient -i -h root_master -a root_master_IP_address -d your.domain name.
```

3. Log on to the designated client as *root* and execute the **nisclient** script to initialize the client.

```
grissom# nisclient -i -h mercury -a 154.50.2.2 -d astro.com.
Initializing client grissom for domain astro.com....
Once initialization is done, you will need to reboot your machine.
Do you want to continue? (Y or N) y
setting up domain information astro.com....
setting up the name service switch information...
Please enter the network password that your administrator gave you.
Please enter the Secure-RPC password for root: rootpasswd
Please enter the login password for root: rootpasswd
Client initialization completed!!
Please reboot your machine for changes to take effect.
```

Setting Up an NIS+ Replica Server

An NIS+ client can be configured as a replica server. In this case, the NIS+ client *grissom* created in the previous sections will be converted to a replica server.

1. Log in as *root* and start the NIS+ service daemon, **rpc.nisd**.

```
grissom# rpc.nisd
```

2. Run the **nisserver** script on the master server to initialize the replica.

```
mercury# /usr/lib./nis/nisserver -R -d astro.com. -h grissom
This script sets up an NIS+ replica server for domain your.domain.name.
Domainname : astro.com.
NIS+ Server : grissom
```

3. You will be prompted to verify the setup.

```
Is this information correct? (Y or N) y
This script will set up machine grissom as an NIS+ replica server for domain as-
tro.com.

In order for grissom to serve this domain, you will need to start up the NIS+ serv-
er daemon (rpc.nisd) with proper options on grissom if they are not already run-
ning.
```

4. Ask once more to verify the setup.

```
Do you want to continue? (Y or N) y
Added grissom.astro.com. to group admin.astro.com.
```

Domain Name Service

The NIS+ name service allows the local administrator to maintain data about the machines in the local domain. The information in the NIS+ tables is only available to hosts within the NIS+ domain. In addition to IP address to hostname mappings, NIS+ stores information about several site specific parameters as described below.

- ☐ **/etc/ethers**—Contains MAC addresses for use by the **tftp** protocol (e.g., diskless client booting).
- ☐ **/etc/inet/hosts**—Local host name to IP address look-up table.
- ☐ **/etc/inet/networks**—List of local networks and aliases.
- ☐ **/etc/inet/netmasks**—List of local netmask values.
- ☐ **/etc/inet/protocols**—List of protocols in use at the local site.
- ☐ **/etc/inet/services**—List of services available at the local site.
- ☐ **/etc/mail/aliases**—List of local e-mail aliases.
- ☐ **/etc/netgroup**—List of hosts allowed to access file systems and services at the local site.

Much of the information stored by NIS+ is useless to systems outside the local network, or worse yet, could be used by hackers to compromise the security of hosts on the local network.

But what happens when a remote host (a host outside the NIS+ domain) wants to communicate with a host within the NIS+ domain? The remote host needs a way to resolve the host name to an IP address, but the NIS+ data are not available to it. Because NIS+ was never intended to service the entire Internet, other name services have been devised to provide this service.

➛ **NOTE:** *The following sections explore the workings of the Domain Name Service (DNS) and provide a limited overview of DNS. Because entire publications have been devoted to DNS, the reader is encouraged to obtain one of the reference books* *listed in the "References" section of this chapter for more information on DNS.*

DNS Terminology

The Domain Name Service (DNS) is the name service used on the Internet. The DNS duplicates some of the information stored in the NIS+ tables, but the DNS information is available to all hosts on the network. The DNS software uses many of the same terms used in NIS+. In particular, the concepts of domains, clients, servers, and databases are found in both name services. However, the operation of DNS is very different from NIS+. Before delving into how DNS operates, understanding how DNS defines these terms is recommended.

Domains

Under NIS+, a domain is defined as an NIS+ server and its collection of NIS+ clients. NIS+ allows for hierarchical domains to a maximum depth of 128 levels. These hierarchical domains allow the corporation to partition the network services in whatever fashion the corporate data management group desires.

Like NIS+, DNS allows hierarchical domains. Unlike NIS, DNS uses the domain name to describe a generic "ownership" of the hosts.

Root Domain

Under DNS the domain name space is very similar to the UNIX file system. One domain at the root of the name space contains all other domains. Known as the the *root domain*, this domain is usually written as a single dot (.) at the end of the domain name.

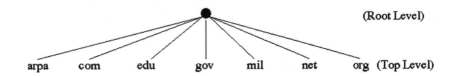

DNS domain name space.

Top Level Domains (U.S.)

DNS defines several "top level" domains. These domains are global in nature. For example, in the United States, the most common top level domains are *edu, com, net, org, gov,* and *mil.* The names of these top level domains make an attempt to describe the type of organizations they contain. For example, the *edu* domain contains educational institutions; *com,* commercial (private sector business) organizations; and *mil,* military institutions.

Under each top level DNS domain there may be any number of subdomains. For example, under the.*edu* domain are subdomains for organizations like The University of Notre Dame (*nd.edu*), Purdue University (*purdue.edu*), Stanford (*stanford.edu*), Cornell (*cornell.edu*), and most other universities. These second level domain names are registered by the Network Information Center (NIC) such that they are unique throughout the Internet. See Chapter 17, "Network Configuration and Management," for information about registering a domain name.

Under each of these second level domains there may be other subdomains. For example, the Computer Science and Engineering Department at The University of Notre Dame is the *cse.nd.edu* domain, while the Engineering Computer Network at Purdue University is the *ecn.purdue.edu* domain. The assignment of third level domain names is delegated (by the NIC) to the "owner" of the registered second level domain. Consequently, the local administrator has some flexibility in assigning subdomain names, and can simplify the operation of the DNS name resolution protocol.

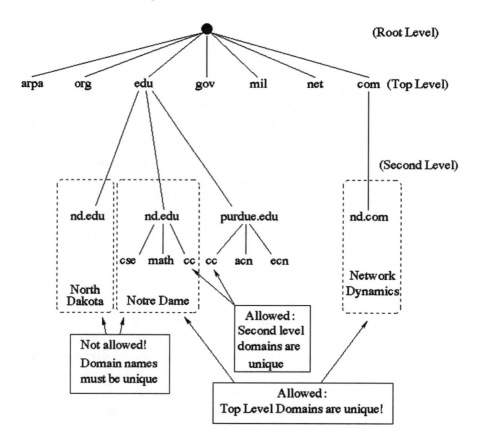

Extended DNS domain space.

Top Level Domains (Outside the U.S.)

Outside the United States, the top level domains usually look a little different than they do within the United States. This difference is an historical artifact that requires a bit of explanation. When the Internet was first put into operation, it only served hosts and organizations within the United States. The top level domains described in the previous section made sense in the context of the network's planned scope.

Over time the Internet expanded to locations outside the United States. The International Standards Organization (ISO) developed a series of standard two-letter country codes (ISO standard 3166). These two-letter domains were inserted under the root domain, with special provisions that the existing U.S. top level domains could remain unchanged.

Most locations outside of the United States subdivide their top level domain in a manner similar to the U.S. top level domains. For instance, *edu* and *ac* are educational or academic organizations, *com* and *co* are for commercial organizations, and so forth. In the United States, the ISO3166 defined second level domains for each U.S. state and territory permitting these entities the flexibility to name further subdomains as they see fit.

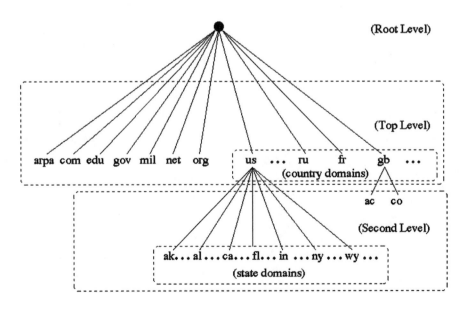

A politically correct rendition of the DNS name space.

Traversing the Domain Tree

Domain names read from left to right, where the leftmost component of the name is the furthest point from the root of the name space. For example, the name *host-name.store21.kmart.com* might describe a host located in store 21 in the *kmart* domain under the *com* top level domain. Similarly, the names *grumpy.cse.nd.edu* and *mischief.ecn.purdue.edu* describe hosts within the *edu* top level domain. The *grumpy* machine is located under the *cse* domain which is under the Notre Dame (*nd*) domain, while the *mischief* machine is located under the *ecn* domain which falls under the Purdue domain.

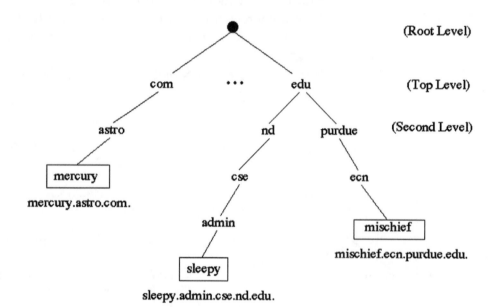

Reading DNS domain names.

To be precise, there should be a dot at the end of any domain/host name (e.g., *grumpy.cse.nd.edu.*). The dot signifies that the domain is fully qualified (specified) all the way to the root domain. When the trailing dot is left off of the name, the name is said to be partially qualified.

DNS Servers

Under DNS, name servers are programs which store information about a domain name space. A name server maintains database files which contain the master information for a domain. The client machines within the domain are configured in such a way that they will query the local name server when name to address mapping is required.

External hosts take a slightly different approach to resolution of name to address mappings. The remote host will query its local name server to get the name to address mapping. The local name server may not have the information required, so it queries another name server in an attempt to resolve the query. This (often recursive) process will continue until the local name server gets a valid mapping (query is resolved), or until all appropriate name servers have been queried and failed to return the requested information (host not found).

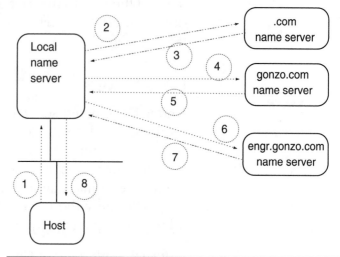

Name resolution example.

Step	Action
1	Host requests address of remote host dilbert.engr.gonzo.com
2	Local name server polls name server in com domain
3	com name server refers local name server to gonzo.com
4	Local name server polls gonzo.com
5	gonzo.com refers local name server to engr.gonzo.com
6	engr.gonzo.com replies with address of dilbert.engr.gonzo.com
7	
8	Local name server passes address of dilbert.engr.gonzo.com to host

In order to ensure that name look-ups will succeed, each domain should have more than one name server configured to answer queries about hosts within that domain. In many cases, organizations may cooperate such that each corporation provides the primary name service for its own domain, but it also agrees to provide backup name service for the other organization's domain.

Primary Name Servers

The primary name server obtains information about the domain from disk files contained on the name server host. The domain may actually have several primary name servers in an attempt to spread the work load around and provide better service. The DNS specification refers to primary name servers as "primary masters." Primary master name servers are defined as the name servers which contain the master information about the hosts within its domain.

Secondary Name Servers

Each domain can also define a number of secondary name servers. The secondary name servers obtain their databases from the primary name servers through a process called "zone transfer." The DNS specification refers to these secondary name servers as "secondary masters." The secondary master name servers are the name servers which provide backup name service for a domain. These secondary name servers are queried in the event the primary name server(s) do not respond to a query.

Caching Name Servers

Caching name servers have no direct access to any authoritative information about the domain. These name servers query primary and secondary name servers with DNS requests, and store the results away in a memory cache for future reference. When these servers spot a DNS look-up request on the network, they reply with the information they have stored in their respective caches. Caching name servers usually contain valid data, but because they do not load information directly from primary name servers, the data can become stale. Caching name servers are not considered authoritative name servers.

Root Name Servers

In order to provide a master list of name servers available on the Internet, the Network Information Center (NIC) maintains a group of root name servers. These name servers provide authoritative information about specific top level domains. When a local name server cannot resolve an address, it queries the root name server for the appropriate domain for information which may allow the local name server to resolve the address.

DNS Clients

DNS clients comprise a series of library calls which issue RPC requests to a DNS server. These library routines are referred to as "resolver code." Whenever a system requires a name to IP address mapping, the system software executes a **gethostbyname()** library call. The resolver code contacts the appropriate name service daemon to resolve the query.

/etc/resolv.conf File

In order to make DNS operate, the administrator must configure the */etc/resolv.conf* file on each host. This file tells the name service resolution routines where to find DNS information. The format of the entries in the */etc/resolv.conf* file is keyword value. Valid keywords appear below.

❑ **nameserver**—Valid values for name servers are IP addresses in standard dot notation. The administrator may specify a maximum of three name servers to query on separate lines.

❑ **domain**—The domain value will be appended to any host name which does not end in a dot. For example, if a user types *telnet mercury*, the resolver code will automatically append *.astro.com* to the request such that the telnet session will actually be issued as *telnet mercury.astro.com*.

DNS Database Files

The DNS service consults a series of database files in order to resolve name service queries. Such database files are often referred to as *db files*. In order to provide scalability and modularity, the database information is split into several files as described below.

❑ /etc/named.boot—Start-up file for the name server. This file contains information specific to the named program. Information in this file determines whether the name server is a primary name server or a secondary name server, where the db files are located, and other local setup information.

❑ db.domainname—This database provides the host name to IP address mapping for the hosts within domain name. There may be multiple *db.domainname* files on a server (one file for each domain for which this server provides name service).

❑ db.address—This database provides the IP address to host name mapping (reverse mapping) for the hosts on network *address*. There may be multiple *db.address* files on the name server (one file for each network within the domain).

Name Service Resource Records

The information stored in the database files is organized into resource records, and the format of these records is determined by the DNS design specifications. Subsequent sections explain the types of records, the format for each record, and the information provided by the record.

Comment Records

As with any programming language, DNS provides a method for embedding comments in the db files. You should place comments in these files in an attempt to explain the local site setup. A comment record begins with a semicolon (;). Everything that follows a semicolon (on a line of the db file) is considered a comment. An example appears below.

```
; The following lines define local host addresses with A records.
```

SOA Records

Start of authority (SOA) records provide information about the management of data within the domain. The SOA record must start in column 1. The format of an SOA record is shown below, followed by an explanation of each field.

```
domain class SOA primary_server rp (
          serial_number              ; comment about serial number
          refresh_value              ; comment about refresh value
          retry_value              ; comment about retry value
          expiration_value            ; comment about expiration value
          TTL )            ; comment about time to live value
```

- ❏ The *domain* field is the domain name (e.g., *astro.com.*).
- ❏ The *class* field allows the administrator to define the class of data. Currently, only one class is used. The *IN* class defines Internet data.

❏ The *SOA* field tells the resolver that this is a start of authority record for *domain.*

❏ The primary server field is the fully qualified host name of the primary name server for this domain. For example, *mercury.astro.com.* would be the name server for the *astro.com.* domain.

❏ The *rp* field gives the fully qualified email address of the person responsible for this domain. Note that the usual at symbol (@) is replaced by a dot (.) in the *rp* record. For example, *curt.mercury.astro.com.* might be the responsible person for the *astro.com.* domain.

❏ The serial number field is used by the secondary name servers. This field is a "counter" which gives the version number for the file. This number should be incremented every time the file is altered. When the information stored on the secondary name servers expires, the servers contact the primary name server to obtain new information. If the serial number of the information obtained from the primary server is larger than the serial number of the current information, a full zone transfer is performed. If the serial number from the primary name server is smaller than the current information, the secondary name server discards the information from the primary name server.

❏ The *refresh_value* field is used by the secondary name servers. The value in this field tells the secondary servers how long (in seconds) to keep the data before they obtain a new copy from the primary name server. If the value is too small, the name server may overload the network with zone transfers. If the value is too large, information may become stagnant and changes may not propagate efficiently from the primary to the secondary name servers.

❏ The *retry_value* field is used by the secondary name servers. The *value* in this field tells the secondary servers how long (in seconds) to wait before attempting to contact a non-responsive primary name server.

❏ The *expiration_value* field is used by the secondary name servers. The *value* in this field tells the secondary servers to expire their information after *value* seconds. Once the secondary server expires its data, it stops responding to name service requests.

❏ The *TTL* field tells name servers how long they can "cache" the response from the name server before they purge the information obtained in response to a query.

An example appears below.

```
astro.com IN SOA mercury.astro.com curt.mercury.astro.com. (
            1997053001          ; Serial - format YYYYMMDD##
            10800                  ; Refresh every 3 hours
            3600                   ; Retry after an hour
            60480                  ; Expire data after 1 week
            86400 )        ; TTL for cached data is 1 day
```

NS Records

The NS records define the name servers within a domain. The NS record starts in column 1 of the file. The format of an NS record follows: *domain class NS fully_qualified_hostname.* An example appears below.

```
astro.com              IN              NS              mercury.astro.com.
```

A Records

The A (Address) records provide the information used to map host names to addresses. The A record must begin in column 1 of the file. The format of an A record is *fully_qualified_hostname class A address*. Examples follow.

```
localhost.astro.com. IN          A          127.0.0.1
glenn.astro.com. IN          A          203.14.5.1
```

PTR Records

The PTR (pointer) records provide the information used for reverse mapping (looking up a host name from an IP address). The PTR record must begin in column 1 of the file. The format of the PTR record is *address class PTR fully_qualified_hostname*.

The PTR record does have one quirk: the address portion appears to be written backwards, and contains information the user typically does not see. The information is presented in this format in order to simplify the look-up procedure, and to maintain the premise that the information at the left of the record is furthest from the root of the domain. An example follows.

```
1.5.14.203.in-addr.arpa. IN PTR glenn.astro.com.
```

MX Records

The MX (Mail eXchanger) records provide a way for remote hosts to determine where e-mail should be delivered for a domain. The MX records must begin in column 1 of the db files. The format of MX records is *fully_qualified_hostname class MX preference fully_qualified_mail_hostname*.

There should be an MX record for each host within a domain. The *preference* field tells the remote host the preferred place to deliver the mail for the domain. This allows e-mail to be delivered in the event that the primary mail server is down. The remote host will try to deliver e-mail to the host with the lowest preference value first. If that fails, the remote host will attempt to deliver the e-mail to the host with the next lowest preference value. Examples appear below.

```
glenn.astro.com. IN MX 1 mail.astro.com.
glenn.astro.com. IN MX 10 www.astro.com.
```

CNAME Records

The CNAME (Canonical NAME) records provide a way to alias host names. For example, the host name *saturn-apollo-13.astro.com* might be a little difficult for some people to type. Users may have an easier time typing *apollo13.astro.com*. To provide this alias mapping, the administrator would have to enter a CNAME record in the appropriate DNS db files.

```
apollo13.astro.com. IN CNAME saturn-apollo-13.astro.com.
```

TXT Records

The TXT records allow the administrator to add text information to the db files giving more information about the domain. The TXT records must start in column 1 of the db file, and the format of TXT records is *hostname class TXT "Random text about hostname in this domain"*. An example follows.

```
mercury IN TXT "Mercury is in room G-101 of Bldg 3 of the astro.com complex"
```

RP Records

The RP (responsible person) records define the person responsible for the domain. The RP records must start

in column 1 of the db files. The format of an RP record
*is hostname class RP fully_qualified_email_address
fully_qualified_hostname.* An example follows.

```
truax IN RP          curt.mercury.astro.com          vonbraun.engr.astro.com.
vonbraun IN TXT          "Engineering Division of Astro.com (900) 555-1212"
```

Setting Up a DNS server

Solaris 2 includes the DNS software as a standard portion of the operating system. This limits the process of
setting up a primary name server to building the database files, then starting the */usr/sbin/in.named* daemon. Prior to building the database files, the
administrator should understand the different pieces
of software which perform the DNS function.

BIND

The BIND software is the Berkeley Internet Name
Daemon. BIND is the current implementation of the
DNS software, and is available for most UNIX ports
currently on the market. BIND consists of the components listed below.

❐ **in.named**—Process which answers name to
address resolution queries.

❐ **nslookup, dig**, and **host**—These utilities allow
users to query the DNS database(s) to determine
what information is available.

❐ DNS **resolver libraries**—Routines which perform the name to address query resolution for
in.named.

Example DNS Configuration

In order to set up a primary name server for a domain, the administrator must create the db files for that domain. The next figure illustrates a sample domain, which will be examined in subsequent sections. The sample domain will not employ subdomains initially, but as the company grows, logical subdomain boundaries might be *engr.astro.com* (engineering), *mkt.astro.com* (marketing), *corp.astro.com* (corporate services), and *adm.astro.com* (administration).

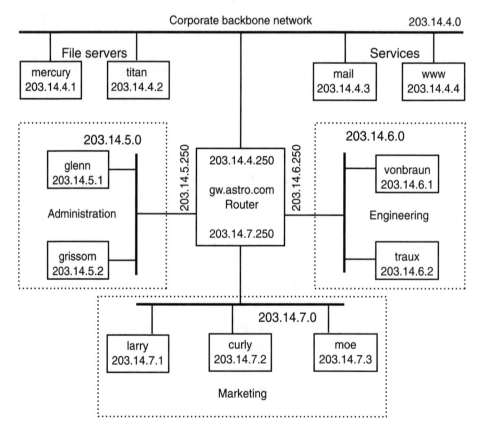

The astro.com DNS domain configuration.

For the example domain, the administrator must create the files listed below.

- ❐ /etc/named.boot
- ❐ db.astro (db.domain)
- ❐ db.203.14.4
- ❐ db.203.14.5
- ❐ db.203.14.6
- ❐ db.203.14.7
- ❐ db.127.0.0
- ❐ db.cache
- ❐ /etc/nsswitch.conf
- ❐ /etc/resolv.conf

/etc/named.boot File

As mentioned previously, the *named.boot* file customizes the DNS information for a particular site. To simplify matters, the *directory* instruction is used to inform the named software to look in */usr/local/named* for the db files for this server. The *primary* instruction tells the named software where to find each individual db file required to resolve DNS queries. The cache instruction tells the named software where to find the file that contains information about the root name servers. For the example network, the *named.boot* file would contain the directives appearing below.

```
directory /usr/local/named
primary astro.com db.astro
primary 4.14.203.in-addr.arpa db.203.14.4
primary 5.14.203.in-addr.arpa db.203.14.5
primary 6.14.203.in-addr.arpa db.203.14.6
primary 7.14.203.in-addr.arpa db.203.14.7
```

```
primary 0.0.127.in-addr.arpa db.127.0.0
cache . db.cache
```

db.domain file

The *db.astro* file contains information mapping host names within the *astro.com* domain to IP addresses for the hosts in the *astro.com* domain. It also contains an SOA record, as well as NS, MX, CNAME, PTR and other records about these hosts. The *db.astro* file for the example would contain the following directives.

```
astro.com.    IN    SOA    mercury.astro.com.    curt.mercury.astro.com.    (
                                  1            ; Serial Number
                                  10800        ; Refresh every 3 hours
                                  3600         ; Retry after 1 hour
                                  604800       ; Expire after 1 week
                                  86400 )      ; TTL is 1 day
;
; Astro.com name servers
;
astro.com.          IN       NS       mercury.astro.com.
astro.com.          IN       NS       www.astro.com.
;
; astro.com A records
;
localhost.astro.com.      IN       A       127.0.0.1
mercury.astro.com.        IN       A       203.14.4.1
titan.astro.com.          IN       A       203.14.4.2
mail.astro.com.           IN       A       203.14.4.3
www.astro.com.            IN       A       203.14.4.4
gw.astro.com.             IN       A       203.14.4.250
glenn.astro.com.          IN       A       203.14.5.1
grissom.astro.com.        IN       A       203.14.5.2
gw.astro.com.             IN       A       203.14.5.250
vonbraun.astro.com.       IN       A       203.14.6.1
```

```
truax.astro.com.          IN        A         203.14.6.2
gw.astro.com.             IN        A         203.14.6.250
larry.astro.com.          IN        A         203.14.7.1
curly.astro.com.          IN        A         203.14.7.2
moe.astro.com.            IN        A         203.14.7.3
gw.astro.com.             IN        A         203.14.7.250
;
; Aliases for astro.com hosts
;
gw-4.mercury.com.         IN        CNAME          gw.astro.com.
gw-5.mercury.com.         IN        CNAME          gw.astro.com.
gw-6.mercury.com.         IN        CNAME          gw.astro.com.
gw-7.mercury.com.         IN        CNAME          gw.astro.com.
;
; MX records for astro.com
;
mercury.astro.com.        IN        MX        1         mail.astro.com.
titan.astro.com.          IN        MX        1         mail.astro.com.
mail.astro.com.           IN        MX        1         mail.astro.com.
www.astro.com.            IN        MX        1         mail.astro.com.
gw.astro.com.             IN        MX        1         mail.astro.com.
glenn.astro.com.          IN        MX        1         mail.astro.com.
grissom.astro.com.        IN        MX        1         mail.astro.com.
vonbraun.astro.com.       IN        MX        1         mail.astro.com.
truax.astro.com.          IN        MX        1         mail.astro.com.
larry.astro.com.          IN        MX        1         mail.astro.com.
curly.astro.com.          IN        MX        1         mail.astro.com.
moe.astro.com.            IN        MX        1         mail.astro.com.
mercury.astro.com.        IN        MX        10        www.astro.com.
titan.astro.com.          IN        MX        10        www.astro.com.
mail.astro.com.           IN        MX        10        www.astro.com.
www.astro.com.            IN        MX        10        www.astro.com.
gw.astro.com.             IN        MX        10        www.astro.com.
glenn.astro.com.          IN        MX        10        www.astro.com.
```

grissom.astro.com.	IN	MX	10	www.astro.com.
vonbraun.astro.com.	IN	MX	10	www.astro.com.
truax.astro.com.	IN	MX	10	www.astro.com.
larry.astro.com.	IN	MX	10	www.astro.com.
curly.astro.com.	IN	MX	10	www.astro.com.
moe.astro.com.	IN	MX	10	www.astro.com.

db.203.14.4 File

The *db.203.14.4* file contains information mapping IP addresses on the 203.14.4.0 network to host names. It also contains an SOA record, as well as NS, PTR and other records about these hosts. The *db.203.14.4* file for the example would contain the directives appearing below.

```
4.14.203.in-addr.arpa.   IN   SOA    mercury.astro.com. curt.mercury.astro.com.  (
                                1              ; Serial Number
                                10800          ; Refresh every 3 hours
                                3600           ; Retry after 1 hour
                                604800         ; Expire after 1 week
                                86400 )        ; TTL is 1 day
;
; Astro.com name servers
;
4.14.203.in-addr.arpa.        IN        NS        mercury.astro.com.
4.14.203.in-addr.arpa.        IN        NS        www.astro.com.
;
; astro.com hosts on 203.14.4 net
;
1.4.14.203.in-addr.arpa.      IN        PTR       mercury.astro.com.
2.4.14.203.in-addr.arpa.      IN        PTR       titan.astro.com.
3.4.14.203.in-addr.arpa.      IN        PTR       mail.astro.com.
4.4.14.203.in-addr.arpa.      IN        PTR       www.astro.com.
250.4.14.203.in-addr.arpa.    IN        PTR       gw.astro.com.
```

db.203.14.5 File

The *db.203.14.5* file contains information mapping IP addresses on the 203.14.5.0 network to host names. It also contains an SOA record, and NS, PTR and other records about these hosts. The *db.203.14.5* file for the example would contain the directives reproduced below.

```
5.14.203.in-addr.arpa. IN   SOA  mercury.astro.com.  curt.mercury.astro.com.(
                        1              ; Serial Number
                        10800          ; Refresh every 3 hours
                        3600           ; Retry after 1 hour
                        604800         ; Expire after 1 week
                        86400 )        ; TTL is 1 day
;
; Astro.com name servers
;
4.14.203.in-addr.arpa.        IN        NS        mercury.astro.com.
4.14.203.in-addr.arpa.        IN        NS        www.astro.com.
;
; astro.com hosts on 203.14.5 net
;
1.5.14.203.in-addr.arpa.      IN        PTR       glenn.astro.com.
2.5.14.203.in-addr.arpa.      IN        PTR       grissom.astro.com.
250.5.14.203.in-addr.arpa.    IN        PTR       gw.astro.com.
```

db.203.14.6 File

The *db.203.14.6* file contains information mapping IP addresses on the 203.14.6.0 network to host names. It also contains an SOA record, and NS, PTR and other records about these hosts. The *db.203.14.6* file for the example would contain the directives below.

```
6.14.203.in-addr.arpa.  IN   SOA   mercury.astro.com. curt.mercury.astro.com.    (
                                  1                ; Serial Number
                              10800                ; Refresh every 3 hours
                               3600                ; Retry after 1 hour
                             604800                ; Expire after 1 week
                              86400 )              ; TTL is 1 day
;
; Astro.com name servers
;
4.14.203.in-addr.arpa.        IN        NS             mercury.astro.com.
4.14.203.in-addr.arpa.        IN        NS             www.astro.com.
;
; astro.com hosts on 203.14.6 net
;
1.6.14.203.in-addr.arpa.      IN        PTR            vonbraun.astro.com.
2.6.14.203.in-addr.arpa.      IN        PTR            truax.astro.com.
250.6.14.203.in-addr.arpa.    IN        PTR            gw.astro.com.
```

db.203.14.7 File

The *db.203.14.7* file contains information mapping
IP addresses on the 203.14.7.0 network to host
names. It also contains an SOA record, and NS, PTR
and other records about these hosts. The *db.203.14.7*
file for the example would contain the following
directives.

```
7.14.203.in-addr.arpa. IN  SOA   mercury.astro.com. curt.mercury.astro.com.      (
                                 1            ; Serial Number
                             10800            ; Refresh every 3 hours
                              3600            ; Retry after 1 hour
                            604800            ; Expire after 1 week
                             86400 )          ; TTL is 1 day
;
; Astro.com name servers
;
```

```
4.14.203.in-addr.arpa.          IN       NS              mercury.astro.com.
4.14.203.in-addr.arpa.          IN       NS              www.astro.com.
;
; astro.com hosts on 203.14.7 net
;
1.7.14.203.in-addr.arpa.        IN       PTR             larry.astro.com.
2.7.14.203.in-addr.arpa.        IN       PTR             curly.astro.com.
3.7.14.203.in-addr.arpa.        IN       PTR             moe.astro.com.
250.7.14.203.in-addr.arpa.      IN       PTR             gw.astro.com.
```

db.127.0.0 File

The *db.127.0.0* file contains information mapping IP addresses on the 127.0.0 network to host names. It also contains an SOA record, and NS, PTR and other records about these hosts. The *db.127.0.0* file for the example would contain the directives appearing below.

```
0.0.127.in-addr.arpa.  IN  SOA  mercury.astro.com. curt.mercury.astro.com.     (
                            1                ; Serial Number
                            10800            ; Refresh every 3 hours
                            3600             ; Retry after 1 hour
                            604800           ; Expire after 1 week
                            86400 )          ; TTL is 1 day
0.0.127.in-addr.arpa.           IN       NS              mercury.astro.com.
0.0.127.in-addr.arpa.           IN       NS              www.astro.com.
1.0.0.127.in-addr.arpa.         IN       PTR             localhost.
```

db.cache File

The *db.cache* file contains hints that the name server can use to contact top level name servers. This file should be the same for every name server on the Internet. A master copy of the *db.cache* file is available via anonymous ftp, as *named.root*, from *ftp.rs.internic.net* (198.41.0.5). The *db.cache* file for the example would contain the following directives.

; This file holds the information on root name servers needed to initialize cache of Internet

; domain name servers (e.g. reference this file in the "cache . <file>"

; configuration file of BIND domain name servers).

; This file is made available by InterNIC registration services under anonymous FTP as

; /domain/named.root on server FTP.RS.INTERNIC.NET

; last update: May 19, 1997 related version of root zone: 1997051700

;

; formerly NS.INTERNIC.NET

. 3600000 IN NS A.ROOT-SERVERS.NET.

A.ROOT-SERVERS.NET. 3600000 A 198.41.0.4

;

; formerly NS1.ISI.EDU

. 3600000 NS B.ROOT-SERVERS.NET.

B.ROOT-SERVERS.NET. 3600000 A 128.9.0.107

;

; formerly C.PSI.NET

. 3600000 NS C.ROOT-SERVERS.NET.

C.ROOT-SERVERS.NET. 3600000 A 192.33.4.12

;

; formerly TERP.UMD.EDU

. 3600000 NS D.ROOT-SERVERS.NET.

D.ROOT-SERVERS.NET. 3600000 A 128.8.10.90

;

; formerly NS.NASA.GOV

. 3600000 NS E.ROOT-SERVERS.NET.

E.ROOT-SERVERS.NET. 3600000 A 192.203.230.10

;

; formerly NS.ISC.ORG

. 3600000 NS F.ROOT-SERVERS.NET.

F.ROOT-SERVERS.NET. 3600000 A 192.5.5.241

;

; formerly NS.NIC.DDN.MIL

```
. 3600000 NS G.ROOT-SERVERS.NET.
G.ROOT-SERVERS.NET. 3600000 A 192.112.36.4
;
; formerly AOS.ARL.ARMY.MIL
. 3600000 NS H.ROOT-SERVERS.NET.
H.ROOT-SERVERS.NET. 3600000 A 128.63.2.53
;
; formerly NIC.NORDU.NET
. 3600000 NS I.ROOT-SERVERS.NET.
I.ROOT-SERVERS.NET. 3600000 A 192.36.148.17
;
; temporarily housed at NSI (InterNIC)
. 3600000 NS J.ROOT-SERVERS.NET.
J.ROOT-SERVERS.NET. 3600000 A 198.41.0.10
;
; housed in LINX, operated by RIPE NCC
. 3600000 NS K.ROOT-SERVERS.NET.
K.ROOT-SERVERS.NET. 3600000 A 193.0.14.129
;
; temporarily housed at ISI (IANA)
. 3600000 NS L.ROOT-SERVERS.NET.
L.ROOT-SERVERS.NET. 3600000 A 198.32.64.12
;
; temporarily housed at ISI (IANA)
. 3600000 NS M.ROOT-SERVERS.NET.
M.ROOT-SERVERS.NET. 3600000 A 198.32.65.12
; End of File
```

/etc/resolv.conf File

The */etc/resolv.conf* file tells the DNS resolver code where to find the DNS name server, and the name of the current domain. The following sample */etc/resolv.conf* file contains two name servers and sets the domain to *astro.com*.

```
domain astro.com
nameserver 203.14.4.1
nameserver 203.14.4.4
```

Starting the DNS Daemon

Once all of the configuration files are set up and checked, the administrator should start the DNS daemon on the name server machine as follows:

```
# /usr/etc/in.named
```

If the configuration files have been set up correctly, the name service will start without a hitch. The operation of the name service can be tested by using the **nslookup** command to query the DNS name server for information on a local host. For example, the command **# nslookup glenn** should return the following information about the *glenn* machine in the *astro.com* domain.

```
Server: mercury.astro.com
Address: 203.14.4.1
Name: glenn.astro.com
Address: 203.14.5.1
```

If **nslookup** does not return the proper information for a host in the database, the name service daemon may not have started correctly. Consult the */var/adm/ messages* file to determine if the named program wrote any error messages to the log file.

Once the **nslookup** command returns correct information for local hosts, the administrator should create start-up files in the appropriate */etc/rc.d* directories such that the */usr/etc/in.named* process starts when the name server is booted. Consult Chapter 5, "Refining System Boot and Shutdown Procedures," for more information on creating start-up scripts.

Using NIS+ with DNS

In order to provide a robust name service environment, it is sometimes necessary to run more than one name service. For example, an Internet connected corporation will need to run DNS for Internet name service, but the local administrator may wish to maintain local network information in NIS+ maps. Still other corporations may have a need to use other name services such as the Federated Name Service (FNS). The following section covers the basics of configuring hosts to use multiple name services. The discussion is limited to NIS+ and DNS name services, but the concepts presented may be extended to other name services.

Configuring Multiple Name Services

Many sites will opt to use the NIS+ name service for local network information, and DNS for external name service. This creates a minor logistics problem for the name resolution libraries. For instance, which name service should be contacted first? In order to facilitate the use of multiple name services, the administrator should configure the system by editing the */etc/nsswitch.conf* file.

/etc/nsswitch.conf File

The */etc/nsswitch.conf* file contains directives that instruct the name service resolution routines on where to look for name resolution information. A few valid options for where the resolver should look are files (search */etc/inet/hosts*, */etc/passwd*, */etc/group*, and so on), DNS (consult a DNS server), and NIS+ (consult an NIS+ server).

The administrator can configure the name service to search a single source, or to search all sources in a specific order. In addition, you can instruct the system to search one source for host name to IP address mappings, another source for user password information, and yet another source for other information (e.g., Ethernet MAC addresses for network boot requests).

In order to provide the most reliable name service for the corporate network, many sites configure the machines to consult NIS+ first. If a resolution is not found from NIS+ information, contact DNS, and if there is still no resolution to the query consult the local files in */etc*.

The following sample */etc/nsswitch.conf* file will implement an NIS+, DNS, and then local files search in order to resolve name service queries. Note that for information other than host name to IP address, or IP address to hostname look-ups, only NIS+ and local files are consulted.

The **[NOTFOUND=return]** directive tells the name service not to return an error if the query cannot be resolved by the name service. If the request cannot be resolved once all of the name service information has been searched, and if the information is not available in local files, an error will be returned for this query.

```
# /etc/nsswitch.conf
# uses NIS+ (NIS Version 3) in conjunction with files.
# "hosts:" and "services:" in this file are used only if the
# /etc/netconfig file has a "-" for nametoaddr_libs of "inet" transports.
# the following two lines obviate the "+" entry in /etc/passwd and /etc/group.
passwd: files nisplus
group: files nisplus
# consult /etc "files" only if nisplus is down.
```

```
# Use DNS and NIS+. Must also set up /etc/resolv.conf file for DNS nameserver
lookup.
hosts: nisplus dns [NOTFOUND=return] files

services: nisplus [NOTFOUND=return] files

networks: nisplus [NOTFOUND=return] files

protocols: nisplus [NOTFOUND=return] files

rpc: nisplus [NOTFOUND=return] files

ethers: nisplus [NOTFOUND=return] files

netmasks: nisplus [NOTFOUND=return] files

bootparams: nisplus [NOTFOUND=return] files

publickey: nisplus

netgroup: nisplus

automount: files nisplus

aliases: files nisplus

sendmailvars: files nisplus
```

References

The following references are listed in an attempt to provide the interested reader with more information on the topics covered in this chapter. The sources listed below are not exhaustive, but should provide the reader with a knowledge base on the topics of computer name services.

❏ Paul Albitz and Cricket Liu, *DNS and BIND*. O'Reilly & Associates, Inc., 1992 (ISBN 1-56592-010-4).

❏ Hal Stern, *Managing NFS and NIS*. O'Reilly & Associates, Inc., 1991 (ISBN 0-937175-75-7).

Summary

This chapter discussed the basic concepts of network name services. Two name services were examined in detail: NIS+, a local name service, and DNS, an Internet-wide name service. Examples outlined the steps required to configure an NIS+ environment in a local area network. In addition, basic step-by-step procedures for creating and initializing an NIS+ master server, NIS+ client, and NIS+ replica server, and the steps required to populate NIS+ tables were presented. Similarly, examples illustrated DNS database information setup, and the configuration of a DNS name server. Finally, the chapter examined ways to make the two name services cooperate in order to provide a robust, reliable name service for corporate network users.

22

WWW Administration

Perhaps no other area of computing holds more interest and promise for the future than the World Wide Web (WWW). Solaris provides an excellent platform for providing WWW services. However, the Web is complex and makes unique demands on the server providing services and on system administrators. The complexity of Web service also adds new challenges in the area of securing the server system.

What is the World Wide Web?

The World Wide Web is a client-server based application originally developed by Tim Berners-Lee at CERN to distribute documentation. Researchers at various locations, notably the National Center for Supercomputer Applications at the University of Illinois, extended the original design to include the distribution of a wide variety of media including graphics, audio, video, and small applications or applets. WWW clients, known as browsers, are used to make requests from WWW servers and display the results in the form of a *page*. These pages may contain the addresses of other pages in the form of links that can be selected. The World Wide Web name derives from the interconnected nature of these page links.

Pages and other resources are referenced using a universal resource locator (URL). The format of a URL is a resource type tag, followed by the name of the system holding the resource, followed by the path to the resource which may include option flags and other data. For example, the URL for the Dilbert comic strip follows. This URL refers to a Web page as noted by *http:* (hypertext transport protocol). The referenced page is stored on a system named *www.unitedmedia.com* in a directory named */comics/dilbert/*.

```
http://www.unitedmedia.com/comics/dilbert/
```

In typical usage, a browser connects to a server to obtain pages as requested by the user. A single page may contain information obtained from more than one server. The communications protocol or language used to communicate between WWW browsers and servers is standardized allowing browsers functioning on different operating systems to access servers operating on different operating systems. The ability to publish information in a form that is viewable independent of operating systems used to serve and view the information is one key to WWW's success.

Another key to the success of the WWW are forms and secure transactions. Web pages can contain fill-in-the-blank forms. The responses entered into these forms can be transmitted back to the server for processing. In addition, pages and the responses from a form can be encrypted and securely transmitted. Web servers have the ability to run programs through a feature known as the Common Gateway Interface. This feature allows for pages to be constructed dynamically and for processing of information received via forms. The combination of these features makes a wide variety of Web-based applications possible, such as on-line product ordering.

Typical WWW page with graphics, text, and underlined links to other pages.

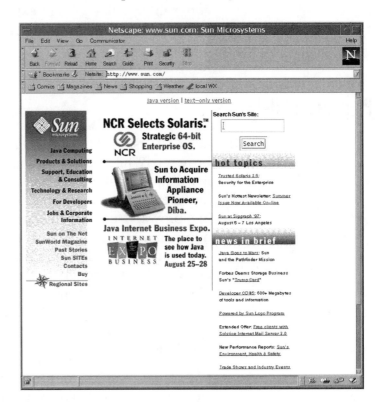

WWW Clients

Administrating WWW clients is primarily a matter of keeping up to date with browser and page content development. At present, leading browsers such as Netscape Communications and Microsoft products are undergoing rapid development. New versions of a browser every six months are common. Likewise, new page content in the form of new media data types are continually being developed. Not all media types are directly viewable by a given browser. Additional software may be needed to view certain content types such as video. Such additions to the browser come in two flavors: (1) extensions to the browser program itself, often called *plug-ins,* or (2) separate applications started under the browser's control, known as *helper applications.*

Plug-ins and Helpers

Solaris version 2.6 will include Sun's HotJava Web browser. While easy to install as part of the Solaris installation, many sites prefer the popular Netscape browser. Netscape Communicator is a suite containing browser, e-mail, calendar, and discussion tools now shipping from Netscape Communications. The suite can be downloaded from Netscape Communications' Web site, installed, and evaluated for 90 days. The browser component of Communicator, called Navigator, has built-in facilities for viewing Web pages and most graphics formats and for running JavaScript and Java applets.

Netscape Communicator installs by way of an installation script provided by Netscape. Once run, the program and its support files are placed in the proper places for use. Additional media formats can be added to Navigator through the *mime.types* and *mailcap* files. Some are added as part of the installation of a plug-in or helper application, while others are added separately. For example, the *xanim* video player is a helper application which requires a line to be added to either the user's *.mailcap* file or the system-wide files.

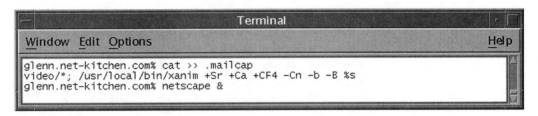

```
glenn.net-kitchen.com% cat >> .mailcap
video/*; /usr/local/bin/xanim +Sr +Ca +CF4 -Cn -b -B %s
glenn.net-kitchen.com% netscape &
```

Adding the xanim player definition to the user's .mailcap file and starting Netscape.

Client Security Issues

Web browsers present several security problems. Problems with downloaded applets and scripts are currently receiving the most attention. Java is a language designed to run within a portable virtual machine. Java applets are downloaded mini-programs that are run inside the browser's Java virtual machine. While the designers of the Java language built in numerous security protections, some implementations of the Java virtual machine and the local libraries used by running Java applets are not secure.

Fortunately, most browsers allow the user to optionally turn off the execution of Java applets. Likewise, the less capable, but more frequently found JavaScript can also be turned off. Turning off these features will disable certain interactive features of Web pages that utilize JavaScript or Java applets while protecting the user from problems caused by bugs in the Java virtual machine or libraries. The desirability of turning off JavaScript and Java applets depends on the specific pages being viewed and the applets they contain. These features may be required for some applications.

In Navigator the controls to deactivate these features can be found in the Preferences dialog box, obtained by selecting the Preferences item from the Edit menu.

The Java language controls are found by clicking on Advanced in the list at the left of the dialog box.

Netscape Navigator's security preferences for languages and cookies.

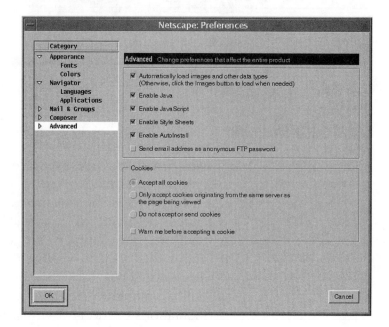

Bugs in the browser itself constitute a more common problem. Browsers are complex, often including their own Java virtual machine as well as internal versions of ftp and other network tools. System managers at sites concerned about security should continually monitor the browser vendor Web pages for updates that address security problems.

Referring page information that some browsers pass along to a Web server as they switch from page to page while following links is a less discussed problem. This information can be used by a Web server to identify the page a user comes from when accessing a new page. The problem derives from the additional information on the referring page that is also passed along. Such additional data may include information believed to be secure if the browser moves from a secure page to an unsecured page. Many Web sites avoid this problem by "wiping the browser's feet" via directing the browser to a blank or unrevealing page after requesting secure information. By default, the Netscape browser will alert users to this problem by posting an alert message when the user moves from a secure page to an unsecured page.

↝ *NOTE: Modern browsers are also capable of storing small pieces of information from Web sites about the user such as a password or usage history. These bits of information are known as "cookies." The same security preferences dialog box mentioned earlier also allows those concerned about cookies to disable them or have the browser announce the delivery of a cookie from the Web site. Turning off cookies will disable password memory and history features of some Web sites. The decision to turn off cookies depends on the user's concerns about her privacy and the Web pages she views most often.*

WWW Servers

Installing and configuring a Web server is a much more involved process. A Web server is a very complex daemon with numerous features controlled by a couple of configuration files. Web servers not only access files containing Web pages, graphics and other media types for distribution to clients, they can also assemble pages from more than one file, run CGI applications, and negotiate secure communications. Basic server configuration issues are discussed in the following sections.

Choosing a Server

Choice of a Web server is dependent on several factors including the list of desired features, cost, and availability. Solaris systems support several different Web servers. Three common choices are listed below.

❏ *Netscape's SuiteSpot.* A commercial Web server with many Web based commerce features built in. This server is frequently updated and supported by Netscape and is intended for production quality Web service.

❏ *Sun's Web Server.* This server is bundled with Solaris 2.6 and is available for free downloading from Sun's Web site. It includes a copy of the Hot-Java browser, and secure Web page transmission tools. This server is new to the market, and its ready availability makes it a good choice for new installations. The server provides Web service on multiple machines or the wherewithal to experiment with Web service.

❏ *Apache.* A free Web server developed by a community of Internet programmers, it is available for many UNIX systems as well as Solaris in source code form. Apache includes the latest features and provides for a broad range of customization and configuration. It is a good choice for sites that require the latest Web server features and possess the required software development tools to compile and maintain the program. The Apache Web server is available on the Web site listed at end of this chapter.

A Web server installation begins by installing the Web server software itself, followed by careful configuration of numerous server options and features. CGI programs run by the server to build Web pages on demand or process information received in forms will be installed and integrated with the Web server as well.

Server Configuration

How a Web server functions is controlled by its configuration file. Configuration options of the major Web servers are similar. In the following discussion, the configuration file locations and options used by the Apache Web server will be used. The Apache Web server's function is controlled via the configuration files described in the next table.

File	Function
httpd.conf	Server configuration file containing the base configuration information.
srm.conf	Resource configuration file which controls local resource information such as the location of icons and image maps.
access.conf	Access configuration file which specifies the actions allowed by specific clients on specific files.
mime.types	Specifies the media types and respective file suffixes.

The Apache server distribution includes a set of samples of the above files which the system administrator can modify. Over 100 configuration options can be applied to control the behavior of the Apache Web server. A few of the most basic options to be examined upon setting up a new Web server are examined in this section.

http.conf

The *httpd.conf* file is the base on which all other configuration files rest. Among the options that can be
controlled by this file are the locations of the httpd
and associated files, server log files, and other configuration files.

At a minimum, the system administrator will want to
modify the User, Group, ServerAdmin, ServerRoot,
and ServerName lines to reflect the local site. The
User and Group lines specify the user id and group id
that the Web server will operate under once started.
The ServerAdmin is an e-mail address to which the
server can send problem reports. The ServerRoot
specifies the installation directory for the server. The
ServerName is the name of the server returned to clients. An example appears below.

```
User nobody
Group nobody
ServerAdmin dwight@astro-corp.com
ServerRoot /usr/local/http
ServerName www.astro-corp.com
```

srm.conf

The Alias lines in the *srm.conf* file may require updating to reflect the location of icons and other local files.
The Alias lines allow Web page designers to use
shortened names for resources such as icons instead
of specifying full paths. An example follows.

```
DocumentRoot /usr/local/http/htdocs
UserDir WWW
Alias /icons/ /usr/local/http/icons/
ScriptAlias /cgi-bin/ /usr/local/http/cgi-bin/
```

Besides making Web page construction easier by providing short names for icons and CGI programs, the *srm.conf* file allows access to users' Web pages. The UserDir line specifies the subdirectory each user can create in his home directory to hold Web pages. This directory, *WWW* in the example, is mapped to the user's log-in name as follows. A user whose log-in name is *bob* has his WWW directory mapped to *http://www.astro-corp.com/~bob*. By default, the Apache Web server will display the *index.html* file in that directory, or a directory listing if the *index.html* file is not found.

access.conf

At a minimum, a new installation of Apache will require changing the *<Directory>* stanzas in *access.conf* to indicate where the server should look for documents to serve and for CGI programs. For example, if the server is installed in */usr/local/http* with the documents and CGI programs in directories under that directory, the following *<Directory >* line may be necessary.

```
<Directory /usr/local/http/htdocs>
```

mime.types

The *mime.types* file includes the mapping from a mime type to a file extension. The most common types are provided in the sample file provided with the Apache distribution. A more detailed description of the myriad available configuration options can be found in the books and Web sites listed at the end of the chapter.

Server Security Issues

Web servers present a difficult security challenge. They must be widely accessible to be useful, but tightly controlled to prevent security breaches. They must be tolerant of any requests submitted to them, including requests specifically constructed to gain unauthorized access to files or to exercise bugs in CGI programs. Basic precautions that should be taken to help prevent problems are discussed below. Additional information on securing a Web server can be found in the Garfinkel and Spafford books listed at the end of the chapter.

File Access Control

The control files which determine the Web server's function as well as the log files it produces should not be accessible to the user ID the Web server runs under. Individuals attempting to gain unauthorized access are thwarted to the extent that they cannot obtain information about the Web server's configuration and function. In addition, Web server options under which Web pages include other files and optionally execute programs should be carefully scrutinized for potential access to files not intended for distribution.

Dangers of CGI

CGI programs are among the biggest potential dangers to Web server security. These programs are run based on a URL passed to the Web server by a client. In normal operations this URL comes from a form or page. However, the URL provided to a CGI program can be given to the Web server by other means and can be carefully constructed to exercise bugs in the CGI program itself.

CGI programs must be carefully constructed to avert potential problems resulting from the input passed to them. One successful method is to use the "tainted" variable facility found in the Perl scripting language. If other languages are used, care must be taken to ensure that *all* possible input characters are properly handled, including shell metacharacters, quotes, asterisks, and braces. Administrators must also be alert to the well-known problem of very large input strings designed to overwrite small input buffers. Security conscious sites should carefully audit CGI programs before putting them into operation.

Log Files

Web servers maintain several log files which can aid in securing the Web server. The next table describes several such log files for the Apache Web server.

File name	Contents
access_log	Listing of each individual request fielded by the Web server.
agent_log	Listing of every program run by the Web server.
error_log	Listing of the errors the server encountered. Errors from CGI programs as well as the server itself are logged to this file.
refer_log	Listing of the previous URL accessed by a given browser.

Of principal interest from a security standpoint are *error_log, agent_log,* and *access_log.* These logs should be reviewed periodically for purposes of identifying CGI program problems and attempts to access files not intended for distribution.

References

The discussion of Web client and server configuration in this chapter has been brief. If you require more detailed information, consider the sources listed below.

Books

Source	Comments
Simson Garfinkel and Gene Spafford, *Web Security and Commerce*. O'Reilly & Associates, June 1997 (ISBN 1-56592-269-7).	Specifically targeted at Web server security and securing transactions carried out over the Web.
Stephen Spainhour and Valerie Querci, *Webmaster in a Nutshell*. O'Reilly & Associates, October 1996 (ISBN 1-56592-229-8).	Covers a wide range of topics including HTML, frames, JavaScript, and server configuration.

Web Sites

Contact information	Comments
http://www.apache.org	Official home page for the Apache Web server. This page includes complete on-line documentation and access to the Web server source code.
http://www-genome.wi.mit.edu/ WWW/faqs/www-security-faq.html	Lincoln D. Stein's WWW Security FAQ. A complete and current discussion of Web server security problems and solutions.
http://hoohoo.ncsa.uiuc.edu/cgi/ security.html	CGI security document of the National Center for Supercomputer Applications. Provides guidelines for writing more secure CGI programs.

Summary

Administering WWW services is an ongoing task. The rapid development of Web based media means that continual updating of Web browsers and Web servers will be required. Administrators running servers should pay close attention to the configuration and security of their systems.

Index

More OnWord Press Titles

NOTE: All prices are subject to change.

Geographic Information Systems (GIS)

GIS: A Visual Approach
$39.95

The GIS Book, 4E
$39.95

GIS Online: Information Retrieval, Mapping, and the Internet
$49.95

Raster Imagery in Geographic Information Systems Includes color inserts
$59.95

INSIDE ArcView GIS, 2E
$44.95 Includes CD-ROM

ArcView GIS Exercise Book, 2E
$49.95 Includes CD-ROM

ArcView GIS/Avenue Developer's Guide, 2E
$49.95 Includes Disk

ArcView GIS/Avenue Programmer's Reference, 2E
$49.95

ArcView GIS/Avenue Scripts: The Disk, 2E
Disk $99.00

ARC/INFO Quick Reference
$24.95

INSIDE ARC/INFO, Revised Edition
$59.95 Includes CD-ROM

Exploring Spatial Analysis in Geographic Information Systems
$49.95

Processing Digital Images in GIS: A Tutorial for ArcView and ARC/INFO
$49.95

Cartographic Design Using ArcView GIS and ARC/INFO: Making Better Maps
$49.95

Focus on GIS Component Software, Featuring ESRI's MapObjects
$49.95

Pro/ENGINEER and Pro/JR.

Automating Design in Pro/ENGINEER with Pro/PROGRAM
$59.95 Includes CD-ROM

INSIDE Pro/ENGINEER, 3E
$49.95 Includes Disk

Pro/ENGINEER Exercise Book, 2E
$39.95 Includes Disk

Pro/ENGINEER Quick Reference, 2E
$24.95

Thinking Pro/ENGINEER
$49.95

Pro/ENGINEER Tips and Techniques
$59.95

INSIDE Pro/JR.
$49.95

INSIDE Pro/SURFACE: Moving from Solid Modeling to Surface Design
$90.00

FEA Made Easy with Pro/MECHANICA
$90.00

OnWord Press Distribution

End Users/User Groups/Corporate Sales

OnWord Press books are available worldwide to end users, user groups, and corporate accounts from local booksellers or from SoftStore Inc. Call toll-free 1-888-SoftStore (1-888-763-8786) or 505-474-5120; fax 505-474-5020; write to SoftStore, Inc., 2530 Camino Entrada, Santa Fe, New Mexico 87505-4835, USA, or e-mail orders@hmp.com. SoftStore, Inc., is a High Mountain Press company.

Wholesale, Including Overseas Distribution

High Mountain Press distributes OnWord Press books internationally. For terms call 1-800-4-ONWORD (1-800-466-9673) or 505-474-5130; fax to 505-474-5030; e-mail to orders@hmp.com; or write to High Mountain Press, 2530 Camino Entrada, Santa Fe, NM 87505-4835, USA.

Comments and Corrections

Your comments can help us make better products. If you find an error, or have a comment or a query for the authors, please write to us at the address below or call us at 1-800-223-6397.

OnWord Press, 2530 Camino Entrada, Santa Fe, NM 87505-4835 USA

On the Internet: http://www.hmp.com